WORLD VEGETABLES

WORLD VEGETABLES

Principles, Production and Nutritive Values

Mas Yamaguchi

Department of Vegetable Crops
University of California at Davis, California

ELLIS HORWOOD LIMITED, PUBLISHERS
Chichester, England

First published in 1983 by

THE AVI PUBLISHING COMPANY, INC.
Westport, Connecticut

United Kingdom Edition published by

ELLIS HORWOOD LIMITED
Market Cross House, Cooper Street, Chichester, West Sussex,
PO19 1EB, England

Distributors:

Australia, New Zealand, South-east Asia:
Jacaranda-Wiley Ltd., Jacaranda Press
JOHN WILEY & SONS INC.
G.P.O. Box 859, Brisbane, Queensland 4001, Australia

Europe, South Africa:
JOHN WILEY & SONS LIMITED
Baffins Lane, Chichester, West Sussex, England

Available elsewhere from
THE AVI PUBLISHING COMPANY, INC.
P.O. Box 831, Westport, Connecticut 06881, USA

Library of Congress Cataloging in Publication Data

Yamaguchi, Mas.
World vegetables.

Includes bibliographies and index.
1. Vegetables. I. Title.
SB320.9.Y25 1983 635 83-6422
0-87055-433-6 (AVI Publishing Company)
0-85312-625-9 (Ellis Horwood Limited, Publishers)

Printed in the United States of America

Contents

Preface

This text and reference book was written because of the urgings of my colleagues here and abroad and also by my students to fill the need for a comprehensive book describing vegetables important as food crops of the world.

The material in this book was first used as outlines for the course on world vegetables taught by the staff of the Department of Vegetable Crops at the University of California at Davis. Over the course of the decade the author taught the course, new topics were added with each succeeding year. In 1978 the handouts were revised and compiled into a syllabus. In addition to the students taking the course, the syllabus was purchased and used by others as a reference book. Word of its existence travelled far; the syllabus is presently used by one other university in the United States and by another in a foreign country.

This edition has been extensively revised; many new materials and references have been added. Included are many photographs taken by the author in his travels abroad.

The book has three parts. The first is introductory and background meant for students who are not familiar with vegetables as food crops. Included is a chapter on toxic constituents and some interesting past and present usage of vegetables. Part II is a very short summary of vegetable physiology, especially the climatic effects on growth and some methods used to grow the crops under adverse conditions. The third and main part covers the world vegetables, divided into starchy crops and succulent vegetables. The appendix is intended to supplement the main parts.

Teaching aids are included. For example, Fig. 8.1 summarizes the exotic root and tuber crops; more detailed accounts are presented in the chapters following. Likewise, the physiology of onions is summarized in

Fig. 17.3, that of cole crops in Table 19.1, and the important legumes used as vegetables are given in Table 21.1.

Although the pests and diseases are only superficially mentioned, they are of major concern in all vegetable growing regions. To adequately cover these aspects would require additional expertise. There is a constant change on recommendations for their control as new advances are made and old ones are discarded.

It is hoped that this book adequately fulfills the need of students and worker in the field.

Mas Yamaguchi

Acknowledgments

The author wishes to acknowledge the use of much materials and information derived from his many colleagues, especially Professors J.F. Harrington, O.A. Lorenz, L.L. Morris, C.M. Rick, P.G. Smith and J.E. Welch, who were among the first lecturers in the course organized in 1967.

There were many people involved in the mechanics of getting the syllabus and the book ready for print. They are Marilynn Berry, Karen Murphy, Moira Tanaka and Marcia Carey for the illustrations; Betty Perry and Kathy Hykonen for typing the manuscript; and Corky Webb, Nancy Folsom and Anne Rundstrom for typing the tables.

Also, the author appreciates and thanks Drs. O.A. Lorenz, V.E. Rubatzky, I.W. Buddenhagen, A. Kader and B.S. Luh, all of this University, for reviewing the various chapters. Their criticisms and suggestions were most helpful.

Last but not least, the author wishes to thank his long time friend and colleague, Dr. Doran L. Hughes, for his most helpful advice and criticisms throughout the years, his expert reproductions of the color slides into black and white photographs and the review of the entire manuscript.

ELLIS HORWOOD SERIES IN FOOD SCIENCE AND TECHNOLOGY

This publishing programme provides an organised coverage on food science for professional technologists, research workers, and students in universities and polytechnics.

Series Editors:
Professor I.D. Morton, Head of Food Science, Queen Elizabeth College, University of London;
Dr. R. Scott, formerly Reading University

FUNDAMENTALS OF FOOD CHEMISTRY
W. Heimann, Director, Institute for Food Chemistry, University of Karlsruhe, Germany

HYGIENIC DESIGN AND OPERATION OF FOOD PLANT
Edited by R. Jowitt, National College of Food Technology, University of Reading

SMALL-SCALE HEATING AND EQUIPMENT SERVICES
D. Kirk, Sheffield City Polytechnic, and A. Milson, The Queen's College, Glasgow

PRINCIPLES OF DESIGN AND OPERATION OF CATERING EQUIPMENT
A. Milson, The Queen's College, Glasgow, and D. Kirk, Sheffield City Polytechnic

TECHNOLOGY OF BISCUITS, CRACKERS AND COOKIES
J.R. Manley, Consultant, Peterborough

SENSORY QUALITY IN FOODS AND BEVERAGES
A.A. Williams and R.K. Atkin, Long Aston Research Station, Bristol

PHYSICAL PROPERTIES OF FOODS AND FOOD PROCESSING SYSTEMS
M.J. Lewis, University of Reading

FLAVOUR OF DISTILLED BEVERAGES
J.R. Piggott, University of Strathclyde

ENERGY MANAGEMENT IN FOODSERVICE
N. Unklesbay and K. Unklesbay, University of Missouri, Columbia, Missouri, USA

THE PSYCHOBIOLOGY OF HUMAN FOOD SELECTION
Edited by L.M. Barker, Baylor University, Waco, Texas, USA

ADVANCED SUGAR CHEMISTRY
R.S. Shallenberger, Cornell University, Geneva, New York, USA

SENSORY EVALUATION OF FOOD: THEORY AND PRACTICE
G. Jellinek, formerly University of Berlin and University of Stuttgart

FISH PROCESSING: Its Science and Technology
S.W. Hanson and W.F.A. Horner, Grimsby College of Technology

MODERN FOOD PROCESSING
J. Lamb, University of Leeds

BATTER AND BREADING TECHNOLOGY
Edited by D.R. Sudermann, Durkee Famous Foods/SCM Corporation, Strongsville, Ohio, USA, and F.E. Cunningham, Kansas State University, Manhattan, Kansas, USA

SUSTAINABLE FOOD SYSTEMS
Edited by D. Knorr, University of Delaware, Newark, Delaware, USA

FOOD OILS AND THEIR USES
T.J. Weiss, Hunt-Wesson Foods Inc., Fullerton, California, USA

WORLD VEGETABLES: Principles, Production and Nutritive Values
M. Yamaguchi, University of California at Davis, California, USA

Part I

Introduction

1

Vegetables and the World Food Supply

WORLD FOOD SITUATION

World population is important in considering the food supply. At the present time the population of the world is increasing at the rate of approximately 2% per year. At this rate it doubles every 30–35 years. It is estimated that the present world population of 4.3 billion (1979), will be over 7 billion in the year 2000 AD and the demand for food will increase accordingly.

The world is categorized by the Food and Agriculture Organization of the United Nations (FAO) into economic classes; the developed countries and the developing countries. Figure 1.1 shows the food situation in the developed and developing regions of the world and Table 1.1 contrasts the two.

With population increases, the land for agricultural purposes declines. Land that is best for growing of most of our economic plants also suits man's living requirements. Prime land, that is, land with deep fertile well-drained soils, is constantly being taken over for nonagricultural purposes. Hence, more and more marginal land will have to be put into use for growing of crops, often with very high costs before it can be used for effective production.

At some point there must be a balance of the number of people and the amount of land needed for food production. Countries of the world vary greatly in population density. Some countries of Asia have already reached the saturation point where agriculture can barely feed the people, and undeveloped land resources are very limited.

Demographers express population density and food producing capabilities as shown in Table 1.2. This is only a gross picture; Canada and Russia have vast areas of extreme cold and Australia has regions of desert; the population is concentrated in the habitable climates. How-

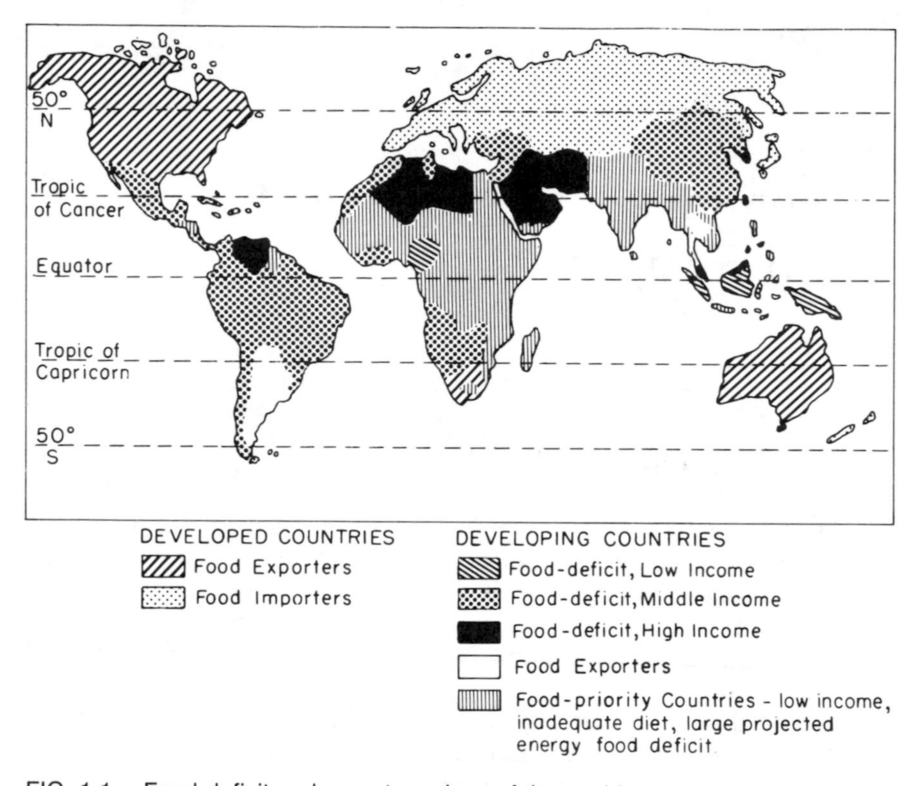

FIG. 1.1. Food deficit and poverty regions of the world.
Redrawn from S. Wortman, Scientific American, Sept. (1976).

ever, arable land per person is a better way of indicating the food producing capacity of a country. Large amounts of arable land per person does not necessarily mean adequate diets as the land is often not

TABLE 1.1. SOME COMPARISONS OF DEVELOPED AND DEVELOPING REGIONS OF THE WORLD

	Developed	Developing
Population	¼ of total	¾ of total
Population growth	1.5% per year	2–3% per year
Population in agriculture	13% (2–48%)	60% (17–93%)
Arable land per person	0.56 ha (1.4 acre)	0.21 ha (0.5 acre)
Food energy:		
Total/person/day	3370 cal	2280 cal
From plant sources	70%	90%
Protein:		
Total/person/day	99 g	58 g
From animal sources	56%	21%

Source: FAO data (1979).

TABLE 1.2. FOOD PRODUCING CAPACITY OF SOME COUNTRIES IN RELATION TO POPULATION DENSITY AND ARABLE LAND IN 1979

Country	Population density (people/km^2)	Arable land[a] (ha/person)
Australia	2	2.97
Canada	3	1.87
Argentina[b]	10	0.94
United States	24	0.86
U.S.S.R.	12	0.86
Congo[b]	5	0.44
Mexico[b]	35	0.32
India[b]	228	0.24
Italy	194	0.17
United Kingdom	233	0.12
Philippines[b]	166	0.11
China[b]	102	0.10
Egypt[b]	41	0.07
Netherlands	413	0.06
Japan	312	0.04

Source: FAO data (1979).
[a] Arable land: (a) If greater than 0.8 ha/person, food supply usually adequate; export of surplus. (b) If in range of 0.4–0.8 ha/person, country is 80% to completely self-sufficient. (c) If less than 0.4 ha/person, diet high in food of plant origin; import of food for adequate diet.
[b] Developing nation.

used or is misused. Thus, protein and vitamin deficiencies occur in Central and East Africa, and in many Asian and Latin American countries.

Often special crops, such as coffee, cocoa, banana, and sugar, are sold in world markets to richer nations. This reduces the food available in developing countries by using the best agricultural lands, fertilizer, and equipment for export crops, leaving little or none for the domestic crops.

Increase of Food Supply

Production

Presently, the increase in food supply is about 2% per year which is just enough to keep up with population growth. About 20% of this increase is from new areas of production expansion into cooler regions, arid lands, and forest regions of the tropics and semitropics. The main increase (80%) is due to technological advances: use of new varieties, improved cultural practices, crop rotation, irrigation, fertilization, optimal plant population, use of growth regulators, and control of weeds, pests, diseases.

Development of New Foods from Other Sources

These can be by-products from present crops' waste and residues, synthetics, use of algae and lower plants, and "farming" of the seas.

Increased Efficiency of Nutrient Production

The food chain is a very important consideration when the food supply is low. Planting of seeds, growing of crops, feeding of the crop to animals, and finally slaughtering the animals for meat take many acres of land and from one to several years of time. This chain of events may be depicted as: plant → animal → man. If we remove the animal from this chain, the efficiency of the land to produce food for man would be increased.

Figure 1.2 shows that the shortest time is taken when the vegetative portions of the plant are harvested for food; fruits, generally, take a longer period to reach harvestable maturity. When animals are raised for meat, additional time is required even though the same land may not be used to rear the animals.

Selection of crops that produce the highest amount of nutrients per unit area of land in the shortest period of time would be the most

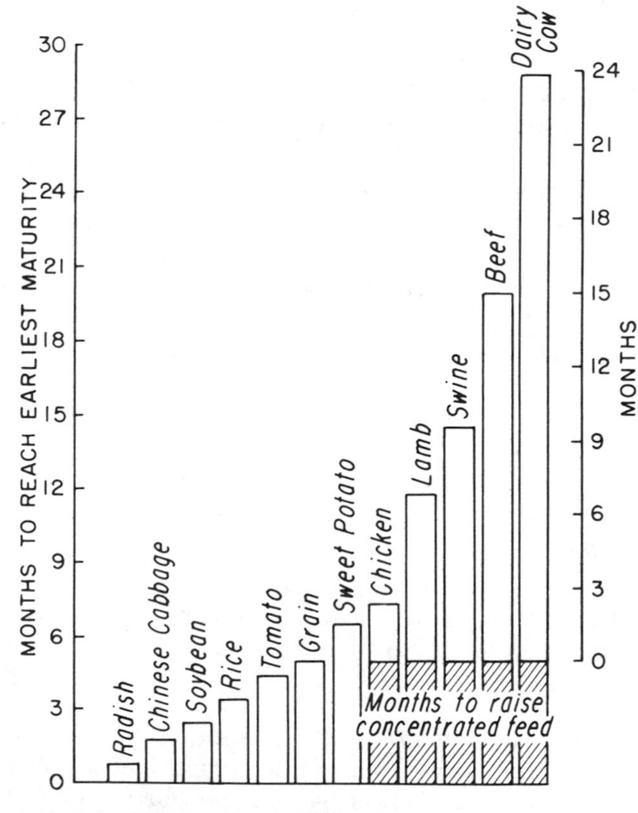

FIG. 1.2. Time required to raise foods.

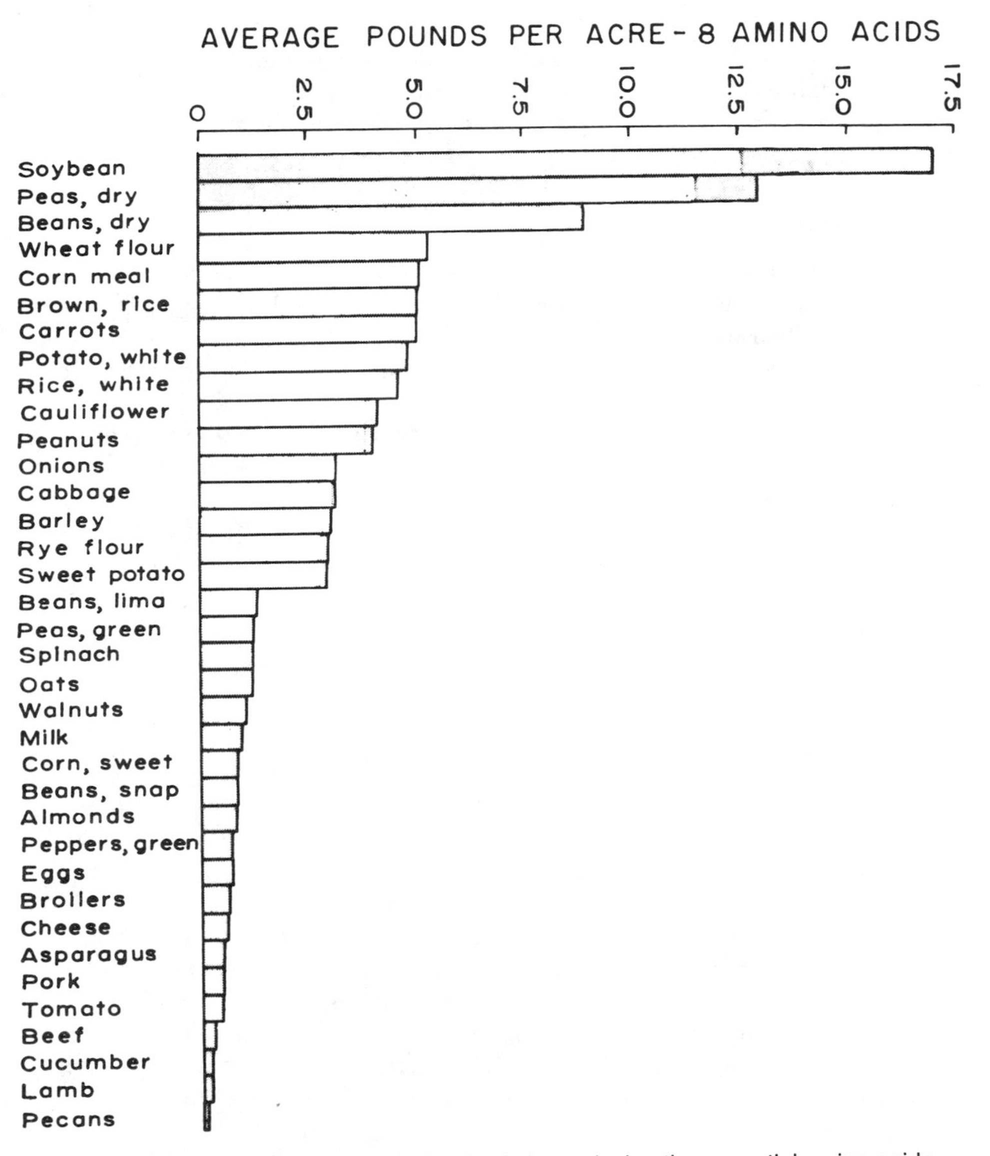

FIG. 1.3. Efficiency of crops and animals in producing the essential amino acids on an area basis.

From MacGillivray and Bosley (1962). Reprinted with permission.

efficient land usage. The efficiency of crops to produce the eight essential amino acids on a unit area of land is depicted in Fig. 1.3. Vegetables rank high among the crops listed. Ideally, the proportion of the eight essential amino acids should be in proportion to the minimum daily requirements (Chapter 4).

Multiple Cropping

Maximum use of arable land can be obtained by using practices in which two or more crops are grown on the same land in a year. This is accomplished by

Sequential cropping: growing of two or more crops in sequence per year.

Ratoon cropping: cultivation of the regrowth of the same crop after harvest, usually by suckers or adventitious shoots, e.g., bananas and sugar cane.

Intercropping: the growing of two or more crops simultaneously on the same land. *Mixed*, *row*, and *strip croppings:* one or more crops planted mixed, in rows or in strips where interaction between the crops can occur.

Relay intercropping: two or more crops grown simultaneously during part of the growing period of each. Usually the second crop is seeded or transplanted after the first crop has reached the reproductive stage or about mid to latter part of the growth period and before the first crop is ready for harvest.

WORLD VEGETABLE PRODUCTION

The kinds of crops grown on arable land are shown in Fig. 1.4. Vegetables including roots and tubers occupy less than 10% of the area. Consumption of vegetables is correlated with per capita income. Those countries with low and middle incomes use more grain products or starchy root crops in their diets. Table 1.3 shows the production and consumption of vegetables by economic and political regions.

Distribution and Conservation of Crops

In many countries often it is economically not feasible to harvest a crop because the price is too low to warrant harvest or transportation to regions of need is too costly or not available.

In some countries over half of the postharvest loss is from pilferage, rodents, infestation of insects, and microbial spoilage. Also, it is esti-

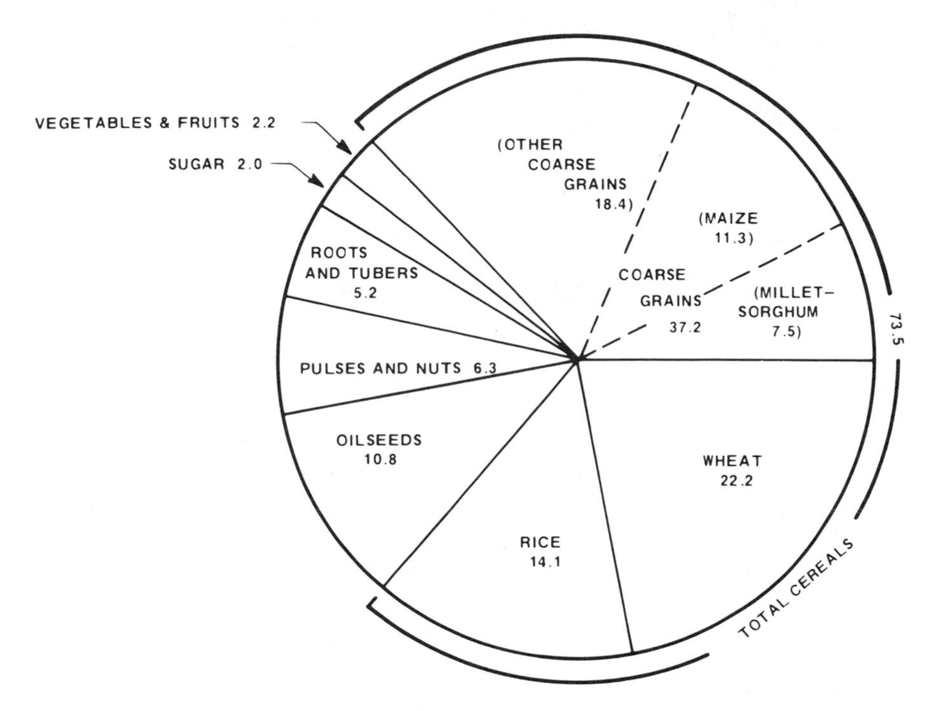

FIG. 1.4. World harvested area of principal crops excluding forage and fodder crops in 1970.
From Univ. of Calif. (1974).

mated that the world's mice and rats consume or make unfit for use enough food annually to feed 130 million people.

Poor handling and storage, lack of refrigeration, and inadequate facilities for processing crops can account for the huge losses. Improper handling and storage of perishable commodities can not only cause physical losses, but also cause losses of nutrients such as vitamins in foods that are still edible.

Increasing Food Supply by Reducing Losses and Waste

The FAO–UN has been endeavoring to remedy losses due to wastage at all stages of production.

The practice of harvesting the largest seedlings in the thinning of Chinese cabbage and marketing them has been standard practice in the Orient for hundreds of years. Such a practice saves on land use and reduces the waste of seeds. The only disadvantage is that since the

TABLE 1.3. ESTIMATED WORLD DAILY VEGETABLE CONSUMPTION AND ANNUAL PRODUCTION

Type of vegetable	Developed countries: Consumption (g/day)	Developed countries: Production ($\times 10^6$ MT)	Developed countries: For sale (%)	Communist dominated countries, production ($\times 10^6$ MT)	Developing countries: Consumption (g/person/day): Africa	America	Southwest/ Central Asia	Rest of Asia	Mean	Developing countries: Production ($\times 10^6$ MT)	Developing countries: For sale (%)	World production ($\times 10^6$ MT)
1. Tomato	34	11.6	85	14.3	9	11	41	12	14	12.5	80	38.4
2. Green leaf (excluding 7, 16, 20)	20	6.8	75	8.4	21	5	11	15	13	11.7	30	26.9
3. Fleshy cucurbits	16	5.5	85	6.7	3	8	9	19	14	12.1	30	24.3
4. Heading cabbage	19	6.5	85	8.0	2	6	12	7	7	5.7	40	20.2
5. Melon, watermelon	18	6.2	85	7.6	3	5	19	4	6	4.8	50	18.6
6. Onions and shallots (dry)	15	5.1	90	6.3	4	2	14	4	5	4.1	80	15.5
7. Lettuce	15	5.1	85	6.3	1	1	3	1	1	1.1	80	12.5
8. Cauliflower, broccoli	14	4.8	85	5.9	0	1	4	1	1	1.0	90	11.7
9. Eggplant	2	.7	90	.8	2	2	7	15	10	8.9	30	10.4
10. Leguminous pods	9	3.1	75	3.8	0	1	2	6	4	3.4	50	10.3
11. Carrot	11	3.8	85	4.6	1	1	6	1	2	1.3	80	9.7
12. Green peas	9	3.1	85	3.8	1	1	5	1	1	1.2	40	8.1
13. Roots, tubers (non-starchy) (excluding 11)	7	2.4	75	2.9	2	2	4	2	2	1.9	40	7.2
14. Shoots, sprouts, flowers, stalks	6	2.1	80	2.5	1	2	2	3	2	2.1	50	6.7
15. Peppers	3	1.0	90	1.3	6	7	6	2	4	3.4	40	5.7
16. Leaf cabbage	3	1.0	80	1.3	1	1	2	4	3	2.5	50	4.8
17. Okra	1	.3	65	.4	6	2	3	4	4	3.4	40	4.1
18. Green beans	3	1.0	85	1.3	2	1	2	2	2	1.7	40	4.0
19. Sweet corn	11	3.8	80	.1	0	0	0	0	0	.0	—	3.9
20. Green onions, leek	3	1.0	85	1.3	0	0	5	2	2	1.5	40	3.8
21a. Garlic	1	.3	95	.4	0	1	2	1	1	.7	70	1.4
21b. Mushrooms	1	.3	95	.4	1	0	1	1	1	.7	30	1.4
22. Various seeds (excluding 12, 18, 23)	0	.0	—	.1	3	0	2	1	1	1.2	30	1.3
23. Dry beans (for sprouting)	0	.0	—	.01	0	0	0	2	1	1.0	40	1.1
Total:	221	75.5	87	88.6	69	60	162	110	101	87.9	47	252.0

Source: G.J.H. Grubben, 1978.

largest and most vigorous plants are removed, the crop takes a few more days to reach harvestable maturity.

Losses of crop yields due to diseases and insect damage can be reduced or prevented by proper pest management and development of resistant and immune cultivars.

It is difficult to increase production on land that has been allowed to erode, where water is wasted in regions of drought, and where diseases and pests are uncontrolled. From erosion alone, it is estimated that 5% of the world's agricultural land [230 million ha (560 million acres) of arable land] has been lost. It will be centuries before enough weathering of the earth's rocks produce fertile soil again.

In India the accumulated losses of all grains in the field, storage, handling, and processing have been estimated to be as high as 50% in some instances; amounting to 22 million metric tons (MT) (24 million tons). The rodent population in this country is estimated at 2.4 billion, four rats for every man, woman, or child. Besides rodents, birds reduce the yield of crops. In Senegal in West Africa, a sparrow-like bird *(Quelea)* has been reported to consume about 800 MT (900 tons) of grain per month.

In many developing countries subsistence crops are almost entirely neglected while valuable exportable cash crops are efficiently cultured, harvested, and graded and protected against rats, insects, and decay. If equal care were given to crops for domestic use, the local food supply should greatly increase.

ENERGY CONSIDERATION IN CROP PRODUCTION

Present farm practices in developed countries depend on large expenditures of energy from fossil fuels. There are two ways in which this energy is used in food production: (1) direct use by tractors and power equipment in tillage, cultivation, water and pesticide applications, and harvest; (2) indirect uses such as manufacture of fertilizer, pesticides, farm equipment and machinery, and transportation of these supplies and materials for crop production.

Tables 1.4 and 1.5 show where the energy is used in the growing of potato and asparagus in North America, respectively. Such calculations are needed for other crops and other essential nutrients in order to determine which crops yield the most nutrients for the least amount of energy expended for production. Recent advances in cropping technology, such as integrated pest management, water economy of plants, and

TABLE 1.4. ENERGY RELATIONSHIPS IN POTATO PRODUCTION (PER HA)[a]

Inputs	Time or quantity	Energy equivalent (10^3 kcal)	% of total
Labor	108 hrs	19	0.3
Tractor and machinery investment[b]		225	3.9
Gasoline[c]	105 liter	862	15.0
Irrigation	3 ha-cm	415	7.0
Pesticides	15.5 kg	380	6.8
Fertilizer (10:10:10)	1335 kg	3832	66.8
Total		5733	100.0

Output = 27,800 kg/ha × 767 kcal/kg = 21,322,600 kcal/ha
Output/input = 3.72/1

Source: Lougheed *et al.* (1975).
[a]Mechanical energy equivalent of 175 kcal/hr.
[b]Proportion of energy cost of manufacture.
[c]Gasoline fuel equivalent, not including energy cost of refining or transport.

TABLE 1.5. ENERGY RELATIONSHIPS IN ASPARAGUS PRODUCTION (PER HA)

Inputs	Time or quantity	Energy equivalent (10^3 kcal)	% of total
Labor	230 hr	58	1.4
Tractor	12 hr	247	5.9
Gasoline	93 liter	988	23.6
Pesticides	3 kg	82	1.9
Fertilizer			
N	112 kg	2075	
P	112 kg	375	64.6
K	112 kg	259	
Herbicide	52 kg	109	2.6
Total		4193	100.0

Output = 1993 kg/ha × 246 kcal/kg = 490,278 kcal/ha
Output/input = 0.12/1

Source: Lougheed *et al.* (1975).

timing and efficiency of fertilizer usage, have reduced production costs in developed countries.

Figure 1.5 shows the food energy yield per unit of energy input in growing of several crops; Fig. 1.6 the protein yield with amount of energy expended on an area–time basis. Vegetables are poor in both categories.

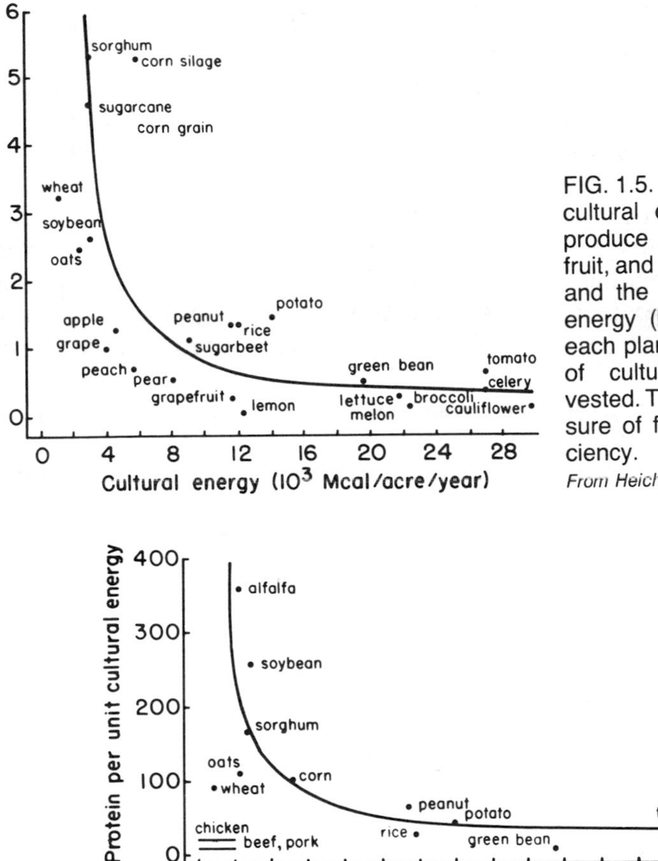

FIG. 1.5. The amount of cultural energy used to produce grain, forage, fruit, and vegetable crops, and the amount of food energy (in calories) that each plant yields per unit of cultural energy invested. The ratio is a measure of food energy efficiency.
From Heichel (1976).

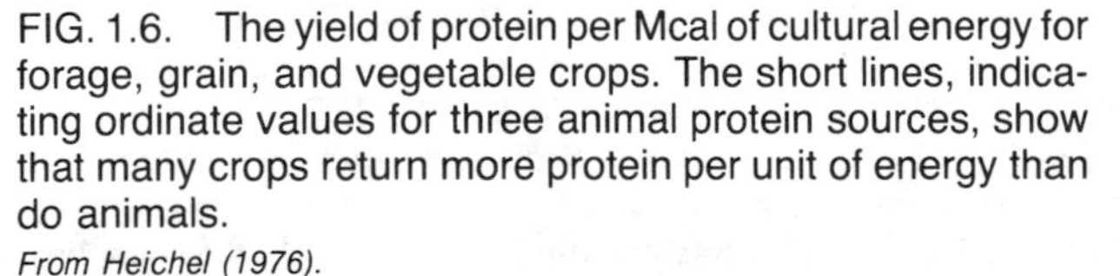

FIG. 1.6. The yield of protein per Mcal of cultural energy for forage, grain, and vegetable crops. The short lines, indicating ordinate values for three animal protein sources, show that many crops return more protein per unit of energy than do animals.
From Heichel (1976).

INTERNATIONAL TRADE OF VEGETABLES

Trading of vegetables, because of the highly perishable nature of the commodity, is very low. Vegetable export from the United States, both fresh and processed, amounted to 2.2% of the total agricultural commodities and the imports amounted to 5.5% in 1973. Much of the export is to Canada, Europe, and Japan, and the imports are from Mexico and other Latin American countries. Middle Eastern and North African countries ship fresh vegetables and fruits to Europe. Export of canned vegetables is important for Taiwan's economy.

Proper handling, especially refrigeration, is necessary to maintain the quality of fresh produce. Improved rapid transport on land, sea, and air and use of controlled atmosphere (CA) storage have steadily increased the trading of vegetables among countries in recent years. There has been increased demand for fresh vegetables by the more affluent countries during the winter months; the tropical and subtropical countries have been able to supply the demand. However, the quality of these imported vegetables sometimes need improvement. Grade standards and strict enforcement of grading of the commodities are desirable.

Although agricultural exports by developing countries can improve their economy, too often little effort is given to growing of crops for home consumption, resulting in nutritional deficiencies in the population.

BIBLIOGRAPHY

CHRISTENSEN, R.P. 1966. Protecting our food. *In* Man's Historic Struggle for Food. Yearbook of Agriculture.

ENNIS, W.B., Jr., DOWLER, W.M., and KLASSEN, W. 1975. Crop protection to increase food supplies. Science *188*, 593–598.

FAO. 1979. FAO Production Yearbook, Vol. 33. Food and Agric. Org., Rome, Italy.

FAO. 1981. Food loss prevention in perishable crops. Agric. Ser. Bull. No. 43. Food and Agric. Org., Rome, Italy.

GRUBBEN, G.J.H. 1978. Tropical vegetables and their genetic resources. International Board for Plant Genetic Resources. FAO-UN, Rome, Italy.

HEGSTED, D.M. 1978. Protein-calorie malnutrition. Am. Sci. *66*, 61–65.

HEICHEL, G.H. 1976. Agricultural production and energy resources. Am. Sci. *64*, 64–72.

HUBER, D.N., WARREN, H.L., NELSON, D.W., and TSAI, C.Y. 1977. Nitrification inhibitors—New tools for food production. Bioscience *27*, 523–529.

LOUGHEED, E.C., PROCTOR, J.T.A., ROWBERRY, R.G., TISSEN, H., SOUTHWELL, R.H., and RIEKELS, J.W. 1975. Fruit and vegetable production and the energy shortage. HortScience *10*, 459–462.

MACGILLIVRAY, J.H. 1956. Factors affecting the world's food supplies. World Crops, August, 303–305.

MACGILLIVRAY, J.H., and BOSLEY, J.B. 1962. Amino acids production by plants and animals. Econ. Bot. *16*, 25–30.

OOMAN, H.A.P.C. 1964. Vegetable greens, a tropical underdevelopment. Chron. Hortic. *4*, 3–5.

STELLY, M. (Editor) 1976. Multiple Cropping. Spec. Publ. No. 27, 378 pp. Am. Soc. Agron., Madison, Wisconsin.

UNIV. OF CALIF. FOOD TASK FORCE. 1974. A Hungry World: The Challenge to Agriculture. Summary Rep., Div. of Agric. Sci., Univ. of Calif., Davis.

WORTMAN, S. 1976. Food and agriculture. Sci. Am. *235* (Sept), 30–39.

2

Origin and Evolution of Vegetables

As of the mid-1800s nothing was known of the origin of our important economic crops. Their origin was said at that time to be an "impenetrable secret." However, Alphonse de Candolle's classic work, "Origin of Cultivated Plants," published in 1886 and, the more recent studies (1932) by Nikolai Vavilov, the Russian botanist, have contributed much in establishing the centers of origin of some of our more important crops.

To understand the origin and evolution of crops, it is necessary to consider the processes that occurred in the development of agriculture.

MAN BEFORE AGRICULTURE

Early man existed on this earth over two million years ago. He gathered all kinds of plants to supplement the food obtained from hunting and fishing. He followed a nomadic or seminomadic existence in his search for food. In areas where food supply was ample, he practiced a somewhat sedentary form of life.

Attempts to cultivate plants started about 8000–10,000 years ago, only in the last 5% of man's developmental history. Why did it take so long? Mainly, he lacked ideas to build on and little time to think about how to grow crops. His time and thoughts were spent mostly on survival. The tools developed were for survival rather than for planting. (Even today some of the primitive people in New Guinea use sharp sticks in harvesting yams and sweet potatoes.)

Conditions Necessary for Domestication of Plants according to Carl Sauer

Fire—The control of fire. Fire was used for cooking, warmth, and protection from predators. It was used to clear land for cultivation. (The "slash and burn" method for clearing land is still used today in the tropics.)

Temperate or subtropical climate with very marked wet and dry seasons (necessary for seed crops; see below).

River valleys were not used as they were subject to periodic and lengthy floods. Grasslands or savannas were too difficult to dig with crude implements, and grasses with rhizomes were impossible to control. Rain forests were too difficult to clear for crops. Burning was used to clear plots for growing. Open woodlands above river bottoms were used for plantings.

Agriculture started in areas with marked diversity of food plants and abundant supply of game. This allowed for permanence of the settlement. With ample food supply, leisure time was available to think and to experiment with growing of crops.

First Types of Cultivated Plants

The first cultivated crops, no doubt, came from plants gathered from the wilds.

Vegeculture

Vegetative propagation. This method of plant propagation is indigenous to the humid tropics and semitropical lowlands of the Americas, Southeast Asia, and Africa.

In the digging of wild starchy roots and tubers with sharp pointed sticks, all the roots or tubers were not gathered in the harvest, and many were left scattered about. In the high rainfall of this region, these started to grow and within several months were ready for harvest again. The area spread with each successive harvest, and thus small plots were unintentionally started.

Another scheme advocated by many anthropologists and plant geographers is: Some of the roots or tubers brought to the family or tribal shelters were dropped or discarded in clearings or in dumps. These rooted in the high moisture of the surroundings and grew rapidly because of the lack of competition and relatively high fertility from discarded foods as well as human wastes. These plants were allowed to grow and were protected from foraging animals. Much less effort was required to gather food from this source than to go into the wilds and search for the same plant.

Seed Culture

Evidence from archeobotanical studies indicates that the origin of cereal in both the Old and the New Worlds occurred in the mountainous regions of the subtropics having marked wet and dry seasons. Starchy seeds (cereals) were gathered for food during the dry season. Seeds were dropped or spilled near the dwellings. With the onset of the wet season, the seeds germinated and grew vigorously, from lack of competition and increased fertility of the area near the dwellings. The plants yielded well because the crop was protected from birds and other invading animals. After thousands of years of such means of harvesting unintentional plantings, there came the practice of deliberately placing vegetative parts or seeds into the soil. The domestication of plants resulted in the selection over the years of plants most desirable as food. For seed culture to be successful, proper time for planting, crop protection, harvesting, and storage were necessary.

Sites of Plant Domestication

Old World

Southeast Asia, probably Thailand, has been designated by Sauer as the "cradle of earliest agriculture" as this area meets the requirements given above. Much archaeological evidence also gives support. Vegeculture came first and seed culture came much later. The techniques developed spread to China, Southwest Asia, Africa, and southern Europe.

New World

Northern South America and southern Central America have been designated by Sauer as the "hearth" region. This occurred independently about 2000 or 3000 years after the start of plant domestication in the Old World. Migration of man to South America came through North America from northeastern Asia. There is no evidence of direct sea travel between the Old and New Worlds during this era.

Changes in Wild Plants during Domestication

Gigantism

Gigantism is nearly always exhibited by domesticated plants. The wild species are smaller and more slender. Cultivated plants, as a rule, have larger, broader, and thicker leaves, sturdier but fewer stems and stalks, larger flowers, and fruits. This increase in size may be from increase in cell size with no increase in cell numbers, or from increase in cell number with no increase in cell size, or an increase in both number and size. The frequent cause of gigantism is polyploidy, the doubling or multiplication of the basic chromosome number.

Seed

The following probably occurred: (1) Increase in size: large seed favors rapid and even germination and vigorous seedlings from ample reserve food supply. (2) Decrease in the number of seeds. (3) Reduced dispersibility: loss of ability to shatter. (4) Loss of dormancy and hard seed coat character. Hard seed coat prevents penetration of water in germination: this causes germination over long periods of time (uneven germination). Dormancy of seeds is a survival character.

Maturity

Early and concentrated maturation of fruits occurred. This character is advantageous to man. Extended ripening is advantageous to the plant in survival of the species.

Response to Temperature and Photoperiod

Biennial habit is desirable when the vegetative parts are harvested as food: flowering occurs only when proper temperature requirements have been met. Photoperiodism is a survival mechanism in which blossoming and, hence, seed production occur in a favorable season. This property considerably impedes the introduction of plants to regions of different day lengths (climate).

Changes in Shape

Perhaps the most dramatic changes in shape occurred in the cole crops (*Brassica oleracea*, cf. Chapter 19), From natural selections of this species evolved the kale, cabbage, cauliflower, broccoli, Brussels sprouts, and kohlrabi. These evolved over the course of several thousand years. Brussels sprouts originated in the nineteenth century.

Another is *Beta vulgaris* (cf. Chapter 24), which includes Swiss chard, table beets, mangels, and sugar beets. Swiss chard, which belong to the Cicla group, is grown for the enlarged leaves; the leaf had broad white or reddish petioles and broad bark green leaf blades. The table beet or beet root is grown mainly for the enlarged dark red usually globular shaped root; the root is mainly hypocotyl and a small portion is root tissue. When the plants are small, the leaves are harvested and used as greens. The sugar beet is grown for the sucrose in the enlarged taproot; the root is mainly root tissue and a small portion of the top part is hypocotyl tissue. Mangel or mangel-wurzel, from which sugar beet was derived, has white fleshy roots, which are used for stock feed.

Reduced or Complete Loss of Survival Ability

Reduction in the number of seeds produced, reduced or nonshattering of seeds, loss of seed dormancy, loss of impermeability of seed coat to water, etc., all contribute to reduction in the survival of the species. Also, selection for characters such as barbless and thornless, or reduction or absence of bitter and/or toxic substances, which are deterrent to foraging animals including man, are factors in plant survival.

Loss of Ancestral Form

Living plants resembling ancestral forms of a number of economic crops have been found in the wilds, but for many species none has been found. Evidence of prehistoric types are often discovered in archeological diggings in arid regions of the world.

Mechanics of Change from Wild to Cultivated Forms

The first cultivated types were from large populations and they were rather heterozygous. Saving of seeds from plants having the desirable qualities obtained by natural hybridization and recombinations in the field resulted in crops possessing desirable traits. It has been stated that this mechanism might be the most important source of variability on which the natural selection can operate.

Mutation is another mechanism in the transformation of wild plants into cultivated forms.

Gene mutations occur infrequently and most of the new characters are disadvantageous to the organism. This leads to their elimination. However, there are alterations which occasionally produce desirable qualities for man's use.

Through such mechanisms and natural crossings, plants gradually accumulated more of the valuable characters. This procedure is slow, taking hundreds to thousands of years, and make up only a minor part of the total variability of a species.

Purseglove's Observations

Plants taken from their original habitat are often much more successful in a new habitat. This is due to (1) lack of diseases and insects that attack the plant, and, (2) lack of competition.

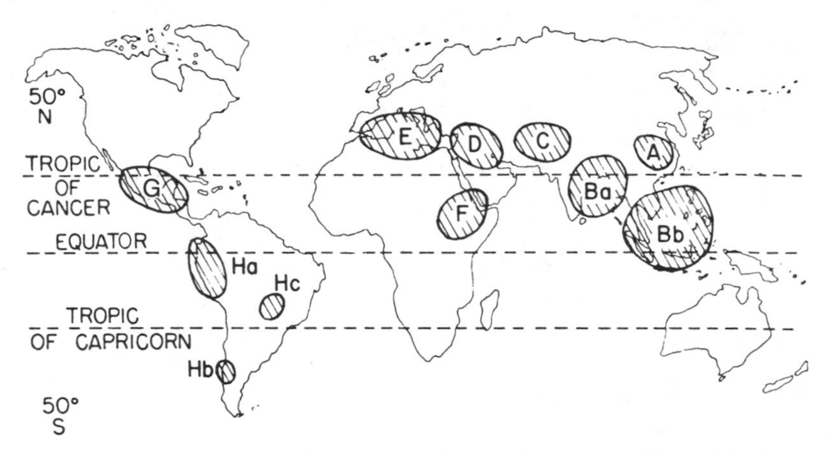

FIG. 2.1. Main centers of origin of cultivated plants according to Vavilov. See text for explanation of letters.

WORLD CENTERS OF ORIGIN OF SOME VEGETABLES ACCORDING TO VAVILOV (FIG. 2.1.)

A. Chinese center—mountians of central and western China and adjacent low lands.
   1. Soybean *(Glycine max)*
   2. Chinese yam *(Dioscorea batatas)*
   3. Radish *(Raphanus sativus)*
   4. Chinese cabbage *(Brassica campestris)* (Chinensis and Pekinensis groups)
   5. Onion (*Allium chinense* and *A. fistulosum*)
   6. Cucumber *(Cucumis sativus)*

B. Indian–Malaysian center
   a. Assam and Burma
      1. Mung bean *(Phaseolus aureus)*
      2. Cowpea *(Vigna sinensis)*
      3. Eggplant *(Solanum melongena)*
      4. Taro *(Colocasia esculenta)*
      5. Cucumber *(Cucumis sativus)*
      6. Yam *(Dioscorea alata)*
   b. Indo-Malayan center (Indo-China and Malay Archipelago)
      1. Banana *(Musa paradisiaca)* (starchy fruit used as vegetable)
      2. Breadfruit *(Artocarpus communis)*

C. Central Asiatic center—northwest India (Punjab and Kasmir), Afghanistan

1. Pea *(Pisum sativum)*
2. Horse bean *(Vicia faba)*
3. Mung bean *(Phaseolus aureus)*
4. Mustard *(Brassica juncea)*
5. Onion *(Allium cepa)*
6. Garlic *(Allium sativum)*
7. Spinach *(Spinacia oleracea)*
8. Carrot *(Daucus carota)*

D. Near-Eastern center—Asia Minor—Transcaucasia, Iran, and Turkmenistan
1. Lentil *(Lens esculenta)*
2. Lupine *(Lupinus albus)*

E. Mediterranean center—includes the borders of the Mediterranean Sea
1. Pea *(Pisum sativum)*
2. Garden beet *(Beta vulgaris)*
3. Cabbage (*Brassica oleracea*, capitata gp.)
4. Turnip (*Brassica campestris*, rapifera gp.)
5. Lettuce *(Lactuca sativa)*
6. Celery *(Apium graveolens)*
7. Chicory *(Chichorium intybus)*
8. Asparagus *(Asparagus officinalis)*
9. Parsnip *(Pastinaca sativa)*
10. Rhubarb *(Rheum officinale)*

F. Ethiopian (Abyssinian) center—Ethiopia and Somali Republic
1. Cowpea *(Vigna sinensis)*
2. Garden cress *(Lepidium sativum)*
3. Okra *(Hibiscus esculentus)*

G. South Mexican and Central American center—southern sections of Mexico, Guatemala, Honduras, and Costa Rica
1. Maize *(Zea mays)*
2. Common bean *(phaseolus vulgaris)*
3. Lima bean *(Phaseolus lunatus)*
4. Malabar gourd *(Cucurbita ficifolia)*
5. Winter pumpkin *(Cucurbita moschata)*
6. Chayote *(Sechium edule)*
7. Sweet potato *(Ipomoea batatas)*
8. Arrowroot *(Maranta arundinacea)*
9. Pepper *(Capsicum annuum)*

H. South American centers
   a. Peru, Ecuador, Bolivia
      1. Andean potato *(Solanum andigenum)*

2. Potato, white *(Solanum tuberosum)* (24 chromosomes)
3. Starchy maize (*Zea mays* amylacea)
4. Lima bean *(Phaseolus lunatus)* secondary center
5. Common bean *(Phaseolus vulgaris)* secondary center
6. Edible canna *(Canna edulus)*
7. Pepino *(Solanum muricatum)*
8. Tomato *(Lycopersicon esculentum)*
9. Ground cherry *(Physalis peruviana)*
10. Pumpkin *(Curcurbita maxima)*
11. Pepper *(Capsicum annuum)*

b. Chiloe center—Island near coast of Chile
   1. Potato, white *(Solanum tuberosum)* (48 chromosomes)

c. Brazilian–Paraguayan center
   1. Cassava *(Manihot esculenta)*

BIBLIOGRAPHY

BAKER, H.G. 1972. Human influences on plant evolution. Econ. Bot. *26*, 32–43.

DECANDOLLE, A. 1885. Origin of Cultivated Plants, Appleton, New York.

HARLAN, J.R. 1975. Crops and Man. Am. Soc. Agron., Crop Sci. Soc. Am., Madison, Wisconsin.

HARRIS, D.R. 1972. The origins of agriculture in the tropics. Am. Sci. *60*(2), 180–193.

HAWKES, J.G. 1970. The origin of agriculture. Econ. Bot. *24*, 131–133.

LOWENBERG, M.E., TODHUNTER, E.N., WILSON, E.O., SAVAGE, J.R., and LUBAWSKI, J.L. 1974. Food and Man, Second Edition, Chapter 1. John Wiley & Sons, New York.

MANGELSDORF, P.C., MACNEISH, R.S., and GALINAT, W.C., 1964. Domestication of corn. *Science 143*, 538–545.

SAUER, C.O. 1950. Cultivated plants of South and Central America. Handbook of South American Indians, Vol 6, pp. 487–433. Smithsonian Inst. Bur. Am. Ethnology Bull. 143. Washington, DC.

SAUER, C.O. 1952. Agricultural origins and dispersals. Am. Geo. Soc., New York.

SAUER, C.O. 1965. Cultural factors in plant domestication in the new world. Euphytica *14*, 301–306.

SCHWANITZ, F. 1966. The Origin of Cultivated Plants. Harvard University Press, Cambridge, Massachusetts. (Translated from German edition Die Entstehung der Kulturpflangen, 1957.)

SIMMONDS, N.W. Editor, 1976. Evolution of crop plants. Longmans-Green, London.

SMITH, C.E., Jr. 1968. The new world centers of origin of cultivated plants and archeological evidence. Econ. Bot. *22*, 253–266.

SMITH, C.E., Jr. 1969. From Vavilov to the present—A Review. Econ. Bot. *23*, 2–19.

VAVILOV, N.I. 1932. The origin, variation, immunity and breeding of cultivated plants. Chron. Bot. *13*, 1949–1950. (English Transl.)

WILSIE, C.P. 1962. Crop adaptation and distribution. W.H. Freeman and Co., San Francisco, California.

ZEVEN, A.C., and ZHUKOVSKY, P.M. 1975. Dictionary of Cultivated Plants and Their Centers of Diversity. Center for Agricultural Publishing and Documentation, Wageningen, Netherlands.

3

Vegetable Classifications

IMPORTANCE OF CLASSIFICATION

Some orderly method of grouping different vegetables is essential to catalog or systemize, to some extent, the voluminous information gathered by man since the dawn of agriculture. Such classification can present this material orderly and eliminate repetition of many of the principles related to culture and storage of the harvested crop.

Vegetables used throughout the world number in the several hundreds. In the United States alone there are over a hundred, including the minor crops. Therefore, some system of classification is essential.

BASIS FOR CLASSIFICATION

Methods that can be used for classification depend on its usefulness. Some of the methods or types used:

1. *Botanical classification* is based according to flower type and structure, and also on genetics and evolution. The groupings of plants are into families, genera, species, and varieties. This classification, based on the botanical relationship, is the most exact system.
2. *Optimum growing temperatures*, e.g., cool and warm season crops; temperate or tropical crops.
3. *Relative resistance of plants to frost or low temperatures.*
4. *Part of plant used for food*, e.g., foliage, stem, roots, flowers, or fruits.
5. *Number of seasons a plant may live*, e.g., annual, biennial, perennial.
6. *Storage temperature* and *storage life.*
7. *Optimum soil conditions*, e.g., soil acidity, and salt tolerance.
8. *Water requirements* to harvestable stage.

TYPES OF CLASSIFICATION OF VEGETABLES

Botanical Classification

All plants belong to one community (plant kingdom or community).

Division:

a. Algae and fungi (Thallophyta)
b. Mosses and liverworts (Bryophyta)
c. Ferns (Pteridophyta)
d. Seed plants (Spermatophyta)

Classes of seed plants (Spermatophyta):

a. Cone-bearing (Gymnosperm)
b. Flowering (Angiosperm)

Subclass of flowering plants (Angiosperm):

a. Monocotyledon (Monocotyledonae)
b. Dicotyledon (Dicotyledonae)

Order:
Family:
Genus:
Species:
Variety (botanical), and *Group:*
Cultivar (horticultural variety):
Strain (horticultural):

Example: Botanical classification of the summer squash cultivar, 'gray zucchini'

Division: Spermatophyta
Class: Angiospermae
Subclass: Dicotyledonae
Order: Cucurbitales
Family: Cucurbitaceae
Genus: Cucurbita
Species: *pepo* L.[1]
Variety: *Melopepo*, Alef.[1]
Cultivar: Zucchini
Strain: Gray

Definitions Used in Vegetable Classification

Group (Botanical Variety—Old Terminology). A population within a species of a cultivated crop which is distinct from the rest of the species

[1] L. is for C. Linnaeus, the person first suggesting the name; Alef. for F. G. C. Alefeld.

forms in one or more clearly defined characteristics; dwarf growth habit, enlarged taproot, etc. Group designation is used for horticultural convenience and has no botanical recognition.

Cultivar (Horticultural Variety). A cultivar denotes an assemblage of cultivated individuals which are distinguished by any character (morphological, physiological, cytological, chemical, etc.) significant for the purpose of agriculture or horticulture and which retain their distinguishing features when reproduced (either sexually or asexually). When naturally occurring populations are also represented by cultivars, the botanical variety name is retained. The cultivar names are set off in single quotation marks ('zucchini').

Strain A strain includes those plants of a given cultivar which possess the general varietal characteristics but differ in some minor characteristics or qualities. A cultivar with disease resistance incorporated, or early maturation, may be considered a strain within the cultivar. Selections within a cultivar for differences in climatic adaption may be considered a strain. The international code for nomenclature does not recognize the term strain. Any selection that shows sufficient differences from the parent is regarded as a distinct cultivar.

Table A.1 lists the botanical classifications of many of the more common vegetables of the world.

Usefulness of Botanical Classification

For biologists to (a) establish relationships and origin, and (b) serve as positive identification, regardless of language.

For horticulturists because (a) climatic requirements of a particular family or genus are usually similar, (b) use of crop for economic purposes is similar, and (c) disease and insect controls are quite often similar for related genera.

CLASSIFICATION ACCORDING TO SEASON GROWN (after J. H. MacGillivray)[2]

1. *Cool Season Crops*—Adapted to mean monthly temperatures of 16°–18°C (60°–65°F).

artichoke	celery	onion
asparagus	chard	parsnip
Brussels sprouts	endive	pea
broccoli	garlic	radish
cabbage	kale	spinach

[2] From J.H. MacGillivray. 1953. Vegetable Production. Used with permission of McGraw-Hill Book Co., New York.

carrot	lettuce	turnip
cauliflower	mustard	white potato

2. *Warm Season Crops*—Adapted to mean monthly temperatures of 18°–30°C (65°–86°F). Intolerant of frost.

cucumber	okra	sweet corn
eggplant	pepper	sweet potato
lima beans	snap bean	tomato
muskmelon	squash and pumpkin	watermelon

This classification is based on temperature zone conditions and even then must be used with some caution. In tropical zones where temperatures are quite uniform, group differences are much less clear.

CLASSIFICATION OF VEGETABLES BASED ON USE, BOTANY, OR A COMBINATION OF BOTH

1. Potherbs or greens

spinach	kale
New Zealand spinach	mustard
chard	collards
dandelion	water convulvulous

2. Salad crops

celery	chicory
lettuce	watercress
endive	

3. Cole crops (all are members of *Brassica oleracea* except Chinese cabbage)

cabbage	Brussels sprouts
cauliflower	kohlrabi
sprouting broccoli	Chinese cabbage

4. Root crops (refers to crops which have a fleshy taproot)

beet	turnip
carrot	rutabaga
parsnip	radish
salsify	celeriac

5. Bulb crops (all species of *Allium*)

onion	garlic
leek	shallot
Welsh onion	chive

6. Pulses

peas
beans (including dry-seeded or agronomic forms)

7. Cucurbits (all members of the Cucurbitaceae)
   - cucumber
   - muskmelon
   - watermelon
   - pumpkin
   - squash
   - several Oriental crops
8. Solanaceous fruits (members of the Solanaceae)
   - tomato
   - pepper
   - eggplant
   - husk tomato
9. white (Irish) potato
10. sweet potato
11. sweet corn

CLASSIFICATION BY EDIBLE PART

1. Root
   a. Enlarged taproot
      - beet
      - carrot
      - radish
      - salsify
      - rutabaga
      - turnip
      - parsnip
      - celeriac

   b. Enlarged lateral root
      - sweet potato
      - winged bean
      - cassava
      - arracacha
2. Stem
   a. Above ground, not starchy
      - asparagus
      - celtuce
      - kohlrabi

   b. Below ground, starchy
      - white or Irish potato
      - yam
      - Jerusalem artichoke
      - taro
3. Leaf
   a. onion group, leaf bases eaten (except chive)
      - onion
      - leek
      - shallot
      - garlic
      - chive

   b. Broad-leaved plants
      1. salad use
         - lettuce
         - cabbage
         - chicory
         - Chinese cabbage
         - celery (petiole only)
         - endive
      2. cooked (may include tender stems in some)
         - spinach
         - edible amaranth

chard
New Zealand spinach
Jew's mallow
dandelion
rhubarb (petiole only)
kale
chicory
Chinese cabbage
mustard
cardoon (petiole only)

4. Immature flower bud
   cauliflower
   broccoli
   broccoli raab
   artichoke
5. Fruit
   a. immature
      pea
      bean, snap
      bean, lima
      bean, broad
      chayote
      summer squash
      cucumber
      zucca melon
      okra
      sweet corn
      eggplant
   b. mature
      i. gourd family (cucurbits)
         pumpkin and winter squash
         muskmelon
         Chinese wax gourd
         watermelon
      ii. potato family
         tomato
         pepper
         pepino
         husk tomato

CLASSIFICATION OF VEGETABLES ACCORDING TO SALT TOLERANCE[3] (RICHARDS, 1954)

Listed from high tolerance (top) to low tolerance (bottom)

7700 ppm (EC $\times 10^3 = 12$)

1. High salt tolerance
   garden beets
   kale
   asparagus
   spinach

6400 ppm (EC $\times 10^3 = 10$)

2. Medium salt tolerance
   tomato

[3] Salt concentration based on saturation extracts of soils; EC=electrical conductivity.

broccoli
cabbage
peppers
cauliflower
lettuce
sweet corn
white potato
carrot
onion
peas
squash
cucumber
cantaloupe

2600 ppm (EC $\times 10^3=4$)

3. Low salt tolerance
   radish
   green beans

1900 ppm (EC $\times 10^3=3$)

CLASSIFICATION OF VEGETABLES ACCORDING TO TOLERANCE TO SOIL ACIDITY

1. Slightly tolerant (pH 6.8 to 6.0)

   asparagus
   beet
   broccoli
   cabbage
   cauliflower
   celery
   spinach
   Chinese cabbage
   leek
   lettuce
   muskmelon
   New Zealand spinach
   okra
   onion
   spinach

2. Moderately tolerant (pH 6.8 to 5.5)

   bean
   Brussels sprouts
   carrot
   cucumber
   eggplant
   garlic
   horseradish
   kohlrabi
   parsley
   pea
   pepper
   pumpkin
   radish
   squash
   tomato
   turnip

3. Very tolerant (pH 6.8 to 5.0)

   chicory
   endive
   potato
   rhubarb
   sweet potato
   watermelon

CLASSIFICATION BY ROOT DEPTH INTO SOIL

1. Shallow < 80 cm (3 ft)

cabbage	potato
lettuce	spinach
onion	sweet corn

2. Medium 80–160 cm (3–6 ft)

beans	eggplant
beets	summer squash
carrot	peas
cucumber	

3. Deep > 160 cm (6 ft)

artichoke	sweet potato
asparagus	tomato
melon	winter squash and pumpkin

CLASSIFICATION BY HABITAT

Hydrophyte (aquatic)	*Mesophyte*	*Xerophyte*
taro	most vegetables grown on soil and requiring moderate amounts of water	cactus, some desert cucurbits (buffalo gourd)
water chestnut		
watercress		
lotus		
water convulvulus		

BIBLIOGRAPHY

BAILEY, L.H. 1949. Manual of Cultivated Plants. Macmillan Co., New York.

BAILEY, L.H., and BAILEY, E.Z. 1976. Hortus Third. Macmillan Co., New York.

GILMOUR, J.S.L. 1969. International Code of Nomenclature of Cultivated Plants. International Bureau of Plant Taxonomy and Nomenclature. Utrecht, Netherlands.

MacGILLIVRAY, J.H. 1953. Vegetable Production. McGraw-Hill Book Co., New York.

RICHARDS, L.A. Editor. 1954. Diagnosis and Improvement of Saline and Alakali Soils. USDA Handb. No. 60, U.S. Dept. Agric., Washington, D.C.

SMITH, P.G., and WELCH, J.E. 1964. Nomenclature of vegetables and condiment herbs grown in the United States. Am. Hort. Sci. Proc. *84*, 535—548.

THOMPSON, H.C. and KELLEY, W.C. 1957. Vegetable Crops. McGraw-Hill Book Co., New York.

4

Importance of Vegetables in Nutrition

Man has and always will be interested in food. The intake of proper nutrients is necessary for his well-being or there will be dire body dysfunctions. Proper nutrition means intake of many different kinds of foods with proper balance and in optimal amounts. The required amounts of nutrients can vary with sex, age, body size, activity, and the genetic and biochemical makeup of the individual. The environmental temperature is important in energy considerations. In the female, pregnancy and lactation can modify the requirements. All foods ingested supply a portion of the nutrients required.

The term *undernourished* is used to designate those people lacking or having inadequate caloric intake. The term *malnourished* is used to designate people lacking minimum daily intake of proteins including essential amino acids, vitamins, and/or essential mineral elements. Undernourished people are likely to be also malnourished.

Marasmus is a nutritional disease caused by partial starvation from chronic calorie deficiency. Symptoms of the disorder are atrophy of limb muscles and emaciated body; a shriveled "old man face" is characteristic of the disease.

Undernourishment and malnourishment can cause increased susceptibility to infection and disease and the stunting of growth in children, not only physical but mental as well since the brain and neurological development is adversely affected. Malnutrition during the development of the fetus and during infancy has a decidedly harmful and, from all indications, an irreversible effect on the brain and neurological development, resulting in lower basic intelligence.

NUTRIENTS IN FOODS

Water

Water, other than some dissolved minerals, may not be considered a nutrient but it is required to maintain the osmotic balance in the cells and to carry the metabolic waste products from the body.

The daily intake of water:

From eaten food—1100 ml (2½ pints)
From liquid drink—1100 ml (2½ pints)
From metabolic water—250 ml (½ pint)

The daily output of water:

From wastes—1350 ml (3 pints)
From evaporative loss (from skin and lungs) 1100 ml (2½ pints).

Metabolic water may be derived from carbohydrates, proteins, and fats:

1 g of starch = 0.60 g of water
1 g of protein = 0.41 g of water
1 g of fat = 1.07 g of water

Caloric or Energy Foods

Generally, all foods ingested have some energy potential as they are broken down in the body and converted into energy. It was established in 1899 by Atwater and Benedict that man required about 2400–2500 kcal/day. The heat of combustion from calorimeter and available metabolic energy differ slightly.

	Heat of combustion (kcal/g)	Metabolic energy (kcal/g)
Carbohydrate	4.1	4.0
Fat	9.4	9.0
Protein	5.6	4.0
Alcohol	7.1	ca 7.1

Depending on the concentration of alcohol in the blood, alcohol can contribute up to about 65% of the basal metabolism.

Carbohydrates and Related Substances

Carbohydrates contain carbon, hydrogen and oxygen, in the proportion of $C(H_2O)$. In plants the main carbohydrates of nutritional importance are starches and sugars, largely, sucrose, glucose, and fructose. There are also small quantities of other sugars, tetra-, tri-, di-, and monosaccharides, but except for few plants, these are less important.

The Compositae (sunflower family) is known to contain large amount of fructosans, which are polymers of fructose. Also, many of the monocots, particularly the alliums (members of onion family), are rich in polymers of fructose.

Inulin, a polymer of fructose is found in some plants as the main storage carbohydrate. Jerusalem artichoke is such an example. This carbohydrate is not metabolized by humans.

Cellulose

Cellulose is a polymer of glucose not utilizable by man for energy. Certain animals, the ruminants, can utilize cellulose because of the presence of microbial flora in their digestive tracts.

Fibers

Fibers are composed of celluloses, hemicelluloses, pentosans, pectins, mixtures of polysaccharides, and lignins, which affect the texture of plant foods. Fibers are considered important as "roughage" in digestion and in regulation of bowel movements.

Lipids: Fats and Oils

Adequate diets should have about 15% of the total calories derived from lipids. Fats are substances that are either liquids or solids at room temperature and are made of glycerol (glycerine) and fatty acids. Whether the fat is a solid or liquid depends on the number of carbon atoms in the fatty acid and degree of saturation with hydrogen atoms.

Fats from vegetable sources are generally oils and are less saturated than animal fats, which are in the solid form at room temperature.

In recent years the blood cholesterol level has been of much concern in human health. High blood pressure and heart disease have been related to excess intake of dietary fats. Fats, particularly the highly unsaturated ones from plant sources, reportedly lowers the blood cholesterol levels while fats from animal sources have the opposite effect.

The controversy of "essential fatty acids," linoleic, linolenic, and arachidonic acids has not been resolved. The significances of these fatty acids in human physiology is not certain. However, they undoubtedly act as precursors of prostaglandins, thought to be active in the control of blood pressure, smooth muscle contraction, and in the induction of enzymes and hormones.

Proteins and Amino Acids

The main function of proteins is to serve as the building blocks of the body cells; also they are part of enzymes necessary to carry out the body functions.

Proteins are digested in the intestinal tract and broken down into amino acids. There are some 36 naturally occurring amino acids and 22 of these are part of the building units of proteins. Of these, eight are essential in man; they are not synthesized by the body cells. Additionally, young adult females require cysteine and tyrosine while infants require supplemental histidine.

It is not sufficient that the quantitative requirements for total proteins be met for adequate nutrition. Ideally, the protein should contain the essential amino acids in the proper proportion so that they are utilized most effectively. This is *protein quality*. The nonessential amino acids can be converted by biochemical processes in the humans to make up the deficit in any of the nonessential ones.

Net protein utilization (NPU) values are used to index the nutritional value of proteins. They are the percentage of amino acids ingested as proteins which are retained and incorporated into body proteins. The NPU of egg protein is over 90%, milk and fish 80%, and meats range from 60 to 70%. Vegetables range from 50 to 70% whereas the mature seeds of legumes range from 30 to 60%.

Another measure of the biological value of proteins is the determination of *protein efficiency ratio* (PER). This is obtained by feeding rats a protein from a test source and measuring the weight gained per unit of protein fed. PER = weight gain/unit of protein fed. The standard is casein, which has a PER of 2.5. Vegetable protein has a PER of 2.0 while meat is rated at 3.0 and hen's egg protein is 3.4.

Protein quality is important in nutrition in that if one essential amino acid is limiting, protein synthesis cannot continue because that particular amino acid is necessary in building of the particular enzyme or structural protein. Protein from animal sources are better in protein quality than those from plant sources. Vegetable protein tend to be low in methionine. Corn is low in lysine. A new high lysine corn variety has been introduced. If one is selective in obtaining protein from various vegetable sources, a good balance of the essential amino acids can be achieved. Table 4.3 on protein quality shows the essential amino acid balance in some vegetables in reference to hen's egg.

Kwashiorkor is a nutritional disorder caused by protein–calorie malnutrition. Infants and children between the ages of 6 months to 6 years can suffer irreparable brain damage from lack of proteins in their diets.

Vitamins

Vitamins are very potent organic substances necessary for normal body functions. They must be supplied as nutrients to the animal or as precursors for synthesis by the animal. The body stores fat-soluble vitamins but water-soluble vitamins are not stored in significant amounts.

Fat-Soluble Vitamins

Vitamin A is essential for vision in dim light. Deficiency causes abnormally dry malformed skin (epithelial cells). Retinol is found in foods of animal origin. α and β-carotene are precursors of vitamin A found in abundance in green plants.

Cryptoxanthin and α- and β-carotene epoxides, found in small amounts in some plants, also have vitamin A activity. One IU of vitamin A = 0.30 μg retinol or 0.60 μg β-carotene or 1.20 μg α-carotene.

Vitamin D is the antirickets vitamin. This vitamin is principally from animal sources. Cereals, fruits and vegetables contain no vitamin D. Exposing ergosterol to ultraviolet light produces vitamin D_2, ergocalciferol. Vitamin D_3, cholecalciferol, is produced by irradiating 7-dehydrocholesterol, a sterol found in animal fats.

The tocopherols (α, β, and γ) all have the same physiological activity as *vitamin E*. They are antioxidants and seem to have control of metabolic rates. Deficiency affects the reproductive mechanisms. The richest sources are the fats of vegetable origin principally in the seed and oils of corn, soybeans, peanut, coconut, and cotton. Cereals and eggs are also high in the tocopherols. Animal fats and meats, and also fruits and vegetables, are low in this vitamin.

Vitamin K is important in the coagulation of blood. It is a naphthoquinone. Vitamin K_1, phytomenadione, is present in fresh dark green vegetables such as kale and spinach; alfalfa is also rich in this vitamin. Foods of animal sources are very poor sources for this vitamin unless they have undergone bacterial decomposition. Vitamin K was first discovered in putrified fish. *Escherichia coli*, the bacterial organism in the colon of humans, synthesizes vitamin K. Except in newborn infants, normally enough is absorbed from this process to supply the needs.

Water-Soluble Vitamins

Vitamin B_1 (thiamin) prevents beriberi. It is a cofactor in carbohydrate metabolism. The lack of this vitamin prevents the normal metabolism of pyruvic acid in the respiratory cycle. Pyruvic acid is an intermediate compound in carbohydrate metabolism. The brain and nerve cells derive their energy mainly from carbohydrates and therefore are first to be affected by B_1 deficiency. Both plant and animal foods are good sources; the germ of seeds are especially rich. All fruits and vegetables contain this vitamin.

Vitamin B_2 (riboflavin) is necessary for energy production in the cells. It is part of the enzyme system in the conversion of foods to chemical energy. Most foods contain varying amounts of this vitamin; there may be some deficiencies in some parts of the world but they are minor. Wheat germ, meats, fish, eggs, cheese, whole grain, pulses, and leafy vegetables are considered high in riboflavin. Highly milled grains, fruits, and most vegetables are considered low.

Niacin (nicotinic acid) prevents pellagra, a disease with disorders of the skin, gastrointestinal tract, and of the central nervous system. Nicotinic acid is part of the coenzyme system in cellular metabolism. It is required in many synthetic reactions. Therefore, all living cells contain nicotinic acid. Wheat bran and germ, meats, and yeasts are very high in this vitamin. Eggs, milk, fruits, and vegetables are low in niacin.

Vitamin C (ascorbic acid) prevents scurvy. Lack of this vitamin can cause anemia, poor wound healing, and connective tissue formation. Citrus fruits and green vegetables are high in vitamin C. Rose hips, acerola fruits, and peppers are very high in ascorbic acid content. Acerola *(Malpighia glabra)*, a native to tropical and subtropical America, is a small cherry-like fruit, referred to often as Barbados, West Indian, or Puerto Rican cherry; it contains over 2000 mg ascorbic acid per 100 g. Tomato juice contains from 15 to 20 mg, which is considered a good source of this vitamin. Although potatoes are rather low in vitamin C, sufficient amounts are eaten in the northern European countries to supply the people's requirement. Synthetic ascorbic acid is very inexpensive and is added as a food supplement.

Vitamin B_6 (pyridoxine) is needed for normal metabolism of tryptophan to nicotinic acid. It has a role in the metabolism in nervous tissue and also in anemia. The compound is a prosthetic group of enzymes. Pyridoxine and related compounds, pyrodixal and pyridoxamine, are widely distributed in both plants and animals. Meats, liver, the bran of cereals, and vegetables are high in this vitamin.

Vitamin B_{12} (cyanocobalamin) is important in anemia prevention. The blood forming tissues of the bone marrow and the gastrointestinal tract are affected by the lack of this vitamin. Vitamin B_{12} is found in animal products; it is also produced by intestinal microorganisms. Plants do not contain this vitamin.

Folic acid (tetrahydrofolic acid) is needed in the synthetic processes of the cell, particularly in nucleic acid and protein syntheses. Liver, oysters, spinach, broccoli, cabbage, and lettuce are good sources. Milk, chicken, pork, and fruits are poor sources.

Pantothenic acid is a constituent of the enzyme system, coenzyme A, present in all living cells. Its distribution is universal so a deficiency of this vitamin is rather unlikely.

Biotin is part of several enzyme systems, particularly in fatty acid synthesis. Liver, kidney, and yeast are good sources. Legume seeds, nuts, and some vegetables (cauliflower) are fair sources. Cereals, meats, and dairy products are poor sources.

Choline has been reported to be a precursor of acetylcholine. All foods containing phospholipids are good sources of choline.

Minerals

Macroelements

Sodium, potassium, and chlorine are part of the system of the blood and tissues of the body, necessary in maintaining the osmotic balance of body fluids.

Calcium, magnesium and phosphorus are important in the blood and skeletal system. Phosphorus is important in many enzyme reactions where transfer of energy is concerned. Strontium may replace Ca in the skeletal system because of its resemblance to Ca.

Sulfur (S) is an element in proteins. It plays an active role in enzyme systems. Iron (Fe) is the metal in hemoglobin of the red blood cells. Iron is also important in the energy transfer system in respiration (cytochromes).

Microelements

Iodine (I) is necessary for thyroid activity.

Copper (Cu) is metal ion necessary for tyrosinase activity; lack of copper impairs iron metabolism, causing anemia.

Zinc is a component of at least eight enzyme systems.

Cobalt is necessary for vitamin B_{12} activity.

Molybdenum is a metal ion in an enzyme, xanthine oxidase. Molybdenum deficiency has not been shown as yet in man.

Manganese is a cofactor in many enzyme reactions.

Fluorine prevents dental decay; excess causes loss of luster and roughness of teeth, also bone disorders.

Selenium is necessary for proper vitamin E function required for glutathione peroxidase activity. It has been reported that an Se deficiency during pregnancy causes cystic fibrosis in infants.

Chromium is involved in glucose and energy transformations.

Recommended Daily Allowances

The U.S. Recommended Daily Allowance for proteins and vitamins are shown in Table 4.1. The amounts of essential amino acids are given in Table 4.2.

FACTORS AFFECTING AMOUNTS OF NUTRIENTS IN VEGETABLES

Cultivar or Genetic Makeup of Plant

If there is a wide variation in the content of any nutrient in a given population grown under the same conditions, then it is possible by breeding and selection to increase the content of that nutrient greater than the mean of the original population.

Environment in Which Crop Is Grown

The following conditions are important:

1. Seasonal factors: temperature, moisture, and light
2. Atmospheric condition and composition
3. Soil factors: soil chemical and physical properties; moisture
4. Cultural practices: fertilization, pests and diseases, competition, and maturity

LOSSES AND STABILITY OF NUTRIENTS

Harvest

Losses during this process can vary. Much depends on the particular vegetable and its resistance to bruising and breaking of cells. With mechanical harvesting more damage is likely to occur than if the crop is harvested by hand. Additional mechanical damage can occur during transport.

Holding and Storage prior to Processing

Time and temperature are important factors. The retention of ascorbic acid, a very labile vitamin, has been used as an index for proper storage conditions.

TABLE 4.1. FOOD AND NUTRITION BOARD, NATIONAL ACADEMY OF SCIENCES–NATIONAL RESEARCH COUNCIL RECOMMENDED DAILY DIETARY ALLOWANCES,[a] Revised 1980

		Weight		Height			Fat-soluble vitamins			Water-soluble vitamins							Minerals					
	Age (years)	(kg)	(lb)	(cm)	(in)	Protein (g)	Vitamin A (μg RE)[b]	Vitamin D (μg)[c]	Vitamin E (mg α-TE)[d]	Vitamin C (mg)	Thiamin (mg)	Riboflavin (mg)	Niacin (mg NE)[e]	Vitamin B_6 (mg)	Folacin[f] (μg)	Vitamin B_{12} (μg)	Calcium (mg)	Phosphorus (mg)	Magnesium (mg)	Iron (mg)	Zinc (mg)	Iodine (μg)
Infants	0.0–0.5	6	13	60	24	kg × 2.2	420	10	3	35	0.3	0.4	6	0.3	30	*0.5*[g]	360	240	50	10	3	40
	0.5–1.0	9	20	71	28	kg × 2.0	400	10	4	35	0.5	0.6	8	0.6	45	1.5	540	360	70	15	5	50
Children	1–3	13	29	90	35	23	400	10	5	45	0.7	0.8	9	0.9	100	2.0	800	800	150	15	10	70
	4–6	20	44	112	44	30	500	10	6	45	0.9	1.0	11	1.3	200	2.5	800	800	200	10	10	90
	7–10	28	62	132	52	34	700	10	7	45	1.2	1.4	16	1.6	300	3.0	800	800	250	10	10	120
Males	11–14	45	99	157	62	45	1000	10	8	50	1.4	1.6	18	1.8	400	3.0	1200	1200	350	18	15	150
	15–18	66	145	176	69	56	1000	10	10	60	1.4	1.7	18	2.0	400	3.0	1200	1200	400	18	15	150
	19–22	70	154	177	70	56	1000	7.5	10	60	1.5	1.7	19	2.2	400	3.0	800	800	350	10	15	150
	23–50	70	154	178	70	56	1000	5	10	60	1.4	1.6	18	2.2	400	3.0	800	800	350	10	15	150
	51 +	70	154	178	70	56	1000	5	10	60	1.2	1.4	16	2.2	400	3.0	800	800	350	10	15	150
Females	11–14	46	101	157	62	46	800	10	8	50	1.1	1.3	15	1.8	400	3.0	1200	1200	300	18	15	150
	15–18	55	120	163	64	46	800	10	8	60	1.1	1.3	14	2.0	400	3.0	1200	1200	300	18	15	150
	19–22	55	120	163	64	44	800	7.5	8	60	1.1	1.3	14	2.0	400	3.0	800	800	300	18	15	150
	23–50	55	120	163	64	44	800	5	8	60	1.0	1.2	13	2.0	400	3.0	800	800	300	18	15	150
	51 +	55	120	163	64	44	800	5	8	60	1.0	1.2	13	2.0	400	3.0	800	800	300	10	15	150
Pregnant						+30	+200	+5	+2	+20	+0.4	+0.3	+2	+0.6	+400	+1.0	+400	+400	+150	[h]	+5	+25
Lactating						+20	+400	+5	+3	+40	+0.5	+0.5	+5	+0.5	+100	+1.0	+400	+400	+150	[h]	+10	+50

[a] Designed for the maintenance of good nutrition of practically all healthy people in the United States. The allowances are intended to provide for individual variations among most normal persons as they live in the United States under usual environmental stresses. Diets should be based on a variety of common foods in order to provide other nutrients for which human requirements have been less well defined.

[b] Retinol equivalents. 1 retinol equivalent = 1 μg retinol or 6 μg β-carotene. Calculation of vitamin A activity of diets as retinol equivalents.

[c] As cholecalciferol, 10 μg cholecalciferol = 400 IU of vitamin D.

[d] α-Tocopherol equivalents. 1 mg *d*-α tocopherol = 1 α-TE

[e] 1 NE (niacin equivalent) is equal to 1 mg of niacin or 60 mg of dietary tryptophan.

[f] The folacin allowances refer to dietary sources as determined by *Lactobacillus casei* assay after treatment with enzymes (conjugases) to make polyglutamyl forms of the vitamin available to the test organism.

[g] The recommended dietary allowance for vitamin B_{12} in infants is based on average concentration of the vitamin in human milk. The allowances after weaning are based on energy intake (as recommended by the American Academy of Pediatrics) and consideration of other factors, such as intestinal absorption.

[h] The increased requirement during pregnancy cannot be met by the iron content of habitual American diets nor by the existing iron stores of many women; therefore the use of 30–60 mg of supplemental iron is recommended. Iron needs during lactation are not substantially different from those of nonpregnant women, but continued supplementation of the mother for 2–3 months after parturition is advisable in order to replenish stores depleted by pregnancy.

TABLE 4.2. MINIMUM REQUIREMENTS OF ESSENTIAL AMINO ACIDS: NATIONAL RESEARCH COUNCIL (g/day)

Amino acid	Young male adult	Young female adult	Infant of 15 lb wt
Leucine[a]	1.10	0.62	1.05
Isoleucine[a]	0.70	0.45	0.83
Lysine[a]	0.80	0.50	0.72
Threonine[a]	0.50	0.31	0.61
Tryptophane[a]	0.25	0.16	0.15
Valine[a]	0.80	0.65	0.74
Methionine[a]	1.10	0.29	0.32
Cysteine		0.22	
Phenylalanine[a]	1.10	0.22	0.63
Tryosine		0.90	
Histidine			0.24

[a] Essential amino acid.

Washing prior to Processing

This procedure has little effect on nutrient retention. If the washing water is recycled and gets warmer and detergents are used, some water-soluble substances can be leached. Contamination with trace metals can occur. Pesticides can be removed by washing.

Peeling and Chopping

Losses of nutrients occur in peeling. Peeling with mechanical equipment or by use of caustic soda may remove nutrients concentrated in the skin. Vitamin C is reported to be concentrated just under the skin of the tomato. Chopping or mincing of plant tissues can cause losses of vitamin C by oxidation; addition of acids to prevent oxidation reduces the loss of this vitamin.

Blanching

This step is used to inactivate enzymes that produce undesirable effects on the product. Blanching is usually by steam or hot water. Significant nutrient losses, especially the water-soluble vitamins as well as minerals, can occur when hot water blanching is used.

Processing

Preservation can be by (1) freezing, (2) dehydration, (3) canning, or (4) pickling. The greatest nutrient loss occurs with thermal processing. A new process of high temperature treatment for a short time has increased vitamin retention.

Packaging and Storage

The container, storage temperature, and time affect nutrient retention. Generally, low or no oxygen (N_2 atmosphere), vacuum, low temperature, and darkness are the best conditions for nutrient retention. Low moisture is important in dehydrated foods. The quality decreases with increases in storage time and temperature.

VEGETABLES AS A SOURCE OF NUTRIENTS

Fresh leafy vegetables alone are not enough to satisfy the daily nutrient requirements. They are high in water and low in dry matter, so that large amounts must be ingested to supply the daily requirement of many nutrients. Carrots, for example, are eaten because they add color to foods and for taste but also, because they are an excellent source of provitamin A. Only 45 g, (slightly under 2 oz.) supply the minimum daily requirement for vitamin A. But, one would have to eat 3 kg (over 6½ lb) of carrots in order to supply the daily protein requirement! However, people who eat nothing but food from plant sources do live and grow just as well as those who eat meat. They can achieve this by eating foods from plant sources containing high amounts of proteins. A proper balance of all nutrients can be obtained from plants alone, but people in the depressed areas and in many developing countries are not informed on how to obtain this balance.

Vegetables can be grouped as sources of nutrients:

High carbohydrates

- white potato
- sweet potato
- cassava
- dry beans
- taro
- yam

High in oils

- legume seeds (soybean)
- mature vegetable seeds contain much oil

High proteins and amino acids

- beans and peas (legumes)
- sweet corn
- most leafy vegetables; crucifers are especially high

High in vitamin A value (carotenes)

- carrot (yellow or orange fleshed)
- sweet potato (orange or yellow fleshed)
- cucurbits (yellow or orange fleshed)
- peppers
- green leafy vegetables
- green beans and green peas

TABLE 4.3. ESSENTIAL AMINO ACID COMPOSITION OF SOME VEGETABLES IN COMPARISON TO HEN'S EGG AND THE LIMITING AMINO ACID DETERMINING PROTEIN SCORE

	(% of FAO hen's egg)							
	Trypto-phan	Threo-nine	Isoleu-cine	Leucine	Lysine	Methio-nine plus cystine	Phenyl-anine plus tryosine	Valine
Collards	88	58	47[a]	64	81	49[a]	71	68
Kale	68	70	51	74	49	34[a]	—	64
Spinach	102	87	70	87	96	67[a]	74	75
Potatoes	67	77	66	57	83	40[a]	62	73
Sweet potatoes	109	92	73	65	74	63[a]	101	102
Common beans	58	85	86	98	116	37[a]	94	83
Broadbeans	58	64	95	99	88	21[a]	69	68
Cowpeas	60	77	73	85	102	52[a]	82	77
Lentils	53	70	80	80	95	28[a]	71	75
Lima beans	59	93	88	94	104	57[a]	85	86
Mung beans	46	61	84	103	107	31[a]	64	81
Peas	67	76	85	94	115	46[a]	91	77
Lima beans (green)	81	88	93	92	98	40[a]	86	89
Peas (green)	52	72	69	71	74	34[a]	63	56
	(g/100 g protein)							
FAO hen's egg[b]	1.6	5.1	6.6	8.8	6.4	5.5	10.0	7.3

Source: Kelley (1972).
[a] Limiting amino acid.
[b] Based on average figures.

TABLE 4.4. ESSENTIAL NUTRIENTS SUPPLIED BY LEGUMINOUS CROPS SUPPLIED IN QUANTITIES SUFFICIENT TO MEET A RECOMMENDED DAILY DIETARY PROTEIN ALLOWANCE

		% recommended daily dietary allowance[a]										
Cooked, drained	wt (g)	P	Fe	Ca	Mg	Vit. A	Vit. C	Vit. B_1	Vit. B_2	Niacin	Niacin equiv.	Energy
Cowpeas (green)	740	110	105	20	115	50	250	185	55	70	70	30
Cowpeas (mature)	1170	115	100	20	270	2	0	155	30	30	64	35
Soybeans (green)	610	115	100	37	—	80	208	160	53	49	—	29
Lima beans (green)	790	95	130	37	108	44	250	120	53	69	86	35
Lima beans (mature)	730	112	150	21	145	0	0	79	29	34	63	40
Mung beans (mature)	710	85	125	29	—	4	0	78	35	43	—	33
Lentils (mature)	770	92	108	19	62	3	0	45	31	31	58	33
Peas (green)	1110	110	133	25	95	120	445	260	82	170	55	31
Peas (mature)	710	85	85	16	128	6	0	153	48	50	70	34
Broadbeans (green)	720	113	105	20	—	32	430	168	82	77	—	30
Broadbeans (mature)	710	97	120	25	—	3	0	104	50	41	62	34
Common beans (mature)	770	114	138	38	131	0	0	157	36	36	62	36

Source: Kelley (1972).

[a] Average daily dietary allowances used were 60 g protein, 1 g phosphorus, 15 mg iron, 1 g calcium, 350 mg magnesium, 5000 IU vitamin A, 50 mg ascorbic acid, 1.2 mg thiamin, 1.5 mg riboflavin, 1.5 mg niacin, 2500 kcal food energy. Niacin equivalent based on 1 mg niacin per 60 mg tryptophan.

TABLE 4.5. POTENTIAL YIELD OF AVERAGE NUTRITIVE VALUE (ANV) FROM SEVERAL TYPES OF VEGETABLES

Vegetable	Yield (MT/ha) Harvest portion	Yield (MT/ha) Edible portion	ANV	ANV per m^2	Duration (days from planting to harvest)	ANV (per m^2 per day)
Fleshy fruits						
Tomato	45	42.3	2.39	101	160	0.63
Eggplant	25	24.0	2.14	51	200	0.27
Sweet peppers	30	26.1	6.61	173	130	1.33
Okra	15	13.5	3.21	43	90	0.48
Cucumber	50	40.0	1.69	68	150	0.45
Pumpkin	20	16.6	2.68	44	150	0.30
Watermelon	40	25.2	.90	23	120	0.19
Leaf vegetables						
Amaranth	30	18.0	11.32	204	50	4.08
Water spinach	80	57.6	7.57	436	270	1.61
Chinese cabbage	30	25.8	6.99	180	90	2.00
Lettuce	20	14.8	5.35	79	50	1.58
White cabbage	40	34.0	3.52	120	90	1.33
Cassava leaves	60	52.2	16.67	870	270	3.22
Leguminous vegetables						
Asparagus bean (pods)	7	6.2	3.74	23.0	150	0.15
Lima bean (fresh)	9	5.1	4.88	25.0	210	0.12
Mung bean (sprouted)[a]	2.5	20.9	2.94	61.5	110	0.56
Hyacinth bean (dry)	3	3.0	14.03	42.1	180	0.23
Bulbs, tubers, roots						
Onion	40	38.4	2.05	78.7	150	0.52
Carrot	20	16.6	6.48	107.6	90	1.20
Taro	20	16.8	2.38	40.0	120	0.33
Turnip	13	10.3	2.03	20.9	80	0.26

Source: G.J.H. Grubben (1978).
[a] 1 kg of dry mungbean seed produces 9 kg of sprouts.

High in vitamin C

- crucifers
- peppers and tomatoes
- melons
- most leafy vegetables
- beans, seeds at immature stage
- bean sprouts
- freshly harvested white potatoes

Minerals

most leafy vegetables, particularly the crucifers and root crops, are particularly rich in minerals.

Many vegetables are low in sulfur-containing amino acids, methionine and cystine (Table 4.3).

Shown in Table 4.4 are the weights of legumes necessary to supply 60 g of protein, the minimum daily requirements for adults, and the percentages of the recommended daily requirement of minerals, vitamins, and energy supplied by those weights.

ASSESSMENT OF NUTRITIVE VALUES OF VEGETABLES AND EFFICIENCY OF NUTRIENT PRODUCTION

Considering the many components that go into human nutrition, it is difficult to assess the overall nutritive value of a crop, but attempts have been made to do this. Rinno (1965) came up with a value, "Essential Nutritive Value" which was renamed "Average Nutritive Value" (ANV) by Grubben. The empirical formula is

ANV per 100 g of edible portion =

$$\frac{\text{g protein}}{5} + \text{g fiber} + \frac{\text{mg Ca}}{100} + \frac{\text{mg Fe}}{2} + \text{mg carotene} + \frac{\text{mg vitamin C}}{40}$$

Adjustments are made for vitamin C if the vegetable is eaten raw and for Ca if the oxalate content is high. The ANV can be used to rank vegetables according to these nutrients considered important and obtained from vegetables, yield per unit area, and efficiency of production on an area-time basis. Grubben (1978) compared some vegetables on these bases in Table 4.5.

BIBLIOGRAPHY

GRUBBEN, G.J.H. 1978. Tropical vegetables and their genetic resources. International Board for Plant Genetic Resources, FAO-UN, Rome, Italy.

HARRIS, R.S., and VON LOESKE, H. 1960. Evaluation of Food Processing. John Wiley & Sons, New York.
HOWARD, F.D., MACGILLIVRAY, J.H., and YAMAGUCHI, M. 1962. Nutrient composition of fresh California-grown vegetables. Bull. No. 788. Agric. Exp. Stn., Univ. of California, Berkeley.
KELLEY, J.F. 1972. Horticultural crops as sources of proteins and amino acids. HortScience 7, 149–151.
LOWENBERG, M.E., TODHUNTER, E.N., WILSON, E.D., SAVAGE, J.R., and LUBOWSKI, J.L. 1974. Food and Man, Second Edition, Chapters 4 & 8, John Wiley & Sons, New York.
OOMEN, H.A.P.C., and GRUBBEN, G.J.H. 1977. Tropical Leaf Vegetables in Human Nutrition. Commun. 69, Dept. of Agric. Res., Koninklejk Institute voor de Tropen, Amsterdam, Netherlands.
RINNO, G. 1965. Die Beurteiling des Ernahrungsphysiologischen Wertes von Gemuse. Arch. Gartenban *13*(5), 415–429.
SCRIMSHAW, N.S., and YOUNG, V.R. 1976. The requirements of human nutrition. Sci. Am. *235*(3), 51–64.
YAMAGUCHI, M., and WU, C.M. 1975. Composition and nutritive value of vegetables for processing. *In* Commercial Vegetable Processing. B.S. Luh and J.G. Woodroof (Editors). AVI Publishing Co., Westport, Connecticut.

5

Toxic Substances and Folk Medicinal Uses of Vegetables

Man's experiences throughout his history have taught him much about avoiding sickness and even death from consumption of natural food products. Much evidence exists of the ancient customs and prejudices handed down generation after generation on food taboos, eating habits, and customs of the various peoples of the world. Some of these customs are practiced even today. Witch doctors, medicine men, high priests, and in more recent times, the physicians of Rome, all have had some knowledge as to the toxins present in plants and the physiological effects on man and animals.

The survival of some plants is probably correlated with the bitter and bad tasting compounds present. If there were no protective mechanisms, a species could become extinct from the foraging of animals. Although most of our present-day food plants have been selected to enhance the desirable characteristics, many still retain their chemical similarity to their wild ancestors.

TYPES OF TOXICANTS IN PLANTS

Enzyme Inhibitors

Protease Inhibitors

Protease inhibitors inhibit proteolytic activity which catalyzes the hydrolysis of proteins to amino acids. Some of them are heat labile and others are heat resistant. Trypsin inhibitor is found in legumes, particularly raw soybeans. Chymotrypsin inhibitor is present in potatoes. Both trypsin and chymotrypsin are proteases present in digestive systems of animals.

Cyanogens

Cyanogens release cyanide of HCN in acidic solutions. Lima beans, bamboo shoots, and cassava (manioc) contain large amounts of cyanogenic glucosides which are enzymatically converted to HCN. Cyanide has inhibitory action on cytochrome oxidase, the respiratory enzyme and other enzymes.

Glucose-6-phosphate Dehydrogenase Inhibitor

Favism is an inherited disorder of certain individuals particularly of southern European origin. These people have NADP-linked glucose-6-phosphate dehydrogenase deficiency when vicine or convicine from fava bean is ingested. Pollen from fava plants in the respiratory tract affects these people also.

Cholinesterase Inhibitors

Cholinesterase inhibitors react with the enzyme which controls nerve impulses.

Alkaloids

The alkaloids generally react with the nervous system. Opium, quinine, and solanine, the bitter substance in white potatoes, are some examples.

Amylase Inhibitors

These substances inhibit starch hydrolysis. Taro corms have salivary amylase inhibitor. Navy beans have pancreatic amylase inhibitor.

Invertase Inhibitors

Invertase inhibitors inhibit the hydrolysis of sucrose. The white potato contains invertase inhibitors.

Other Enzyme Inhibitors

The heat-sensitive phallolysin, a glycoprotein from the poisonous mushrooms *Amanita phalloides*, affects the action of phospholipase enzymes of cell membranes.

Physiological Disorganizors and Irritants

Hemagglutinins

Hemagglutinins are proteins in nature. They cause low food absorption and low nitrogen retention, resulting in reduced growth. They are present in soybean, common beans, and castor beans.

Saponins

Saponins cause bloat in ruminants because of foaming and trapping of gasses; this is said to inhibit enzyme action because it reduces surface tension. Primitive people used plants containing large amounts of saponins from certain legumes to catch fish since these substances disable the breathing mechanism. Apparently, there are no ill effects from eating fish caught by the method.

Lathrogens

Chick peas and chickling vetch contain γ-glutyamylamino proprionitrile. This substance blocks the formation of cross-links in the collagen molecule of the connective tissue causing paralysis of arm and leg muscles. Also *Lathyrus* species and *Vicia* species have the amino acid, β-cyanoalanine, which causes rigidity, weakness, and paralysis of muscles.

NO_3^- and NO_2^-

These nitrogenous compounds occur with excessive nitrogen fertilization. NO_3^- gets into the blood and converts the hemoglobin to methemoglobin causing cyanosis, especially in infants. Concentration from 64 to 140 ppm NO_3^- in H_2O is reported to be toxic to babies. However, NO_2^- at 200 ppm is considered not hazardous.

Oxalates

For many years oxalates were thought to interfere with calcium uptake. Poisoning from rhubarb leaves is not from oxalates but from other poisons present in the leaves. However, oxalate crystals are thought to cause irritations.

Irritant Oils

Terpenes and oils from certain plants cause irritation; poison oak and ivy are examples of such plants.

Allergens

Allergens are usually proteinaceous and widespread throughout the plant kingdom. Proteins ingested or pollen on mucous membrances cause physiological reactions with production of histamines or histamine-like substances. In many cases cooking denatures the protein and lessens the degree of allergic reactions. Legumes (peanuts, lentils, and peas) have heat-stable allergens.

Alterations of Hormonal Actions

Goitrogens

These substances cause hypothyroidism and enlargement of the thyroid gland. They are largely plants belonging to the mustard family and contain the compound, L-5-vinyl-2-thiooxazolidone:

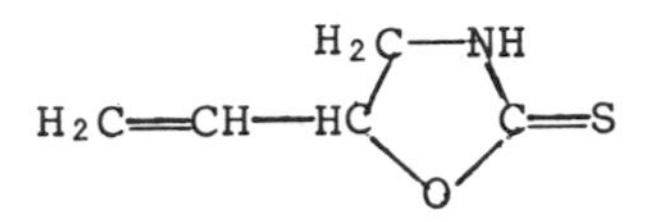

This compound is inactivated by heat. Unfortunately, it is transferred to humans by the drinking of milk from cows feeding on plants containing this substance.

Estrogen

Yam family plants have steroidal substances resembling hormones. Ergot, the fungus on grain, has hormone-like activity.

Antimetabolites

Toxic Amino Acids

Hypoglycine A causes hypoglycemia. Ingestion could cause the blood sugar level to decrease to 20 mg/100 ml (normal 80–100 mg). Fruit of *Blighia sapida*, known in Jamaica as "ackee" and in Nigeria as "isin," contains *methylene cyclopropyl alanine*. Litchi seeds *(Litchi chinensis)* contain *methylene cyclopropyl glycine*. Both compounds cause hypoglycemia.

Creeping indigo causes liver damage. The active compound is indospicine, L-α-amino-ε-amindinocaproic acid:

$$\underset{\displaystyle NH}{\overset{}{NH_2\!-\!\underset{\|}{C}}}\!-\!(CH_2)_4\!-\!\underset{\underset{\displaystyle NH_2}{|}}{C}H\!-\!COOH$$

Indospicine

Mimosine is found in the legume *Leucaena glauca*, called "Koa haole" in Hawaii. Ingestion of the substance causes loss in weight and loss of hair. Indonesian people who eat this legume often lose their hair.

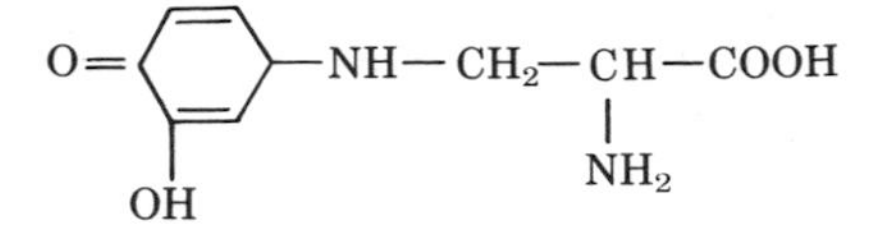

Mimosine

Mimosine is an analogue of tryosine, an essential amino acid; their structures are quite similar.

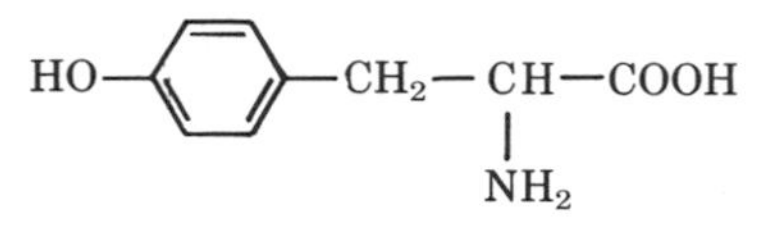

Tyrosine

Djenkolic acid is a toxin found in bean from leguminose tree, *Pithecolobium labatum*; it is reported to cause kidney failure. The toxin appears as needlelike cluster in bloody urine.

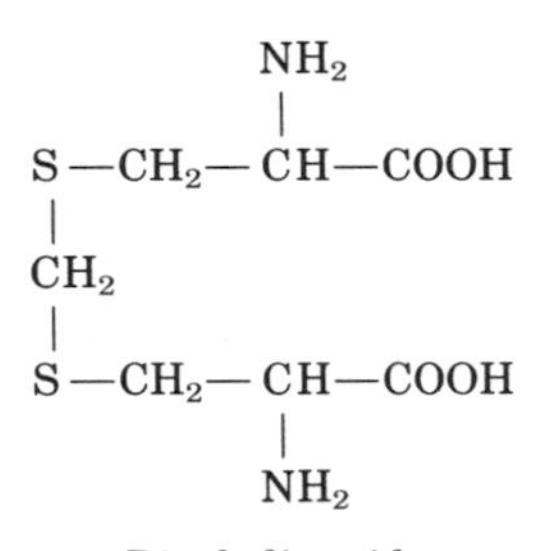

Djenkolic acid

Dihydroxyphenylalanine (dopa). This compound may also be associated with *favism* (fava beans). It lowers the gluthathione level due to the reaction with glucose-6-phosphate dehydrogenase.

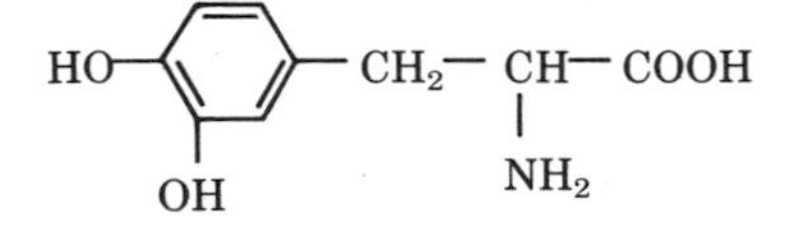

Dopa

α-Amino-β-methylaminopropionic acid. This compound is found in cycads and it causes neurotoxicity.

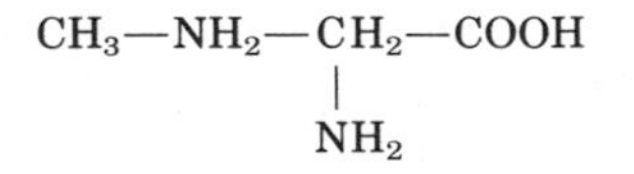

α-Amino-β-methylamino propionic acid

The action of this substance is similar to Lathyrism which is caused by the chemical β-cyanoalanine:

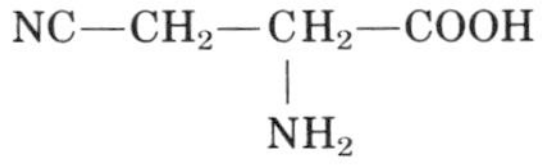

$$\begin{array}{c} NC-CH_2-\underset{\displaystyle NH_2}{\underset{|}{CH_2}}-COOH \end{array}$$

β-Cyanoalanine

Antivitamins

Antivitamin A

Lipoxidase in soybeans oxidizes and destroys carotenes (vitamin A precursors). Citral in orange oil competes with retinene in the metabolism of endothelial cells.

Antivitamin B_1 (Antithiamine)

Antivitamin B_1 is found in bracken fern *(Pteridium aquilinium)*.

Antivitamin B_2 (Antiriboflavin)

There are evidences that hypoglycine A is an antimetabolite of riboflavin. Hypoglycine A is found in akee plants *(Blighia sapida)*.

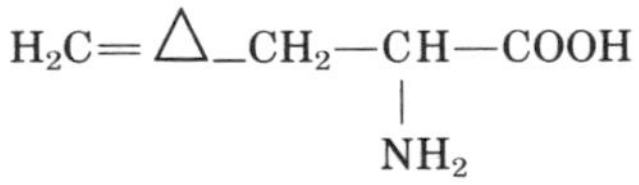

$$H_2C=\triangle-CH_2-\underset{\displaystyle NH_2}{\underset{|}{CH}}-COOH$$

Hypoglycine A

Antiniacin

Sorghum is reported to contain substances that produce niacin deficiency in dogs (black tongue disorder).

Antipyridoxine

Flax seed contains linatine, the precursor of 1-amino-D-proline, which is an antagonist of pyridoxine. Amino-D-proline forms a complex with pyridoxine.

Antivitamin D

Uncooked soybean has antivitamin D properties. The rickitogenic compound has metal-binding properties; it ties up Ca^{2+}.

Antivitamin E (anti-α-tocopherol)

Peas, alfalfa, and beans have compounds that interfere with vitamin E utilization.

Antivitamin K

Sweet clover has dicumerol, which lowers the prothrombin or blood clotting time.

Carcinogens and Tumorigens

These are natural substances in plants that cause malignant and benign tumors.

Cycasin

Cycasin from cycads has been reported as carcinogenic toxin.

$$CH_3—N{=}N—CH_2—O—\text{glucose} \xrightarrow{\text{glucosidase} + H_2O} CH_3—N{=}N—CH_2OH + \text{glucose}$$

The product causes liver damage.

Aflatoxin

The fungus *Aspergillus flavus*, growing on foods, produce aflatoxins that cause carcinomas of the liver in rats, ducks, and trout. These carcinogenic compounds are heat stable. Eating spoiled food caused by this organism is a potential hazard. The structures of four aflatoxins are shown below:

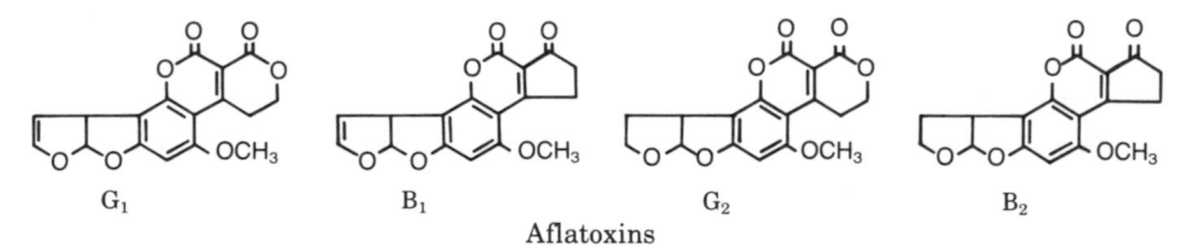

Aflatoxins

TOXICANTS AND FOLK MEDICINAL USES OF SOME VEGETABLES

Fungi

The poisonous genus *Amanita* causes 90% of the deaths from eating of mushrooms. *Amanita phalloides*, *A. verna*, and *A. virosa* are the more common poisonous species.
There are two types of toxin present in fungi:

1. Phallin, a hemolytic (red cell-destroying) glucoside is inactivated by heating or cooking. It is also destroyed in the stomach.

2. From *Amanita* two deadly toxins, amanitine and phalloidine, are present; they affect the liver, kidney, and heart. These are not inactivated by cooking or by heating. Symptoms of poisoning appear 10 to 15 hr after eating.

In recent years a treatment has been discovered for *Amanita* poisoning. It was first successfully used in Czechoslovakia and more recently in the United States. Intravenous injections (75 mg) of thioctic acid (α-lipoic acid) are given every 6 hr for 7 days.

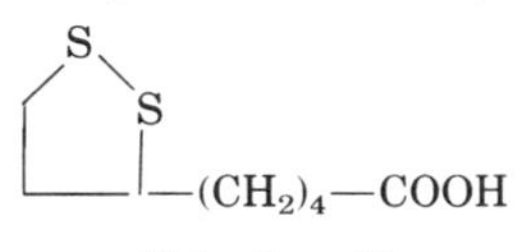

Thioctic acid

Thioctic acid is an acetate replacing and a pyruvate oxidation factor. It is also a prosthetic group of a coenzyne in metabolism.

With *Muscaria*, another poisonous species, the symptoms occur very soon after ingestion.

When gathering wild mushrooms, the only way to be sure is to examine each and every mushroom as belonging to the edible species. Quite often more than one species, poisonous and edible, grow in the same area.

The edible mushroom, coprinus or inky cap, has been reported to contain disulfiram, a substance that interferes with the metabolism of alcohol. Laplanders use the common toadstool to kill pain; they burn the mushroom on the skin near the location of the pain.

Polypodiaceae

Pteridium aquilinum, bracken fern or tree fern, has been known to cause bone marrow damage and intestinal mucosa disorder. It also contains an enzyme, thiaminase, which hydrolyzes thiamine making it inactive as a vitamin.

Gymnosperms

Cycadaceae

These plants called cycads contain glycosides, which when eaten in large quantities can sometimes be lethal. The paralytic condition has been noted among people and animals fed cycads. The incidence of amyotropic lateral sclerosis, a neurological disease, was found to be significantly greater (100 times more) among natives eating cycads.

Natives of a number of southwest Pacific islands use cycads for curing skin ulcers. In India the seeds of cycads *(Cycas circenalis)* are used as a laxative.

Gynkgoaceae

Fruits of the ginkgo were immersed in oil by the ancient Chinese herbalists. This was used for treatment of tuberculosis. It is also used in Chinese folk medicine to dissipate phlegm, and to expel intestinal worms.

Angiosperms: Monocotyledons

Gramineae

Bamboo shoots contain a cyanogenic glucoside, which, when ingested, is poisonous because of cyanide released. Parboiling in water leaches the compound to render the shoots nontoxic.

Araceae

Taro contains oxalates which cause acridity. Thorough cooking removes the acridity. Dasheen has low oxalic acid content.

Liliaceae

Asparagus shoots were given for jaundice and to "cleanse the bowels." It was also recommended for disorders of the breast. Methyl mercaptan is found in the urine after ingestion of asparagus and is thought to give the distinctive odor. Some state that asparagine aminosuccinic acid monoamide is the odor-causing substance.

Amaryllidaceae

Both onion and garlic have been used for centuries as remedies for man's diseases. Garlic was used as a confluent for smallpox. This was accomplished by applying cut pieces of the clove to the feet about 8 days following the onset of the disease. It has been shown that the antibiotic effects of alliums is due to the sulfur compounds.

Dioscoreaceae

Certain species of yams contain a poisonous substance called diosgenin. The edible species contain little or none. Some of the wild types contain as much as 10% diosegenin on a dry weight basis. Diosgenin in yams is used to manufacture cortisone and other sex hormones.

Angiosperms: Dicotyledons

Polygonaceae

Rhubarb contains large amounts of oxalates: oxalates tie up the calcium (combines to form insoluble calcium oxalate) preventing absorption of the essential element. During World War I rhubarb leaves were recommended as a substitute for green vegetables. Many cases of acute poisoning resulted; death occurred in some cases. Rhubarb contains a highly irritant anthraquinone, a glycoside. This compound rather than the oxalate is the suspect. Foods containing oxalates are not considered much of a dietary hazard because they do not ordinarily constitute a sufficient part of the diet. Rhubarb has been used for medicinal purposes since before Christ. Dioscorides, who was physician to Anthony and Cleopatra, recommended rhubarb for diseases of the liver and weaknesses of the stomach. He used it to cure ringworm by applications of a concoction of roots in vinegar.

Chenopodiaceae

Beets, chard, and spinach all contain oxalates. The ancients used spinach for various medicinal purposes. The country people in England crushed spinach and placed it to cleanse wounds and to cure warts.

Aizoaceae

New Zealand spinach contains oxalates. Captain Cook, in his 1769 visit to New Zealand, harvested this plant and fed it to his crew to guard against scurvy (vitamin C deficiency).

Cruciferae

This family comprises cabbage, cauliflower, broccoli, radish, rape, rutabaga, etc. Plants of this family contain goitrogenic substances, the thioglycosides. The seeds have a higher content of this substance than the vegetative tissues; hence, the meal of rape seeds, after oil extraction, can be quite toxic when fed to animals.

Brassica

Plants of this genus contain *S*-methylcysteine sulfoxide, a compound known as the "kale anemia factor." Cattle and sheep are often poisoned in the winter feeding of rape, cabbage, Brussels sprouts, and rutabaga tops and roots. This same compound has been shown to lower blood cholesterol levels.

Cabbage was considered a very important drug by the Romans. The only medicine given for many diseases was cabbage. The Greeks recommended the juice of cabbage as an antidote for eating poisonous mushrooms.

Leguminosae

Plants belonging to this family contain saponins (soybeans), lathrogens (chick peas), glycosides that cause favism (seeds of *Vicia* species), and cyanogenic glycosides (lima beans). Some also contain protease inhibitors (soybeans, lima beans, peanuts, and fava beans) as well as hemagglutinins (castor bean, soybean, and the common beans).

Roman physicians used peas boiled in seawater to cure erysipelas (a bacterial infection of the subcutaneous tissue). A superstition among the ancients was that if one ate peas too freely, one could contract leprosy. The Romans used chick peas and rosemary mixtures for jaundice and edema. Pythagoras, the Greek philosopher (582 BC), forbade his disciples to eat beans because he supposed that beans came from the same putrid matter that man was formed. Beans were used in older times to gather votes of the people in elections. In a trial a white bean signified acquittal and a black one condemnation.

Euphorbiaceae

The bitter variety of cassava or manioc contains cyanogens, which must be removed before using. The sweet variety contain little of the toxic substance, therefore, need no treatment. The castor bean is a member of this family and contains the poisonous alkaloid, ricin.

Araliaceae

The ginseng root for thousands of years has been used as a drug plant in the Orient. It has been and is still being used for all ailments, ranging from heart disease and fever, to insomnia, as well as a love potion. Recent physiological experiments indicate that extract from this plant have a mild stimulating effect on the vital organs.

Umbelliferae

Plants of this family contain alkaloids. The poison hemlock contains the alkaloid coiine. In ancient times, prisoners were forced to eat this plant in order to carry out death sentences.

Carrots contain carotoxin, not an alkaloid, which is toxic to mice. The wild carrot, Queen Anne's lace, has been a suspect in causing mild intoxications of horses and cattle. Recent reports state that extracts of carrot seeds prevented pregnancy by inhibiting the implantation of the fertilized egg in the uterus.

Convolvulaceae

The sweet potato contains a bitter substance called ipomeamarone. In New Zealand the natives use an infusion of the sweet potato to lower fever, and apply the crushed leaf tissues for diseases of the skin.

Solanaceae

Most of the plants in the nightshade family contain some alkaloids.

Potato tubers ordinarily contain little or no solanine, a bitter and toxic alkaloid. However, on exposure to light, the alkaloid readily forms. Also, the formation of chlorophyll occurs with light exposure. Hence, green potatoes and solanine presence are correlated. The sprouts from potato tubers are extremely poisonous, as are the tops. The Maori of New Zealand apply the extract from boiled potato tubers to cure pimples and soothe burns.

The bitter taste in overmature eggplants and other solonaceous fruits is probably caused by alkaloids.

Tomato leaves, stems, and immature fruits contain an alkaloid, tomatine. With maturation of the fruit, an enzyme degrades the compound into a nontoxic form. Wild tomatoes contain more of this insect repellent alkaloid than the commercial cultivars. Perhaps when the tomato was first introduced into the Old World, it did have more tomatine present and therefore was considered poisonous.

The mandrake root *(Mandragoia officinarum)* from ancient times has the reputation of producing sleep. This was probably the first use of anesthesia before surgical operations. Too much of this drug caused mental disease and even death. The active ingredient of mandrake roots is a pain-deadening alkaloid, hyoscine or scopolamine.

The natives of Australia chew the leaves of the genus *Diboisia*, which contains hyoscine for relief from fatigue, hunger, and thirst. Also, they use the leaves to stupify and catch fish and also to capture the emu, which comes to drink in the treated waterhole.

Cucurbitaceae

Members of the gourd family contain bitter glycosides. Cucumbers contain glycosides called curcurbitacins; some varieties contain more than others. The summer squash, 'zucchini' has been reported to contain cucurbitacins. Reports indicate that these substances are very toxic to humans as well as to animals. Some wild watermelon in Africa are reported to be quite bitter. Perhaps the custom of plugging and tasting watermelons originated because of the bitter fruits. It has been reported that some people (inherited trait) cannot taste this bitterness.

Pumpkins and squashes contain cholinesterase inhibitors.

Balsam pear or bitter melons are grown in the Asian countries as a vegetable. This vegetable is prepared by parboiling to remove the bitter toxic substances. In the Orient the sliced, dried fruit is sold by herbalists as medicine for hemorrhoids, gout, rheumatism, and vermifuge. Scientists have isolated a hypoglycemic ingredient, cheratin, in the fruit. This substance causes the abnormal lowering of blood sugars.

Compositae

A bitter taste of lettuce is probably due to an aromatic compound called lactucopicrin. The dried latex from German lettuce *(Lactuca virosa)*, called lettuce opium for many centuries, has been used as a hypnotic drug. Greek and Roman physicians used extracts of wild lettuce leaves to combat fever.

Boiled dandelion leaves were frequently prescribed for fevers and as a diuretic. The English gave the dandelion plant a bad name because children, who ate the plant for supper, wet their beds.

BIBLIOGRAPHY

KEELER, R.F., VAN KAMPER, K.R., and JAMES, L.F. 1978. Effects of Poisonous Plants on Livestock. Academic Press, New York.

KINGMAN, A.D., Editor. 1977. Toxic Plants, Columbia Univ. Press, New York.

LIENER, I.E. 1969. Toxic Constituents of Plant Foodstuffs. Academic Press, New York.

MARTIN, F.W., and RUBERTE, R.M. 1975. Tropical leaves that are poisonous. *In* Edible Leaves of the Tropics. Mayaguez Inst. Trop. Agric., Sci. Ed. Admin., U.S. Dept. Agric., Mayaguez, Puerto Rico.

NAS. 1973. Toxicants occurring naturally in foods. Publ. No. 2117 Nat. Acad. Sci., Washington, DC.

Part II

Principles of Growing Vegetables

6

Environmental Factors Influencing the Growth of Vegetables

Growth and development of plants depend on climatic and soil factors. Weather is the state of the atmosphere with respect to temperature, moisture, air movement, solar radiation, or any meteorological phenomena over a short period of time. Climate is the average course or condition of weather at a location over a period of many years; it is the integrated effects of temperature, precipitation, humidity, sunlight, and wind. The soil condition in many ways is dependent on climate.

The kinds or species of plants that can be grown in a given region is controlled by the climate. However, the growth of a plant is dependent on the weather during its life cycle.

CLIMATE FACTORS

Temperature

Temperature has profound effects on all living organisms; it limits growth and distribution of both plant and animal life. The source of this energy is the sun. About 2.0 g cal/cm^2 (2.0 Langleys) come to the earth every minute.

World Distribution of Temperature

Although the world is a rotating sphere and each latitude receives its portion of the insolation, the temperature of any particular point on the earth is governed by its proximity to other land masses and to the air and ocean currents. Thus, the world map of temperature is not a uniform cord around a sphere, but is somewhat skewed; still, the equatorial zone is the hottest and the polar zones the coldest. Due to the inclination of the earth (66½°) to the plane of orbit around the sun, all portions of the earth receive a regular increase and decrease of insolation according to the time of the year and latitude. At the higher latitudes, the difference in insolation is very large, resulting in the seasons of winter, spring, summer, and fall. If it were not for the tilt of the earth, there would be no seasons: temperature and climate would vary little from day to day, and every day would be 12 hr long.

The seasonal temperature variation can change as little as 3°C (5°F) all year round in the tropics and as much as 45°C (80°F) in the polar zones. The range of the variation is much larger over the continents than over or near the oceans or seas because of the larger heat capacity of water as compared to soil.

Diurnal fluctuations occur due to the rotation of the earth as solar radiation is cut off at night and from radiational losses which occur from the surface at night. Thus, each day has a temperature maximum which occurs shortly after noon and a minimum which occurs just at sunrise.

Temperature decreases with increases in altitude; a drop of about 6°C occurs with each 1000 meters in elevation (3.5°F per 1000 ft).

Temperature Effects on Plants

Cardinal Temperatures

Cardinal temperatures are (1) *minimum* value where growth ceases, (2) *maximum* value where growth ceases, and (3) *optimum* value where growth occurs most rapidly. The cardinal temperatures are not the same for all plants but vary with different families, genera, and species. Plant breeders select lines for tolerances to low temperature in cooler regions and for high temperature tolerance in the tropics.

Van't Hoff's Law

For every 10°C rise in temperature, the rate of dry matter production (growth) doubles. This is valid only in the range of about 5° to about 35°C depending on the organism. This is also called Q_{10}.

Length of Growing Season

One of the oldest means of evaluating or predicting where crops can be grown is the *number of frost-free days*. This is the average period between the last killing frost in the spring and the first killing frost in the fall, that is, the frost-free season.

Heat Units

The concepts of heat units is based on the theory that plant growth is dependent on the total amount of heat to which it is subjected during its lifetime. The unit is *degree-days,* which is usually calculated by subtracting the minimum threshold temperature from the average temperature (maximum + minimum)/2 each day. The degree-days are summed for a desired period, usually from planting to harvest. The minimum threshold temperature varies with the crop. With cool season crops, it is 10°C (50°F). If the average temperature for the day is equal to or less than the threshold temperature, the degree-day is zero. Figure 6.1 shows the degree-days for the United States from March 1 to July 27, 1974, using a 50°F base. On the Celsius scale the heat unit would be 5/9 that for Fahrenheit scale.

Diurnal Change (Thermoperiodicity)

Generally, a large diurnal range is favorable for net photosynthesis. Particular night temperatures are important for some crops (e.g., tomato) and particular day temperatures are optimal to crops such as peas. Generally, very high night temperatures are not beneficial as respira-

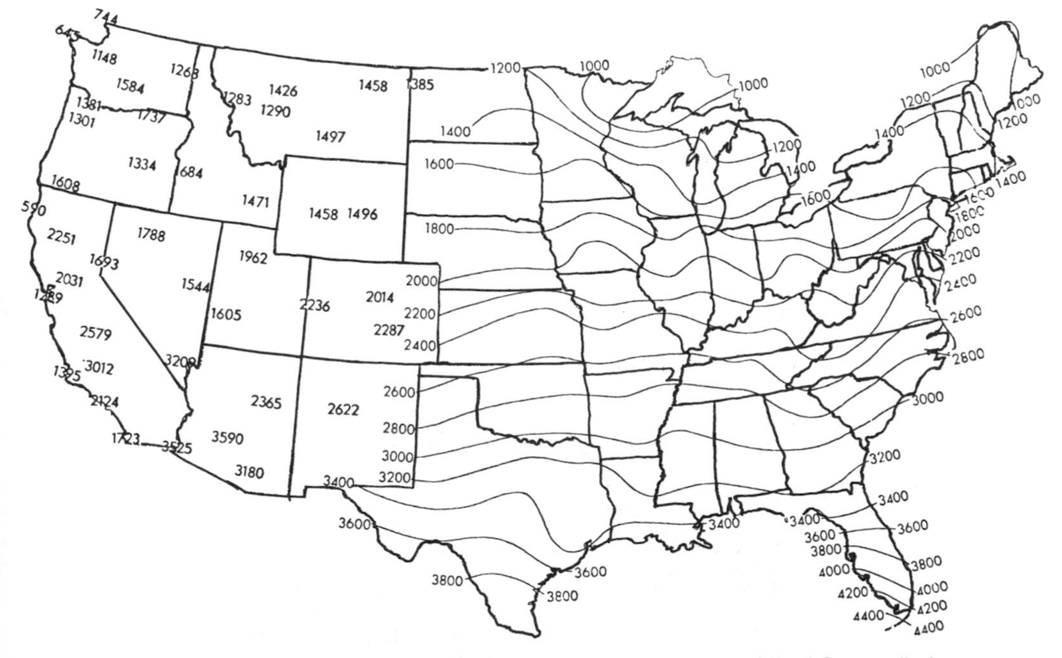

FIG. 6.1. Total degree-days from March 1 to July 27, 1974, United States; limits: 50°–86°F, base 50°F.
Courtesy National Weather Service, NOAA.

tion rates increase using up the photosynthates produced during the day.

Vernalization

The exposure of certain plants to low temperatures induces or accelerates flowering (bolting). This is vernalization. The required length of low temperature exposure varies with species.

With some species, devernalization (reversal of vernalization) can occur if the plant is exposed immediately to high temperatures (30°C) (86°F) or higher, following the exposure to low temperatures. Figure 6.2 shows the phases of growth of a biennial plant in the temperate region.

In some species, seedling and young plants still in the juvenile stage are insensitive to conditions that promote flowering in older plants. In other species, such as winter wheat and winter rye, the soaked partially germinated seeds can be vernalized. The seeds must have sufficient water to allow the vernalizing process to occur but still low enough to hinder radicle growth. Holding at 15°–18°C (59°–64°F) (nonvernalizing temperature) for 10–24 hr is sufficient to start germination processes. The seeds are then vernalized at temperatures of 1°–6°C (34°–43°F) for 15–60 days depending on the variety. The spring grains do not require vernalization for flowering to occur.

Certain deciduous trees (cherry, peach, almond, etc.) require chilling temperatures before dormant buds become active. Certain tubers, corms, and bulbs require low temperatures following moderately high temperatures before growth resumes.

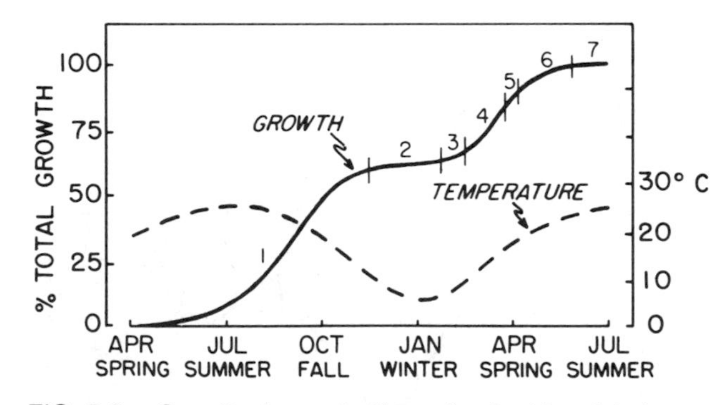

FIG. 6.2. Growth phases in life cycle of a biennial plant. 1, Vegetative growth; 2, vernalization; 3, floral initiation; 4, seed stalk emergence and elongation; 5, flowering and fertilization; 6, seed maturation; 7, mature seed.

Freezing and Chilling Injury

Many cold-climate plants may be frozen at low temperatures without injury. Conifers in the subarctic survive temperatures of −60°C (−80°F). Most vegetables are injured at temperatures at or slightly below freezing. Tropical or subtropical plants may be killed or damaged by the cold at temperatures below 10°C (50°F) but above freezing. This latter type of damage is called *chilling injury.* Vegetables of tropical origin such as cucumber, tomatoes, and sweet potatoes, when exposed to prolonged periods of temperature below 10°C (50°F), become injured. A few hours slightly below this temperature does little or no harm if the temperature rises high enough immediately following the cold exposure; there is a time–temperature relationship.

Susceptibility to cold damage varies with different species and there may be differences among varieties of the same species. The susceptibility to cold damage varies with stages of plant development. Plants are more sensitive to cold temperatures particularly shortly before flowering through a few weeks after anthesis.

Hardening of Plants

Many plants can be made somewhat resistant (hardened) to cold temperatures by subjecting the plant successively to lower and lower temperatures. This hardening process causes an adaptation of the protoplasm to low temperatures. Such a process takes place in the fall of the year. Cabbage can be hardened experimentally by subjecting plants to 0°C (32°F) a few hours each day for several consecutive days. Hardening also occurs when plants are subjected to gradual water stress.

High Temperature Injury

High temperature in many semiarid and arid regions may be a limiting factor in the production of economic crops. Under high isolation and high humidity, leaf temperatures can reach 8°C (15°F) above air temperatures. When temperatures rise too high, heat destruction of the protoplasm results in death of cells. This occurs in the range of 45°–50°C (113°–122°F).

At Davis, California, in mid-August when the temperature maximums are above 38°C (100°F) in tomato fields, temperature of fruits exposed on the vine can reach temperatures of 49°–52°C (120°–125°F). If green fruits are at these temperatures for an hour or more, they become sunburned; and ripe fruits become scalded.

As with cold resistance, the plant cells can become gradually acclimated, to a certain extent, to heat by slowly raising the temperature and lengthening the exposure daily.

Transpiration (evaporation of water) from leaf stomata helps cool leaves. It has been calculated that transpiration can reduce heating about 15 to 25%.

Rest and Dormancy in Seeds and Organs

Most mesophytic plants go through a phase of little or no growth at some stage in the life cycle. Seeds, buds, tuberous roots, tubers, rhizomes, bulbs, and corms often show this phenomenon. Under natural conditions this period usually coincides with unfavorable environmental conditions such as low or high temperature and/or lack of or excesses of moisture; sometimes photoperiod is also involved.

Seeds and vegetatively reproductive organs are said to be at *rest* when the organ shows no signs of growth resumption even when the environmental conditions are favorable. This is also known as *rest period*, *internal*, or *innate dormancy*.

Seeds and organs that have the potential to germinate or resume growth but do not because of unfavorable environmental conditions are termed *dormant*. Other terms used to describe this state are: *external* or *imposed dormancy* and *quiescence*.

Actually the phases of rest and dormancy are not abrupt, but the phases occur gradually as changes in inhibitor(s) and/or promotor(s) concentrations occur over a period of time; the rate of the changes is definitely affected by temperature. Some plant physiologists describe several stages such as early rest (predormancy) → rest (middormancy) → after rest (postdormancy). At early rest the organ may resume growth after such treatment as high temperature. However, when the organ is at rest, it may be extremely difficult to induce resumption of growth. Following rest, there is a transition period where it becomes progressively easier to induce resumption of growth. Temperatures at which the organ is held can influence the rate of these changes; at high temperatures each stage is shorter in duration than at low temperatures.

Thermoclassification of Vegetables

As given under the topic of classification, vegetables can be grouped on a climate basis: cool weather or warm weather for proper growth and development. This classification groups vegetables according to their average monthly temperature requirements and tolerance for certain minimum and maximum temperatures. Table 6.1 shows the groupings of cool and warm season crops, the temperature range for seed germination, and the time in days required from seeding to harvest. The data are valid for temperate climates; in the tropics and subtropics, the days to maturity may deviate from the figures given.

TABLE 6.1. CLIMATIC RANGES OF VEGETABLES

Crop	Germination (soil temperature)	Days to maturity	Notes
I. Cool season crops			
A. Optimum 16°–18°C (60°–65°F), intolerant of monthly mean above 24°C (75°F), some tolerance to freezing			
Spinach	4°–16°C (40°–60°F)	35–40	Long day and high temperature induces bolting
Beet	10°–30°C (50°–85°F)	50–70	Below 10°C (50°F) bolting induced; high temperature, zoning of red pigment
Parsnip	10°–24°C (50°–75°F)	100–130	Below 10°C (50°F) bolting induced
Turnip	10°–35°C (50°–95°F)	40–60	
Cabbage	10°–30°C (50°–85°F)	30[a] + 60–110	Below 7°C (45°F) bolting induced
Radish	10°–30°C (50°–85°F)	21–30 summer 56–60 winter	
Broccoli	10°–30°C (50°–85°F)	30[a] + 60–110	
Also: Brussels sprouts, kale, rutabaga, collard, kohlrabi, salsify, watercress, broad bean, rhubarb			
B. Optimum 16°–18°C (60°–65°F), intolerant of monthly mean above 24°C (75°F), damaged by freezing weather (more easily near maturity)			
Cauliflower	10°–30°C (50°–85°F)	30[a] + 55–150	
Pea	10°–30°C (50°–85°F)	55–75	
Swiss chard	10°–30°C (50°–85°F)	50–60	
White potato (tubers)	7°–27°C (45°–75°F)	90–110	Short day, low temperature induces tubers
Celery	10°–24°C (50°–75°F)	80[a] + 80–130	Below 10°C (50°F) bolting induced
Carrot	10°–30°C (50°–85°F)	60–85	Below 10°C (50°F) bolting induced
Lettuce	4°–27°C (40°–80°F)	40–100	High temperature and possibly long day induce bolting
Also: endive, globe artichoke, cardoon, celeriac, chicory, Chinese cabbage, fennel, parsley, mustard			
C. Optimum 18°–30°C (65°–86°F), tolerant of frost			
Onion	10°–30°C (50°–86°F)	90–250	Bulbing day length sensitive below 10°–16°C (50°–60°F) bolting induced
Asparagus	16°–30°C (60°–86°F)	Perennial	Tips of shoots are injured at temperatures below freezing
II. Warm season crops			
A. Optimum 18°–30°C (65°–85°F), intolerant of frost			
Sweet corn	16°–35°C (60°–95°F)	65–100	
Snap bean	16°–30°C (60°–86°F)	45–75	Seedlings subject to chilling injury below 10°C (50°F)
Lima bean	21°–30°C (70°–86°F)	65–90	Seedlings subject to chilling injury below 10°C (50°F), no fruit set in low humidity
Tomato	16°–30°C (60°–86°F)	50[a] + 60–100	No fruit set in low humidity

(Continued)

TABLE 6.1. *(continued)*

Crop	Germination (soil temperature)	Days to maturity	Notes
Pepper	16°–35°C (60°–95°F)	50[a] + 60–85	Seedling and fruit subject to chilling injury below 10°C (50°F); no fruit set in low humidity
Cucumber	16°–35°C (60°–95°F)	50–75	Seedlings and fruit subject to chilling injury below 10°C (50°F)
Muskmelon	21°–32°C (70°–90°F)	80–120	Seedlings subject to chilling injury below 10°C (50°F)
Also: summer squash, winter squash, pumpkin, chayote, New Zealand spinach, roselle			
B. Long season, will not thrive below 21°C (70°F)			
Watermelon	21°–35°C (70°–95°F)	70–95	
Sweet potato (roots)	21°–32°C (70°–90°F)	40[a] + 150	Chilling injury below 10°C (50°F)
Eggplant	21°–35°C (70°–95°F)	50[a] + 75–85	
Okra	21°–35°C (70°–95°F)	50–60	

Source: Revision of MacGillivray (1953).
[a] Days to grow transplants in nursery or in plant beds.

MOISTURE

Precipitation is equally important in climate as is temperature since water is a prime necessity for life.

Hydrologic Cycle

Evaporation of water from land and water surfaces and transpiration from plants as water vapor are the main sources of water that fall to the earth as rain, snow, hail, fog, dew, or frost.

Humidity

Moisture in the atmosphere is measured as absolute humidity or as relative humidity. *Absolute humidity* is the amount of water present in a unit volume of air. *Relative humidity* (RH) is the amount of water present in air as a precentage of what could be held at saturation at the same temperature and pressure. RH is more useful in horticulture. *Dew point* is the temperature of the air at which the water vapor is at the saturation point. Dew point varies with the amount of water vapor in the air.

High humidity increases the incidence of many diseases and increases insect population. However, it can be beneficial to the water economy as transpiration is decreased. Also pollen viability increases in certain crops as humidity increases.

World Distribution of Precipitation

The average total precipitation over the entire earth is about 1000 mm (39 in.). It varies from a high of 11,500 mm (450 in.) per year in Cherrapungi, India, to a low of 0.5 mm (0.02 in.) in Arica in northern Chile.

The amount of rainfall is closely related to air pressure and winds. Heavy rain falls along the equatorial low-pressure belt which is due to abundant moisture from inflowing ocean air. The greatest rainfall occurs near the coasts or where mountains favor the ascent of air masses which result in cooling of moisture laden air.

Heavy rainfalls occur in the tropics between 20°N and 20°S latitudes. The high pressures are between 20° and 40° in both the Northern and Southern Hemispheres and are known as the horse latitudes, regions of low rainfall. Sahara, Ababia, Turkenstan, Gobi Deserts, southwestern United States and northern Mexico, Australia, and parts of Argentina are arid. Between 40° and 60° latitudes, precipitation is fairly heavy. Near the poles high pressures create conditions for low precipitation.

The low rainfall regions such as lower California, Peru, northern Chile, and Senegal and Angola in southwest Africa are caused by cold ocean currents, which allow little evaporation from the surface. Thus, the air from the cold waters contains little moisture.

The summer monsoons bring in moist ocean air into India and Southeast Asian countries. This is due to the low pressure in the interior of Asia.

It is not only the amount of rainfall but also the distribution of rainfall that is important. Approximately 500 mm (20 in.) of rain falls in south–central Canada. This is sufficient for a good crop without irrigation because a large portion of it occurs during the summer growing season. In contrast, about the same amount falls in northern California but nearly all of it is in winter; thus, irrigation is necessary to grow summer crops.

Classification of Regions according to Amount of Precipitation

Arid	<250 mm (10 in.) per year
Semiarid	250–500 mm (10–20 in.) per year
Subhumid	500–1000 mm (20–40 in.) per year

Humid	1000–1500 (40–60 in.) per year
Wet	>1500 mm (60 in.) per year

Precipitation from Fog and Dew

Minute hygroscopic particles are effective in the condensation of water vapor into water droplets called fog. Fine salt particles and smoke are important in fog formation as nuclei for condensation of water vapors. When air masses of high moisture close to the earth are cooled, fog can form. Also, fog forms when cold air comes in contact with warm waters. Water from fog can collect in appreciable amounts on the leaves of plants. This is especially important along the coastal region where in certain seasons rainfall is low or nil, and plants often obtain enough moisture from fog to survive. In the coastal valleys of California, fog is important in the conservation of moisture by plants.

Dew is the result of heat lost from a surface (ground or leaf) by radiation at night. The air adjacent to the ground is cooled to the point of saturation and the water vapor condenses on the cooled surface. The temperature of the air when this occurs is called the dew point. If the dew point is above 0°C (32°F), the condensed water vapor is liquid (dew). However, if the temperature of the surface is below 0°C (32°F), the condensed water vapor is ice (frost).

Dew is very important in arid regions of the world. In Israel as much as 25 mm (1 in.) of dew is deposited in a year. The amount of "dewfall" can be as high as 0.3 mm per night. This is important because water can be directly absorbed by the leaves. Under dry conditions, dew on the leaves increases the turgor of the cells necessary for growth, especially at night. Excess dew drips from the plant and is absorbed by the soil adjacent to the plant. At or near freezing temperatures, the condensation of the water as dew releases heat which prevents the plant from freezing. When water vapor condenses to liquid, 540 calories of energy are released from each gram or milliliter of water.

There are disadvantages in having dew. Dew increases the infection caused by many pathogenic fungi. Dew and rain promote the germination of spores such as late blight of potato *(Phythophthora infestans)*.

Ecological Classification of Plants according to Water Relations

Hydrophytes

Hydrophytes are water-loving, aquatic plants that grow normally in water and swamps. There are many hydrophytic vegetables: taro, water convolvulus, water chestnuts, lotus roots, watercress, etc. Most of these are tropical or semitropical plants.

Mesophytes

Mesophytes are most common of the terrestrial plants; they grow in well-drained soils. These species wilt after losing about 25% of their total water content; in some plants wilting of leaves occurs at much lower percentages.

Xerophytes

Xerophytes are capable of enduring long periods of drought without injury. They can lose from 50 to 75% of their water without wilting. New Zealand spinach and cactus pads are examples of vegetables in this group.

Physiology of Water in Plants

Water is the major constituent in plant tissues; it is the medium in which the metabolic processes occur in cells, and the medium for transport between cells in tissues, organs and the whole plant. Little water is required in the above processes; it is estimated that less than 1% of the water that passes through the plants is used in the photosynthetic process. However, in plants under water stress, photosynthesis and growth are very much reduced. The photosynthetic rate is reduced well before the available soil moisture is depleted.

Most of the water loss from plants occurs by the process called *transpiration*, through openings in the leaf called stomata. By transpiration, minerals absorbed by the roots are transported to the upper part of the plant. Transpiration has a cooling effect on leaves; this is especially important in climate where the air temperatures exceed 40°C (104°F). A rapidly transpiring leaf can lower the leaf temperature as much as 8°C (15°F) as compared to a nonevaporating surface with a similar exposure.

Evapotranspiration is a term used to describe the combined evaporation from the soil surface and transpiration plus cuticular losses of water from the plant. It is expressed as rate of water loss from a given area and is useful in estimating the water requirements of crops growing in the field. About 85 mm (0.33 in.) of water per day evaporates from an open surface of water on a dry still clear day. Table 6.2 shows the water requirement of some vegetable crops.

The critical moisture sensitive stage of some of the vegetables are given in Table 6.3. For most vegetable crops little or no moisture stress during the entire growth period generally gives high yields and good

TABLE 6.2. WATER REQUIREMENT OF SOME VEGETABLES[a]

Shallow rooted	cm	in.	Medium rooted	cm	in.	Deep rooted	cm	in.
Cabbage	30	12	Bean	30–45	12–18	Artichoke	30	12
Lettuce	45	18	Beet	45	18	Asparagus	50	20
Onion	40–60	15–24	Carrot	40	15	Melon	60	24
Spinach	25	10	Cucumber	45	18	Tomato	60	24
Sweet corn	45	18	Summer squash	45	18	Watermelon	45–60	18–25
			Pea	45	18	Sweet potato	45	18
			Pepper	45	18	Winter squash	45	18

Source: Doneen and MacGillivray (1943).
[a] Water required to raise crop to maturity.

TABLE 6.3. CRITICAL MOISTURE SENSITIVE STAGE OF SOME VEGETABLE CROPS[a]

Crop	Moisture sensitive stage(s)
Broccoli	From flower bud development through harvest
Cabbage	From head formation to harvest
Cauliflower	Sufficient soil moisture at all stages
Radish	During period of root enlargement
Turnip	From root enlargement to harvest
Lettuce	At heading stage to harvest; low soil moisture can cause tip burn
Onion	During period of bulb formation
Peas	At flowering through pod enlargement to harvest
Potato, white	From tuber initiation through tuber enlargement
Snap beans	During flowering and pod elongation
Soybean	During plant growth and flowering
Sweet corn	During period of silking and ear development

Source: Some data from Chang (1968).
[a] For most vegetable crops little or no moisture stress during entire growth period generally gives high yields and good quality. Flooding is to be avoided.

quality. Except for hydrophytes flooding is to be avoided as this condition restricts oxygen to the roots.

LIGHT

Light is another important component of the environment necessary for plant life. Light is necessary for photosynthesis. The length of day (length of dark period) is important in inducing morphological changes in some plants.

Light Intensity

Solar radiation at noon varies from about 1.75 g cal/cm^2/min on high mountain tops [12,000 foot-candles (fc) or 130,000 lux] to 1.50 g cal/cm^2/min at sea level (10,000 fc or 108,000 lux). Atmospheric conditions, smoke, dust, gases, and clouds can reduce the amount of energy reaching the earth's surface. Dust (ashes) from volcanic eruptions has affected the atmosphere.

At 4.3 lux (0.4 fc), photosynthesis is negligible. About 1080 lux (100 fc) is the compensation point for many plants. Compensation point is the light intensity at which the rate of photosynthesis equals the rate of respiration. Figure 6.3 shows the National Weather Service Map of the percent of possible sunshine in the United States for the month of June, 1974.

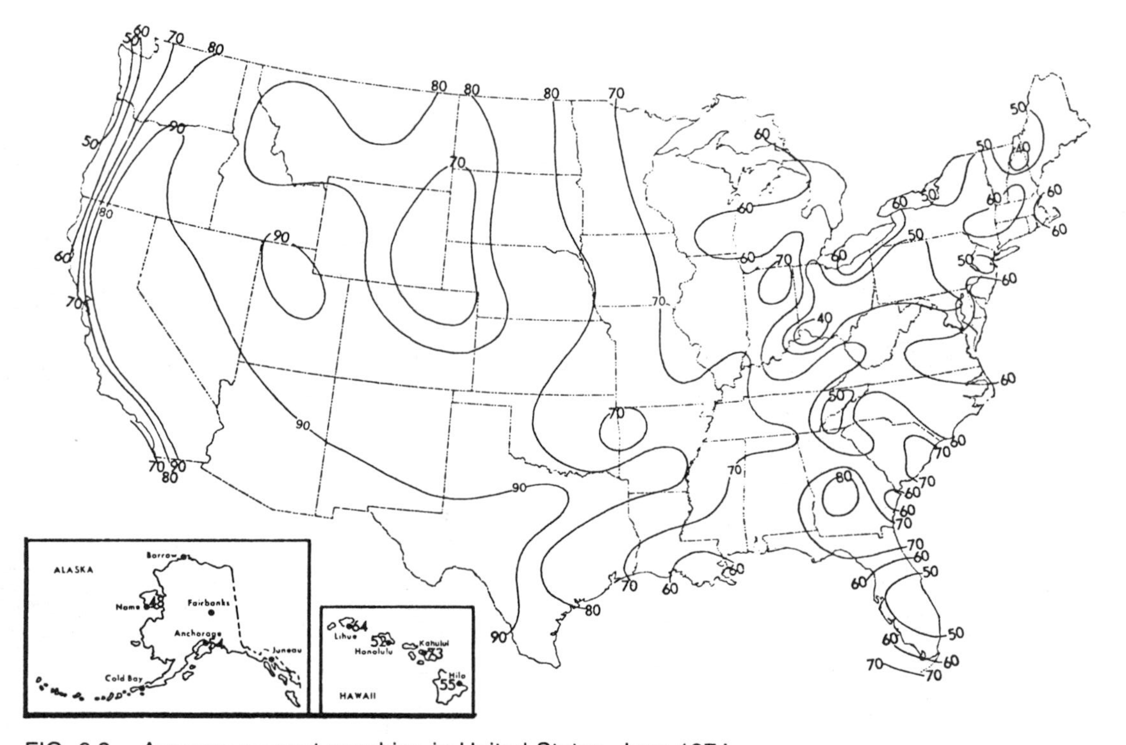

FIG. 6.3. Average percent sunshine in United States, June 1974.

From U.S. Dept. Commerce: NOAA, USDA Stat. Report. Service. Weekly Weather and Crop Bulletin.

Light Quality

White light (sunlight) is composed of all the colors, from violet to deep red (from 400 to 700 nm) the sensitive range of the human eye. Plants respond to a slightly wider range, from about 350 to 780 nm.

Physiological Responses of Plants to Certain Wavelengths of Light

Response	Wavelength (nm)	Remarks
Stem elongation	1000–720 (far red)	
Inhibition of germination of certain seeds	1000–720	'Grand Rapids' lettuce
Stimulation bulbing of onion	1000–720	
Supress bulbing in onion	690–650 (red)	
Red pigment (lycopene) synthesis	690–650	Tomato fruits—43 lux (4 fc)
Stimulate flowering of long day plants	690–650	
Inhibit flowering of short day plants	690–650	
Promote germination of certain seeds	690–650	'Grand Rapids' lettuce
Promote red color formation (anthocyanins)	690–650	Red cabbage color
Photosynthesis	700–400	
Chlorophyll formation	650–400	
Phototropism	500–350	

Duration of Light

Because the earth revolves on its axis, we have day and night, and because the earth is tilted at 66½° and it orbits around the sun, the length of the light period varies from 0 to 24 hr according to the season of the year at the poles (90° latitude) and very little variation at the equator (0° latitude). At the equinox (March 21, September 22), the length of day is the same (12 hr) everywhere on the earth. Figure 6.4 illustrates how the photoperiod changes with seasons at different latitudes. The length of the twilight varies considerably at the high latitudes and must be taken into account in plant responses to light.

The flowering response of plants to the relative length of day or night is *photoperiodism*. Plants that develop and reproduce normally only

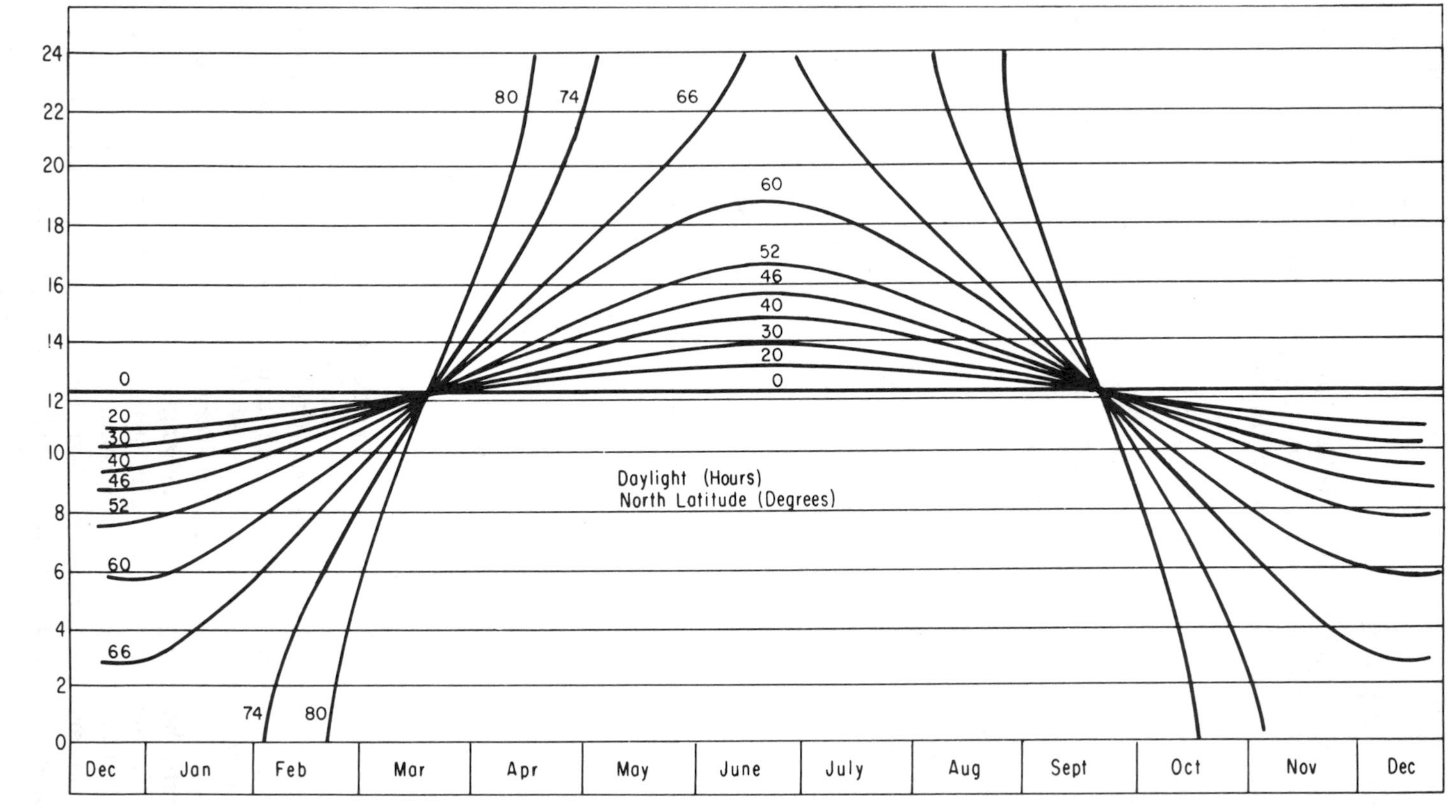

FIG. 6.4. Length of photoperiod with relation to season and latitude.

when the photoperiod is less than a critical maximum are called *short-day plants* and those that flower only when the photoperiod is greater than the critical minimum are *long-day plants*. For short-day plants the *duration of the dark period* is the critical condition, rather than the duration of the light period. *Phytochromes*, which absorb red and far red lights, is the pigment necessary for the response in photoperiodism. In *day-neutral* plants flowering is not affected by photoperiods. Besides flowering, there are other responses of plants to photoperiods. Some responses of vegetable plants to light are listed below:

Flowering Response

Long-day vegetables:	spinach, radish, Chinese cabbage
Short-day vegetables:	soybean, chayote, roselle, sweet potato, chrysantheumum, winged bean, amaranth
Day-neutral vegetables:	tomatoes, early peas, squashes, beans, peppers, eggplant, most cucurbits

Growth Response Other than Flowering

Long days for bulbing:	Onion
Short days for tuber initiation:	White potato, Jerusalem artichoke, yam
Short days for root enlargement:	Cassava, sweet potato

Leaf Area Index

A knowledge of the photosynthetic efficiency of the leaf canopy of a plant is an important consideration in the evaluation of the dry matter production of a crop. Leaf canopy or leaf foliage density is expressed as leaf area index (LAI). LAI is calculated as the total leaf area (leaf blades) subtended per unit area of land.

Most crops have an LAI ranging from 2 to 6; some monocotyledonous plants may have a leaf area index of 9 or even as high as 12. Crops with a vertical foliage have a high LAI. The optimum LAI is not necessarily the maximum LAI. The lowest leaves (least exposure to light) should be barely above the compensation point. If the lower leaves are below the compensation point, the leaves will lose weight or will have to be supported by the upper leaves, resulting in decreased efficiency in dry matter production by the plant.

The optimum LAI will vary with the intensity of solar radiation; it is larger for higher intensity of isolation (Fig. 6.5).

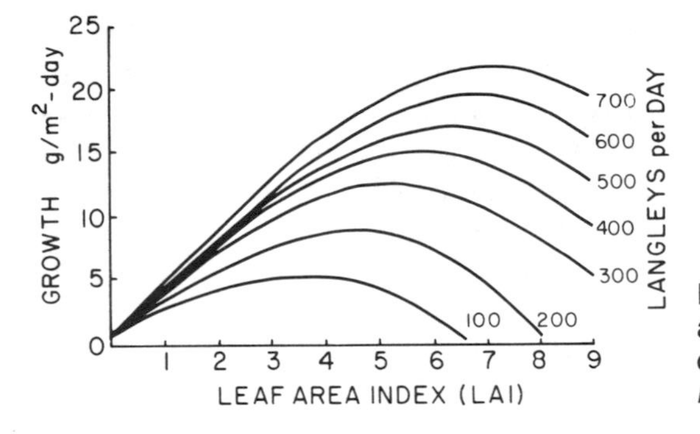

FIG. 6.5. Relationship of leaf area index and solar radiation of subterranean clover.
From Black (1963).

WINDS

Winds are caused by differences in air pressure; temperature differences produce pressure gradients which give rise to air motions. Air masses move from high to low pressure areas. Wind speeds and types of damage to plants are given in Table 6.4. Wind affects plant growth in several ways.

Transpiration

Winds affect the humidity in the atmosphere; they can bring in humid or dry air. They can remove the humid air adjacent to leaf surfaces, increasing transpiration rates and decreasing temperatures.

TABLE 6.4. WIND SPEEDS AND EFFECTS ON PLANTS

Wind speed	mph	km/hr	Remarks
Calm	<1	<2	Smoke rises vertically
Light air	1–3	2–6	Smoke drifts
Breezes			
Light	4–7	7–12	Leaf movement
Gentle	8–12	13–19	Small twigs in motion
Moderate	13–18	20–29	Raises dust, small branches move
Fresh	19–24	30–39	Small trees with leaves begin to sway
Strong	25–31	40–50	Large branches move, whistling occurs
Gales	32–63	51–100	Low speeds: leaves blown off, whole trees in motion; moderate speeds, twigs and branches break; high speeds, trees uprooted
Storm	64–75	101–120	Extensive damage, trees, uprooted
Hurricane, Typhoon	>75	>120	Extensive damage, trees uprooted

CO_2 Concentration

When photosynthesis is rapid under a heavy canopy, CO_2 can become limiting. Winds can effectively replenish the CO_2 supply to leaves deep within the canopy.

Mechanical Damage

High winds can injure or break the aboveground portion of the plants.

Aid in Reproductive Processes

Wind aids in pollen transport of certain species; also, it disperses seeds and spores. In certain species in which the vegetative parts can easily root, pieces of stems or shoots are carried by winds to other favorable habitats. Water currents can serve in the same capacity for hydrophytes.

CLASSIFICATION OF CLIMATES USEFUL IN CROP ECOLOGY

Climate is the average course or condition of weather at a location over a period of years. It is not just the average weather as the variations from the mean are just as important as the mean value itself. The components of weather, i.e., temperature, precipitation, light, and wind, are quite variable from day to day and from season to season. Controls limiting the extent of these variations are the latitude (the maximum height of the sun with respect to the zenith), the distribution of land and water, the semipermanent low and high pressure locations responsible for the wind patterns and the storms, the altitude, mountain barriers, and ocean currents.

The earliest and simplest classification is *tropic*, *temperate*, and *frigid* zones. However, this is not sufficient to study crop ecology. We need to know in much finer details about the climate of a particular region. Many geographers and climatologists have attempted to do this. Wladimir Köppen, an Austrian geographer, has classified the world's climate based on temperature, precipitation, seasonal characteristics, and natural vegetation. Trewartha modified Köppen's classification to relate climate to crop adaptation and distribution. The world climates according to this scheme is shown in Appendix Fig. 1.

Köppen–Trewartha Classification of Climate

A = Tropical rainy—coldest month greater than 18°C (64°F)
B = Dry—evaporation exceeds precipitation—arid and semiarid
C = Humid, mild winter temperate—coldest month between 18°C (64°F) and 0°C (32°F) (mesothermal forests)
D = Humid, severe winter temperate—coldest month below 0°C (32°F), warmest month above 10°C (50°F) (mesothermal forests)
E = Polar—warmest month below 10°C (50°F)

Each major climate is divided further into several subclimates (see Appendix, Table A.1).

Thornthwaite's System

Another modification of Köppen's, includes the consideration of precipitation and evapotranspiration. Evapotranspiration is the reverse of precipitation, the return of moisture to the air from the land and water surfaces and transpiration of plants. This is very important in the consideration of available moisture for growing of crops.

Climate Analogues

Another method of studying crop ecology is the use of climate analogues: the comparison of the similarities of climate of one region of the world with another. By such procedures, one can predict whether a certain crop successful in one region can be grown satisfactorily in a similar climate in a different part of the world. Such extensive studies have been made by the American Institute of Crop Ecology in Washington, DC. They have many publications on agroclimatic analogues. Many detailed studies have been made on small areas; plant climatology maps have been of help to determine where crops can be grown.

The use of infrared aerial and satellite photos reveal much about the vegetation and topography of the earth. These have been used to study present conditions and potentials for crop production.

PHYSIOGRAPHIC AND EDAPHIC FACTORS

Physiographic factors include the topography or relief (elevations and slope) of the land. This affects silting and erosion processes. Topography produces a marked effect on local climates: summits of mountains, slopes, and valleys each have different climates, all within a relatively short distance. High altitude locations have low air and soil temperatures and have greater wind velocities and exposure than valleys.

Slopes may get more or less rainfall than valleys or at very high summits and also more erosion. Slopes have better drainage than level areas. Southern slopes are much warmer than the northern slopes in the Northern Hemisphere.

Edaphic factors include the entire soil environment: the atmosphere, the physical and chemical properties, and organisms.

Plants are dependent on the soil for anchorage, water, and nutrients. The character of the soil is of greater importance to plants under natural conditions than when under cultivation. The edaphic factors are altered by cultivation, fertilization, irrigation, drainage, and cropping.

Soil Composition

Soil is composed mainly of material derived from parent rock and has developed largely through interaction of the substratum with climate and living organisms, i.e., soil is composed of finely divided particles of modified parent material mixed with varying amounts of organic matter ranging from 0 to 100%. Soil is classified according to its texture or its makeup. It is composed of sand (2.0–0.02 mm diameter), silt (0.02–0.002 mm), and clay (<0.002 mm). Figure 6.6 shows the percentage of the three mineral fractions for different soil classifications.

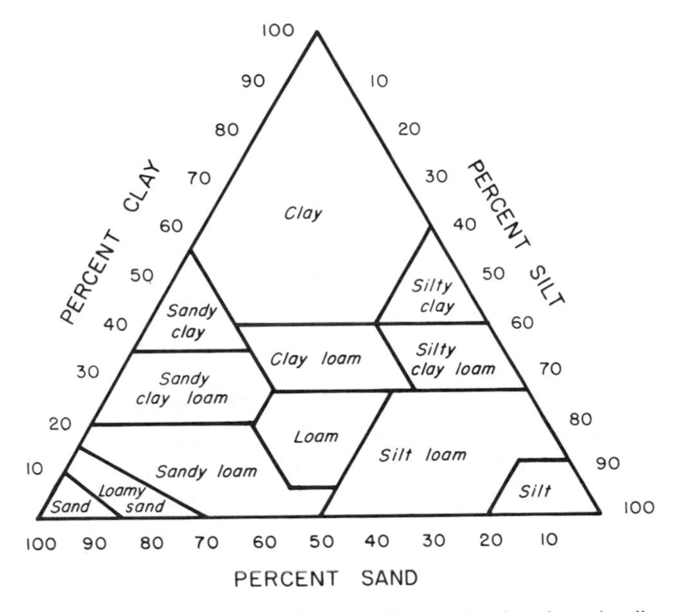

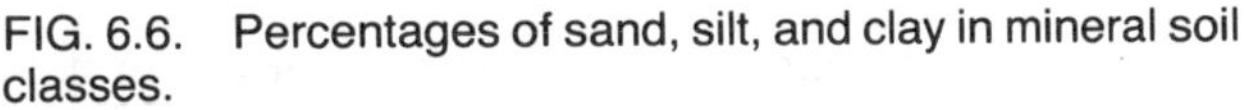
FIG. 6.6. Percentages of sand, silt, and clay in mineral soil classes.

Diagram furnished by O.A. Lorenz, University of California, Davis.

Sand has very low moisture holding capacity and is low in plant nutrients, whereas clay has very high moisture holding capacity and usually high mineral availability. Sand is important because it increases the pore space, which improves the aeration of the soil. Generally, sandy loams are more easily manageable than soils with high clay content. Clay soils tend to drain slowly or poorly; hence, "waterlogging" can occur.

Soil Structure

Soil structure is the arrangement or aggregation of soil particles. Soil aeration depends on soil structure. Good soil structure allows adequate exchange of gases (CO_2 and O_2) needed for root activity. Moderately coarse textured soils have relatively large interstitial spaces.

Compacted soils, resulting from poor soil management, have very little air space and have reduced rates of water penetration. Even with adequate water, plants grow poorly because roots do not develop well due to mechanical impedence and poor aeration.

Mineral Nutrients

Besides water, the soil is the main source of nutrients for the plant. Macronutrients: N, P, K, Ca, Mg, and S; C, H, and O from CO_2 and H_2O can also be included. Micronutrients: Fe, Cu, Mn, Zn, B, Co, Mo, and Cl.

Excessive amounts can cause toxicity and insufficient amounts can cause poor and/or abnormal growth; proper balances of these nutrients are necessary for optimal growth and reproductive processes.

Soil pH

Another important factor is soil acidity or alkalinity (pH). Most plants grow within a range of pH 5.8 to slightly above 7.5. There are "acid-loving" plants and only a few cultivated plants that can tolerate alkaline soil conditions. High rainfall tends to form acid soil, and in low rainfall regions (arid conditions), the soil usually contains carbonates, which create alkaline conditions. Depending on the soil pH, certain elements become unavailable or highly available, which can cause mineral deficiency or toxicity, e.g., in alkaline soils Fe and Mn become deficient, in acid soils these same elements can be available in excessive amounts.

Soil Temperature

Soil temperature is dependent upon air temperature; it lags and is generally lower than the air temperature in the spring and summer and the opposite is true in the fall and winter. Soil temperature is important in seed germination, root growth, and development of underground storage organs. For example, when the soil temperature is above 30°C (86°F), there is poor tuber formation in white potatoes. It is possible for the temperature of dry soil surfaces to be greater than the air temperature.

Soil Moisture

The water in the soil includes vapor, free, capillary, hygroscopic, and crystalline waters. Plant roots can absorb water vapor, free water, and capillary water, but cannot utilize hygroscopic and crystalline water because it is held very tightly to the colloidal particles of the soil. In dry soils some hygroscopic water can volatilize and become water vapor, but the amount is too small to be of significance for plant use. Crystalline water is held very strongly by chemical bonds of the soil minerals and cannot be utilized by plants.

When all the pores are filled with water, the soil is saturated. Free water percolates through the soil as there are no forces except gravity acting on it. After free water has drained, capillary water remains; the soil is said to be at *field capacity.* When most of the capillary water is depleted, the soil is at the *permanent wilting point* at which point plants cannot extract the remaining water from the soil; this is at about 15 atmospheres *diffusion pressure deficit* (DPD) or *moisture tension.* The three states described above are diagrammed in Fig. 6.7.

Sandy soils hold very little water but most of the water is available to plants for use. Clay soils have a large water-holding capacity. Depending on the size of the clay colloids, a fair percentage of this is hygroscopic water (varies from 20 to 40% of the total amount of water) and is unavailable to the plant. The different kinds of water in the soil and the force needed to remove them is depicted in Fig. 6.8. Table 6.5 shows the moist conditions required to germinate some vegetable seeds.

Excessive soil moisture (saturated soils) for prolonged periods, poorly drained soils, and compacted soils can cause damage to plants because of insufficient oxygen for the cells in the roots. Compacted soils are apt to have higher concentrations of ethylene, a plant growth regulator, than noncompacted soils. For active ion uptake and water absorption, root hairs need to be active. It is possible for plants to wilt in spite of the available free water.

Plant growth and photosynthesis are very much reduced before the permanent wilting point is reached in the soil. Irrigation water is supplied to the crop by furrow or sprinkler irrigation systems. Recently, the "drip irrigation" method has been coming into prominent use in the

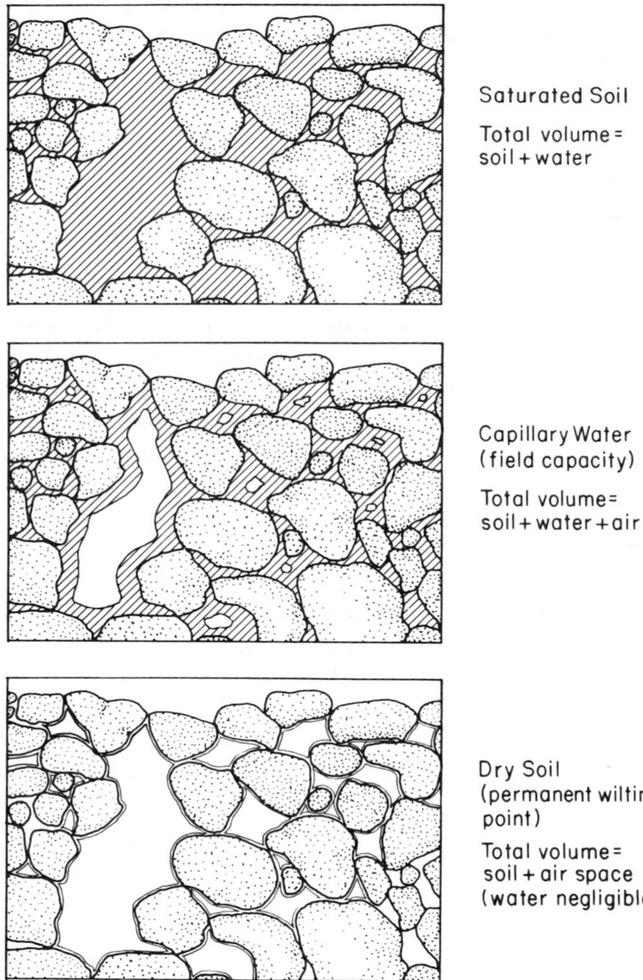

FIG. 6.7. Diagram showing soil water at different moisture contents.

growing of horticultural crops. The system supplies water constant and continuous to each plant; the amount varies with the kind of plant, size, soil type and salinity, and the quality of the water applied. Plant nutrients can be supplied in the applied water. With use of drip irrigation some crops grow faster and yields are higher with less water usage than other methods of irrigation.

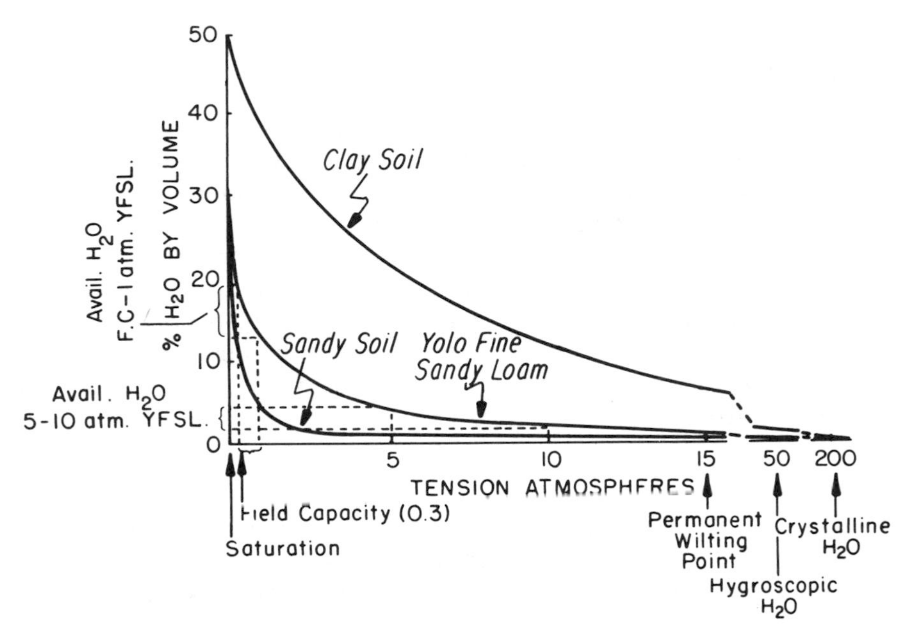

FIG. 6.8. Water content and energy required for water extraction from three types of soil. YFSL, Yolo fine sandy loam.

TABLE 6.5. EFFECT OF SOIL MOISTURE ON GERMINATION OF VEGETABLE SEEDS[a]

Group A.	Germinate over full range of available soil moisture (from field capacity (FC) to permanent wilting point (PWP)).		
	cabbage	summer squash	tomato
	radish	winter squash	pepper
	turnip	muskmelon	
	sweet corn	watermelon	
Group B.	Germinate poor near PWP; good over rest of available moisture range.		
	cucumber	spinach	snap beans
	onion	carrot	New Zealand spinach
Group C.	Germinate well in moister half of available soil moisture range.		
	lettuce	peas	lima beans
	beets		
Group D.	Germinate best at soil moisture near or above FC.		
	celery		

[a] From data of Doneen and MacGillivray (1943). In soils drier than half available moisture, the time lapse from seeding to emergence increases with decreasing soil moisture.

MAN'S ROLE IN AFFECTING THE ENVIRONMENT

Man changes the environment in whatever he does and this is inevitable so long as he inhabits the earth. The objective should be to make only slight changes so as not to greatly disturb the ecological balance.

Atmosphere

Man's use of fossil fuels has increased the CO_2 content of the atmosphere. Whether the increase in CO_2 will increase photosynthesis in plants so that a balance is reached one can only speculate. Increases in atmospheric CO_2 increase the air temperature.

Increased irrigation of arid land has decreased summer air temperatures in certain areas. Air pollution (smog) causes injury not only to man's health but also to plants; spinach, endive, and beets are among the vegetables sensitive to smog. Particles in the air can reduce the energy reaching the surface of the earth, affecting photosynthesis and air temperatures.

Water

Irrigation and drainage of land by man can cause profound changes in local climates. The leaching of arid land can cause problems of excessive salts to users of ground and downstream waters. Also, use of fertilizers can increase the nitrate content of ground and surface waters.

Soil

Unless properly managed, clearing of new land for agriculture can create problems of soil erosion. Usage of heavy farming equipment can cause soil compaction, which can result in poor water drainage and poor soil aeration, leading to reduced yield of crops. Use of agricultural land for urban development, roads and buildings, has changed the local environment. Because of the heat generated in these areas, the temperature is higher by two or three degrees than the surrounding countrysides.

In tropical rain forests where "slash and burn" agriculture is practiced, the soils may become excessively leached. Such soils puddle easily (poor drainage) and become very difficult to manage. Land is abandoned after 4 or 5 years and new land is cleared for growing of crops. The jungle regrows in the abandoned field and after 15 or 20 years the soil can return to a manageable condition. Soil erosion usually occurs in this cycle. These tropical rain forests play an important role in maintaining the water, heat, and O_2/CO_2 balance that affect the overall climate on earth.

BIBLIOGRAPHY

BLACK, J.N. 1963. Relationships of growth, LAI and solar radiation of subterrain clover. Aust. J. Agric. Res. *14*, 20–38.

CHANG, J. 1968. Climate and Agriculture, Adline Publishing Co., Chicago, Illinois.

DONEEN, L.D., and MACGILLIVRAY, J.H. 1943. Germination (emergence) of vegetable seed as affected by different moisture conditions. Plant Physiol. *18*, 524–529.

DONEEN, L.D., and MACGILLIVRAY, J.H. 1943. Suggestions on irrigating commercial truck crops. Univ. Calif., Agric Exp. Sta. Lithoprint Ser. (7686). Berkeley, California.

ENGLISH, J.E., and MAYNARD, D.N. 1978. A key to nutrient disorders of vegetable plants. HortScience *13*, 28–29.

FINCH, V.C., TREWARTHA, G.T., ROBINSON A.H., and HAMMOND, E.H. 1957. Elements of Geography. McGraw-Hill Book Co., New York.

HARRINGTON, J.F., and MINGES, P.A. 1954. Vegetable Seed Germination. Agric. Ext. Serv. Leaflet, Univ. of California, Davis.

MACGILLIVRAY, J.H. 1953. Vegetable Production. McGraw-Hill Book Co., New York.

WILSIE, C.P. 1961. Crop Adaptation and Distribution. W.H. Freeman & Co., San Francisco, California.

7

Devices and Means of Controlling Climate for Vegetable Production in Adverse Climates and During Off Seasons

The production of crops at times other than during normal seasons has been practiced only in recent times. In the tropics and subtropics, many kinds of fruits and vegetables can be grown in abundance practically all year round. However, in the northern and southern temperate climates many crops are almost impossible to grow from late fall through winter due to cold temperatures. It is in these latter regions where farmers have made progress in the growing of crops during "off seasons."

Presently, glass and plastic houses, plastic tunnel structures and plastic mulches have been developed into specialized industries in which thousands of hectares of land are devoted to the culture of vegetables during the winter season in northern Europe, northern United States, Canada, and Japan.

TEMPERATURE CONTROL

Local (Meso-) and Microclimates

Topographical Features, Use of Hillsides

The south slopes on hillsides in the north temperate zone receive more radiation during the day than on level surfaces, and conversely, the north slopes are cooler. Thus, the southern slopes are used for growing crops during cool weather and the northern slopes during hot weather. At night cool air, being denser, drains off the slope of hills, so the slopes are warmer than the valley below.

Shape of Plant Beds

The same kind of effect as planting on hillside slopes is achieved by planting on sides of mounded beds rather than on leveled or flat portions of beds on rows running east–west. Figure 7.1 is a diagrammatic representation of how the radiation is intercepted on the slope and top of the bed in the northern hemisphere. Temperature differences at the different places on the bed are shown in Fig. 7.2 for Davis, California, in mid-April. In the northern hemisphere a 5° angle of slope of the bed has the same effect as moving 50 km (30 miles) to the south.

Soil Temperature

Color

Light colored soils are cooler than dark colored soil.

Texture

Coarse soils tend to warm up faster than fine-textured soils; also, soils high in organic matter do not warm up as fast as mineral soils.

Moisture

Moist soils warm up slower than dry soils because water has a higher heat capacity than soil. Conversely, moist soils cool off slower than dry soils. The cooling effect due to evaporation of water from the soil complicates the soil–temperature relationships.

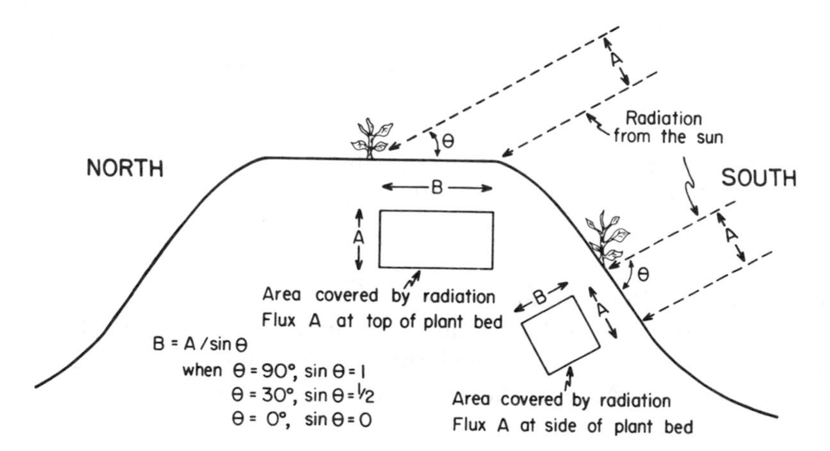

FIG. 7.1. Amount of radiation at the top and side of bed, running east-west, Northern Hemisphere, where θ is the angle of the rays to the surface of the bed.

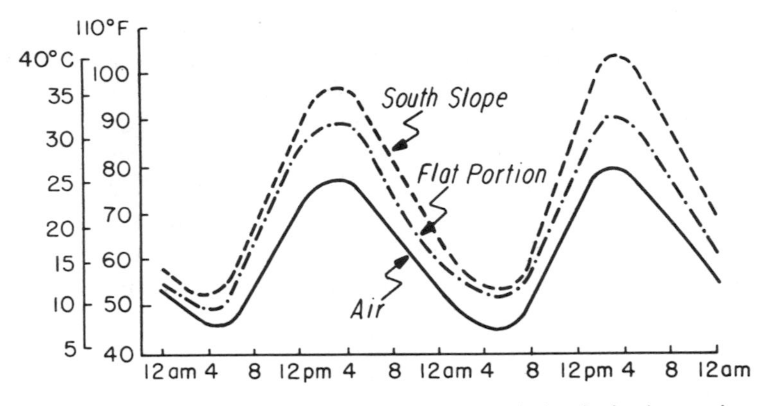

FIG. 7.2. Soil temperature at 12 mm (1/2 in.) depth; beds running east–west; Davis, California in April.
From M.P. Zobel, D.M. Holmberg, and P.A. Minges, unpublished data, 1950.

Mulches

Pulverized Soil

The air between soil particles acts as insulation. Porous soils conduct less heat than compacted soils.

Straw Mulch

Straw and similar materials have a greater insulating effect than pulverized soil mulch. On hot days soil temperatures under straw can be as much as 17°C (30°F) lower than without mulch. Also, it will keep the soil warmer than the ambient air temperature on cold days.

Plastic Mulch

Clear plastic warms the soil more than black plastic. Plastic prevents losses of moisture from the soil. Black plastic mulch is effective in controlling weeds (Table 7.1).

Asphalt or Petroleum Mulch

Water emulsion of asphalt is as effective as clear polyethylene in warming the soil. It is also a good anticrusting agent (Table 7.1). Asphalt mulch was practical when petroleum was inexpensive.

Aluminum Mulch

Thin aluminum with biodegradable backing has been reported to be effective as aphid repellent in viral disease control.

TABLE 7.1. EFFECT OF ASPHALT AND PLASTIC MULCHES ON SOIL TEMPERATURE

Depth		Average temperature 11 AM to 3 PM [a]			
cm	in.	Nonmulched	Asphalt	Clear poly	Black poly
0	0	27.2 (81)	31.1 (88)	31.1 (88)	27.8 (82)
2.0	3/4	25.0 (77)	28.3 (82)	28.9 (84)	26.7 (80)
3.8	1½	23.3 (74)	25.0 (77)	24.4 (76)	20.0 (75)
7.5	3	19.4 (67)	21.7 (71)	21.1 (70)	20.0 (68)
15.0	6	15.0 (59)	16.7 (62)	16.1 (61)	15.6 (60)
30.0	12	13.0 (56)	14.4 (58)	13.9 (57)	13.9 (57)
		Moisture after 24 days[b]			
0–10	0–4	5.2%	9.9%	7.9%	9.3%

Source: Takatori *et al.* (1964).
[a]Air temperature 23.3 C (74 F); 30 cm (12 in.) width mulch.
[b]Initial moisture 19.8%.

Frost Protection

Fogging, Water Sprinkling

Fogging or sprinkling with water is effective in prevention of freezing. Because water has a high heat capacity (1 cal/g), it is effective in frost prevention. Generally, the temperature of the water is higher than the air temperature on frosty nights as heat is released when water cools. This is useful when the temperature drop is rapid. Application should be continuous when air temperature is below freezing. If ice forms, a heavy layer can cause damage to plants from weight of ice alone. When a gram of water freezes, 80 calories of heat are released to the surroundings preventing the temperature from dropping lower.

Water in Irrigation Furrows

The same principle applies as with sprinkling alone.

Heaters and Smudge Pots

Heaters and smudge pots warm up the air. The smoke, as with clouds, hinder radiation (loss of heat) from the ground to the sky. Heaters are most effective when the inversion layer is 9–15 m (30–50 ft) above ground as the volume of air to be heated is relatively small.

Fans, Wind Machines, and Light Aircraft

Large volumes of air are moved by such devices; the cold air near the ground mixes with the warm air above. When heaters and wind machines are used together, more protection is achieved. Currently, helicopters are being used where wind machines are not installed. These devices are effective only when temperature inversion exists. Temperature inversion is a condition in which there is an increase of air temperature with height, so that warmer air overlies a colder air mass, contrary to normal conditions. Even under strong temperature inversions, the gain in ground temperature is small, usually less than 3°C (5°F). If the temperature is down to −3° to −6°C (27°−21°F), wind machines are not usually effective.

Plant Protective Structures

Brushing and Use of Barriers

Bushes or paper on the north side (in the Northern Hemisphere) of plantings is used for protection. These barriers prevent cold north winds from cooling the soil and helps to trap the sun's heat during the day and reduce the radiation losses at night.

Hot Caps

Plastic or glassine paper, a miniature greenhouse, is used to cover the plant. Protection can be for one or more plants in special structures.

Plastic Tunnels

Plastic sheets are supported over rows by wood, wire, or bamboo; these structures are like miniature greenhouses.

Cloches

Cloches are individual plant or row greenhouses of glass. The structure is braced by wires or wood and is very rigid; it can withstand strong winds.

Cold Frames

Cold frames are structures made of wood or concrete with plastic or glass covered tops. Glass traps heat from the sun, making the air several degrees higher than the outside. Cloth or straw mats may be used to cover the glass to reduce heat losses at night.

Hot Beds

Hot beds are cold frames with provision made for heating. Heat sources may be from decomposing manures, flues warmed by hot air, hot water or steam through pipes, or electric heat cables.

Greenhouses

Greenhouses may be glass or plastic; structure is large enough for a man to maneuver. Heating can be performed as with hot beds. Also, the interior can be cooled with evaporative water coolers, whitewash, and the sunlight controlled by louvered slats. Plastic structures are often supported by a large blower creating a positive pressure on the inside.

Baskets and Shade Boards

Baskets and shade boards are used to protect plants from heat and excess light from the sun during the summer in hot climates. They are used to protect transplants.

Lath- and Screenhouses

Lathhouses are used to reduce light intensity and heat from the sun. Screenhouses are used to keep insects away, but they may get hotter than the ambient temperature because air circulation is reduced.

Wind Barriers and Shelter Belts

Winds can injure crops by whipping or blasting from sand and dust. Wind barriers and shelter belts can help alleviate damage from winds. Shelter belts are barriers of living plants (trees and shrubs) of sufficient height to present an obstruction to winds. A good shelter belt can reduce the wind velocity 50% at a distance of 10 times the height of the barrier. The temperatures behind these barriers can be increased from 1.5° to 4°C (3° to 7°F) depending on the type of barrier and soil and moisture conditions. At night, however, more heat can be lost by radiation, particularly if the barrier is dense and causes the air to be stilled. Significant increases in yields and early maturation of crops can be obtained by use of windbreaks.

TRANSPLANTING AND HARDENING

Plants can be started in cold frames, hot beds, or greenhouses early in the season and transplanted when the outdoor conditions become favorable.

Transplanting of bare-rooted plants can be a severe shock to rapidly growing seedlings. Hardening of plants before transplanting can be accomplished by withholding of water and/or by cooler temperature treatment. Growers use peat pots to reduce root damage and transplant shock. Hardened plants can withstand unfavorable conditions better than rapidly growing succulent transplants. Unhardened cabbage seedlings can be injured at –2°C (28°F) but hardened ones are not injured at –6°C (22°F).

OTHER REGULATION OF PLANT GROWTH

Light Control

Intensity

Full sunlight is over 54,000 lux (5000 fc). Compensation point is 1100–3200 lux (100–300 fc), the point at which the plant will maintain itself but will not grow.

When light intensity is low, supplemental lighting is often necessary. Light intensities of at least 8600–11,000 lux (800–1000 fc) are required to grow most plants to maturity.

Maximum efficiency for photosynthesis of plants varies with temperature and light intensity as shown in Fig. 7.3. Also, photosynthetic efficiency can be increased by raising the CO_2 concentration.

Photoperiod

Day length affects physiological processes such as flowering. Plants respond to light intensities of less than 540 lux (50 fc). A long day of 18 hr or more can be achieved by a 3 hr light period in the middle of the dark period or by a flashing light for 4 sec every minute during the dark period.

Day lengths can be reduced by covering the plant with a dark cloth or moving it into a dark chamber. Crops will mature earlier under long

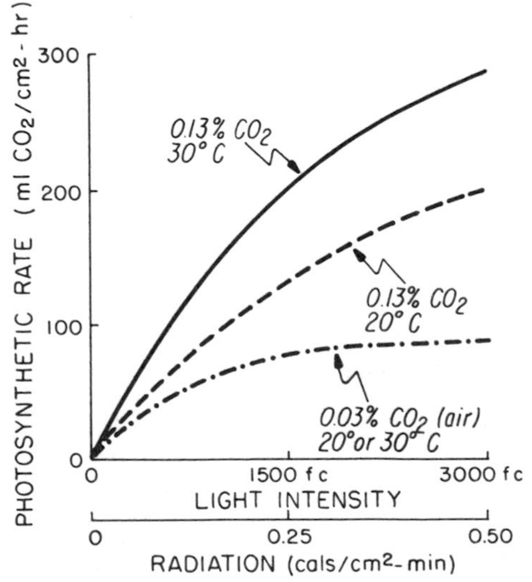

FIG. 7.3. Photosynthetic rate of cucumber leaves under increased CO_2 at 20° and 30°C.
Redrawn from Gaastra, 1963.

days because of longer time for photosynthesis than when grown under short-day conditions.

Application of Growth Regulating Chemicals

p-Chlorophenoxyacetic (PCPA) and β-naphthaleneacetic acid (BNA) are used for setting of tomato fruits under cool temperatures. PCPA in combination with gibberellic acid is used to obtain firm nonpuffy fruits.

Ethephon, 2-Chloroethylphosphonic acid, an ethylene releasing chemical, is used to hasten ripening of tomatoes and red peppers; this chemical is also used to increase perfect or female flowers in cucurbits.

Use of Insects for Pollination

Colonies of bees are used for pollination of cucurbits and other plants requiring insect pollination.

Flies are used for pollination of Umbellifers and Alliums in plant breeding. The pupae, kept dormant under refrigeration until need, are allowed to hatch into adults in cages containing plants.

Mechanical Devices in Pollination

Vibrators and blowers are used in greenhouses for pollination of vegetables.

Anticrusting Agents

These susbtances facilitate emergence of seedlings through soil crusts caused by drying following rains. They also help aeration of soils and prevent root diseases. Gypsum is applied as a band on the seedbeds to prevent crusting.

Vermiculite, which is expanded mica, can be used alone, but is more often used with the asphalt mulch in bands to immobilize the particles on the soil surface over the planted seeds.

Controlled Germination

The use of seeds coated by substances with certain properties which defer germination until conditions are favorable for plant growth is in the experimental stage. Partially germinated seeds are experimentally used to get earlier and more uniform germination and hence uniform stands of seedlings.

BIBLIOGRAPHY

GAASTRA, P. 1963. Photosynthesis of crop plants as influenced by light, carbon dioxide, temperature and stomatal diffusion resistance. Meded. Landbouwhogesch. Wageningen, *59*, 1–68.

HALL, B.J., and BESEMER, S.T. 1972. Plastics in California. HortScience *1*, 373-378.

LIPPERT, L.F., TAKATORI, F.H., and WHITING, F.L. 1964. Soil moisture under bands of petroleum and polyethylene mulches. Proc. Am. Soc. Hort. Sci. *85*, 541.

TAKATORI, F.H., LIPPERT, L.F., and LYONS, J.M. 1961. Petroleum mulch studies for row crops in California. Bull. No. 849, California Agric. Exp. Stn., Univ. of California, Berkeley.

TAKATORI, F.H., LIPPERT, L.F., and WHITING, F.L. 1964. The effect of petroleum mulch and polyethylene films on soil temperature and plant growth. Proc. Am. Soc. Hort. Sci. *85*, 532.

Part III

World Vegetables

8

Global View of Vegetable Usage

The history of man has been the struggle for food (cf. Chapter 2). Throughout the ages the food patterns have been influenced by the availability of the food supply and the social and religious customs of the people. Foods of the common people of nations or regions of the world have not changed much in the past century. Closely followed are the food patterns of our ancestors. Generally people consume, if possible, the same kinds of food that were eaten from childhood. Circumstances such as movement from one region to another where certain food materials are not available, ethnic background of persons preparing food, etc., affect the kinds of foods consumed.

In pre-Columbian times the crops in Europe and Asia were fairly well established. Some interchanges of crops between regions and continents of the Old World took place via traders, itinerant monks, and invading armies. Since the discovery of America, many new species from the New World have been introduced into the Old World and vice versa. With colonization by Europeans, crops from the Old World were established in the Americas, Australia, New Zealand, and parts of Africa as these were the preferred foods of the immigrants.

The mixing of cultures, through intermarriages, etc., brought about additional regional changes in food patterns and the establishment of a mixture of crops, endemic and introduced. An example of such a changing pattern is Brazil, where endemic vegetables, such as cassava, corn, and beans, important to the natives, are mixed with many of the vegetables, such as onions, garlic, carrot, and brassicas, brought by the Portuguese from Europe and with okra, jilo, and watermelon brought by the African slaves; and in more recent times, vegetables of the Orient have been introduced by the Japanese immigrants.

With greater mobility of people and ease of shipment of food commodities in recent years, the food patterns in some parts of the world have changed at accelerated rates.

SOURCES OF VEGETABLES USED FOR FOOD

The sources of vegetables used for food, according to G.J.H. Grubben[1] can be categorized into five types:

1. Wild plants collected from spontaneous vegetation: an estimated 1500 species are used. This type is important in the more primitive areas of developing countries.
2. Wild vegetables growing as spontaneous "weeds" in food crops or in protected compounds. There are about 500 species that are natural selections of primitive varieties. A small percentage of this type comprises vegetables in developing countries.
3. Home gardens and mixed croppings in fields: about 200 species of which many are primitive cultivars. This type comprises about 40% of the vegetables in developing countries and 15% in developed countries.
4. Small scale, labor intensive market production, usually grown in monoculture but sometimes in a multiple cropping system (cf. Chapter 1) where land is scarce. In developing countries this type comprises about 40% and in developed countries less than 10% of the total vegetables produced.
5. Cultivated in highly intensive production system: about 20 species and a small number of cultivars of each species are grown. A small percentage of the total production is of this type in developing countries and over 75% in developed countries.

It is beyond the intent of this book to list plants of types 1 and 2 above. Publications such as Martin and Ruberte (1975) and Knott and Deanon (1967) list many wild plants used as vegetables.

Many of the vegetables of types 3, 4, and 5 are listed in the Appendix, (Table A.1). Besides the scientific and common names, the table gives the climatic regions, according to Köppen–Trewartha's designations, where the vegetable can be grown. The world distribution of climatic types are shown in the Appendix (Fig. A.1). Also, the countries and regions where the crop is of importance are listed. Table A.1 is not inclusive; many of the very minor vegetables used in some regions of the world are not listed.

[1] G.J.H. Grubben, Royal Tropical Institute, Amsterdam, Netherlands. Personal communications, Dec. 1976.

STARCHY ROOTS, TUBERS, AND FRUITS

Starchy vegetables are very important in supplying the energy needs of many of the tropical and semitropical regions of the world. Table 8.1 shows that the developing countries in Africa derive over half of their food production from these crops. Countries of Europe and mainland China obtain from one-fourth to one-third of the energy from root and tuber crops.

The root crops are principally the sweet potato and cassava (manioc; the tuberous crops are the white potato and the yam; the corm crops include taro, dasheen, yautia; the rhizome crops include arrowroot and canna. The types of these storage organs are shown in Table 8.2. Starchy fruits include sweet corn, plantain, and breadfruit.

TABLE 8.1. WORLD PRODUCTION OF ROOTS AND TUBERS AND CEREALS

	Roots and tubers			Cereals
	Area (10^3 ha)	Production (10^3 MT)	Yield (MT/ha)	production (10^3 MT)
World	49,855	547,501	11.0	1,533,076
Continent				
Africa	12,130	82,049	6.8	66,480
North and Central America	1,121	22,393	20.0	356,703
South America	3,952	42,904	10.9	63,602
Asia	19,623	185,190	9.4	629,984
Europe	5,810	122,031	21.0	239,984
Oceania	249	2,634	10.6	24,312
U.S.S.R.	6,970	90,300	13.0	172,011
Developed countries	13,821	239,004	17.3	799,577
Developing countries	36,034	308,497	8.6	753,499

Source: FAO (1979).

TABLE 8.2. TYPES OF STARCHY UNDERGROUND STORAGE ORGANS

	Modified roots	Modified stems
Accepted names	Storage root, tuberous root	Tuber, rhizome, corm
Vascular anatomy	Root	Stem
Lateral branches	Adventitious roots arising from interior (pericycle)	Nodes (eyes) produce adventitious shoots and roots
Vegetative propagation	Crown divisions, plantlets from roots, stem sections with one or more nodes	Whole tuber or pieces of tuber with eyes, rhizome, corm, cormels
Examples	Sweet potato, cassava, arracacha	Tuber: white potato, yam, Jerusalem artichoke; Rhizome: edible canna, arrowroot; Corm: taro, yautia

Among these starchy roots, tubers and fruits, the white or Irish potato is, by far, the most important, the world production in 1979 being 284 million MT (Table 9.1) followed by sweet potatoes with 114 million MT (Table 10.1) and cassava with 117 million MT (Table 11.1). Each of these three crops is important in different parts of the world; the white potato in the temperate regions of North America and practically all of Europe, the sweet potato in central and southern Asia, and the cassava in central Africa and South America (Brazil). The yam and the aroids are confined to specific regions, the world production being 20 million MT and 4¼ million MT, respectively. However, these latter two crops are very important in supplying the energy needs of certain areas of the tropics.

The world production of sweet corn is very difficult to estimate. Sweet corn is important in the United States; 2.2 million MT was processed in 1975 and 0.6 million MT was produced for the fresh market. The world field corn production in 1975 was 313 million MT. A small percentage of this was probably harvested at the "milk stage" (immature) as fresh corn.

Banana production in 1979 was 39.1 million MT and plantain was 20.6 million MT. Unripe bananas are used in many countries as starchy cooked vegetables. In southern Brazil very little plantain is grown but it is very important in the northern South American countries, Columbia in particular.

The efficiency of energy production of the major energy food crops of the world is shown in Table 8.3. Yam and cassava outproduce the grain crops in the energy production on area and time-growing bases. With respect to labor requirements in the tropics, cassava yields the highest per acre and man-day bases and sweet potatoes ranks second with yams a close third (Table 8.4).

TABLE 8.3. COMPARISON OF SOME TROPICAL STAPLE CROPS, MAXIMUM POTENTIAL YIELDS, AND MAXIMUM POTENTIAL ENERGY-PRODUCING CAPACITY

Crop	Yield (Tons/acre/year)	Energy produced (10^3 calories/acre/day)
Rice	10.5	71
Wheat	4.7	45
Maize	8.3	81
Sorghum	5.3	42
Cassava	29.0	100
Sweet potato	26.4	73
Yam	38.0	108
Taro	19.0	59

Source: Coursey and Haynes (1970).

TABLE 8.4. LABOR REQUIREMENTS OF SELECTED TROPICAL CROPS

Crop	Man-days/acre					Yield[b] (lb)	
	Preparation of land[a]	Establishment of crop	After cultivation	Harvesting, threshing, store work, etc.	Total	Per acre	Per man-day
Maize	15	9	13	19	56	1,600	27
Sweet potatoes	16	19	16	21	72	5,430	75
Yams	27	17	61	39	144	8,750	67
Cassava	6	19	63	38	126	12,650	109
Large millets	6	3	11	11	31	852	27

Source: Raeburn *et al.* (1950).
[a] Costs of opening land from bush or grass are not included.
[b] Yields are expressed in terms of threshed grain and fresh roots per acre.

To many people in the Western world, the cassava, yam and the aroids, covered in Chapters 11, 12, and 13 are new and exotic crops, and it is difficult to distinguish the differences among them. Table 8.5 summarizes some of the important similarities and differences amongst these crops.

The nutritive values of the starchy vegetables are compared in Table 8.6 and 8.7, and the essential amino acids contents are listed in the Appendix (Table A.2).

SUCCULENT ROOTS, BULBS, TOPS, AND FRUITS

Most of the vegetables belong to this category. They are relatively high in water content, low to intermediate in caloric content, but high in nutrients such as vitamins A and/or C, minerals, and fibers; some, such as green leafy and leguminous vegetables, are good sources of proteins.

To this group belong hundreds of vegetables, some of which are used by peoples throughout the world because of their wide adaptability to climate, ease of growing, and acceptability, whereas others are restricted to use by few people in localized regions. Table A.1 lists many of them. The total production of vegetables and melons in 1979 is shown in Table 8.8; the consumption of vegetables in developed countries was double that in developing countries. The chapters following starchy vegetables cover crops of major importance as well as the minor ones. Each chapter is concerned with vegetables within a family because of

TABLE 8.5. SOME COMPARISONS OF TROPICAL ROOT AND TUBER CROPS

Crop	Cassava (manioc)	Yam	Taro (dasheen)	Yautia
Classification	Dicot.—Euphorbiaceae *Manihot esculenta*	Monocot.—Dioscoraceae *Dioscorea* Many spp.	Monocotyledon *Colocasia esculenta*	Araceae *Xanthosoma sagittifolium*
Other names	Yuca, mandioca, tapioca plant	Igname (Fr.), Name (Span.), Yampi (Lat.-Am.)	Eddo, "old" coco yam, curcas, gabi (Philippines)	Tannia, "new" coco yam, malanga
Botany of leaf	Deeply palmate, many lobed. (about 1/4 to 1/6 size)	Simple cordate, each has accurate primary veins. (about 2/3 to 1/2 size)	Leaves entire and peltate. (about 1/10 to 1/20 size)	Leaves entire and sagittate, prominent marginal vein. (about 1/10 to 1/20 size)
Storage organ used as food; other edible parts	Roots; tender leaves	Tubers	Corms and cormels; immature leaf blade and petioles	Corms
Types	*Bitter varieties.*—Contain cyanogenic glucosides. Must be leached before use. Present in industrial varieties for starch. *Sweet varieties.*—Not necessary to pretreat. Some bitterness in skin.	Some species contain alkaloid, *dioscorine*. Many contain *saponins*. Those with high saponin content are specially grown for use in cortisone manufacture.	Both taro and yautia (corm and leaf tissues) contain acrid substances, mainly *oxalates*. These are removed by boiling and leaching. Dasheen is a nonacrid variety, low in oxalates.	
Regions grown	Tropics and semitropics 30°N to 30°S latitudes. Can be grown at elevations of 1800 m.	Most are grown in the tropics and semitropics. A few species can be grown in temperate zone during the summer.	Grown mainly in the tropical and semitropical regions of high rainfall. Taro and dasheen are more popular in the southeastern Asian countries, while Yautia in African tropics and in West Indies and tropical America. Crops grown up to 1500 m in the tropics.	
Growth requirements				
Temperature	Grows best at means above 18°C. Growth stops at 10°C and frost kills plant. It can withstand means of over 35°C.	Below 18°C the crop grows poorly; optimum temperature is about a mean of 30°C.	Minimum temperature for growth 10°C; below this chilling can occur.	
Moisture	Grown in areas of 500–5000 mm of annual rainfall. It can withstand periods of drought by dropping its leaves. Roots penetrate 50–100 cm to get moisture.	A crop can be grown with only 600 mm but yields are highest in areas with rainfalls of 1300–5000 mm. Crop can withstand periods of drought.	Adapted to grow in very moist locations. Taro can withstand water logging. In southeastern Asia taro is grown as with rice in paddies; it is also grown in uplands. Crop requires adequate rainfall or water supplied by irrigation. Yautia is grown only in upland culture.	

Soil	Best on sandy loam but can be grown on any type; provided it is not water-logged. Crop can be grown on exhausted soil as last crop of rotation in tropical clearing.	Requires loose, deep soil with good drainage. Waterlogged soil causes tuber rot. Yams have a weak rooting system; therefore, they need good soil for high yields.	Rich, sandy loam is most desirable. This gives well-formed, high quality corms. Crop is shallow rooted, requiring care in cultivation.
Culture			
Propagation	Stems cut from previous crop are used. 30 cm sections stuck vertically in the ground give high yields.	Small tubers or pieces of tuber showing sprouts are planted. Rooted stem sections are also used. Bulbils or aerial tubers. About 1/5 of previous crop needed for propagation.	Mother corms or cormels which are sprouting used for propagation. About 8–10% of the previous crop is saved for propagation. Taro shoot sections (setts) can be used for propagation immediately after harvest.
Vine support	None required.	Staking required; bushes or trees left when clearing ground for support of yam vines.	None required.
Growing season	6–9 months minimum. 12–18 months normally. For starch (tapioca) plants are grown for 30–40 months.	6–8 months minimum. Some are grown for 24 months.	Crop matures in 7–11 months after planting.
Harvest and yield (yields vary tremendously according to region and varieties)	Can be mechanically harvested or hand harvested. 13 to 22 MT/ha.	Tubers easily bruise. Must be carefully dug. 18–27 MT/ha.	Soil drained and allowed to dry out. Corms plowed out or dug by hand. Corms separated from roots and soil after lifting from ground. 13 to 20 MT/ha.
Storage	Roots very perishable after harvest. Cannot be stored more than a few days at 0°–3°C and at 85–90% relative humidity. Roots are "stored" in the ground, i.e., harvested as needed.	Special storage structures: well-ventilated, shaded yam houses. There is little loss after 2 months but losses increase to 50% in 6 months.	Stored at 10°C with good ventilation. Prolonged low temperature causes chilling injury. Can be stored for 6 months under optimum conditions.
Nutritional value	Roots: high carbohydrate, high vitamin C.	Tuber: high carbohydrates.	Corms: high carbohydrates.
	Leaves, cooked as greens: protein and vitamins A and C.		Leaves: good source of proteins and vitamins A and C. Oxalates must be removed by boiling and subsequent leaching.
Special uses	Africa: Cassava, yam, or yautia is pounded in a large wooden mortar to a stiff doughy mass called *fufu*.		
	Tapioca and starch.	Some species contain diosgenin used to make steroid drugs.	Hawaii: *Poi* is made of taro as with fufu but is fermented for a few days.

TABLE 8.6. NUTRITIVE VALUE OF STARCHY ROOTS AND TUBERS IN 100g EDIBLE PORTION

			Macroconstituents					Vitamins									
				g					mg				Minerals (mg)				
Crop	Edible part	% refuse	Energy (cal)	Water	Protein	Fat	CHO	A (IU)	B_1	B_2	Niacin	C	Ca	Fe	Mg	P	Ref.[a]
White potato	Tuber	5	71–75	78–80	2.0–2.3	0.1	16–19	Trace	0.09–0.13	0.01–0.02	0.03–0.06	6–36	9–19	0.7–1.3	20–30	38–65	1
Sweet potato	Root	15	91–115	67–70	1.4–2.0	0.2	22–27	Trace–14,000[b]	0.10–0.20	0.02–0.04	0.02–0.08	20–35	35–60	0.7–0.9	28–34	60–70	1,2
Sweet potato	Leaves	—	49	83	4.6	0.2	10	5,900	0.10	0.28	0.9	70	158	6.2	—	84	1
Cassava	Root	—	132	65	1.0	0.4	33	Trace	0.05	0.04	0.6	19	40	1.4	—	34	2
Cassava	Leaves	—	80	77	6.8	1.4	12.8	30	0.12	0.27	1.7	290	206	2.0	—	86	2
Yam	Tuber	20	85–100	65–75	1.8–2.0	1.4	20–29	Trace	0.08	0.01	0.4	6	23	0.6	29	65	1,3
Taro	Corm	—	85	77	2.5	0.5	19.1	Trace	0.18	0.04	0.9	10	32	0.8	25–46	64	
Taro	Leaf	—	69	80	4.4	0.2	12.2	20,400	0.10	0.33	2.0	142	268	4.3	—	78	
Taro	Petiole	—	19	94	0.2	1.8	4.6	340	0.01	0.02	0.2	8	57	1.4	—	23	
Yautia	Corm	—	132	66	1.7	0.2	31	10	0.13	0.03	0.7	5	14	0.8	—	56	2
Jerusalem artichoke	Tuber	—	65	80	2.2	0.3	17	5	0.20	0.10	—	6	32	0.6	—	88	
Edible canna	Rhizome	—	130	67	0.9	0.1	31	Trace	0.03	0.01	0.4	7	15	1.4	—	63	2
Arrowroot	Rhizome	—	157	57	2.4	0.1	39	0	0.08	0.03	0.7	9	20	3.2	—	24	2
Arracacha	Tuber	—	104	73	0.8	0.2	25	Trace–60[b]	0.06	0.04	3.4	28	29	1.2	—	58	2
Oca	Tuber	—	63	84	1.0	0.6	14	Trace	0.05	0.07	0.4	37	4	0.8	—	34	2
Anu	Tuber	—	52	86	1.6	0.6	11	15	0.06	0.08	0.6	67	7	1.2	—	42	2
Ullucca	Tuber	—	51	86	1.0	0	12.5	0	0.04	0.02	0.3	23	3	0.8	—	35	2

[a] Key to references: (1) Howard *et al.* (1962); (2) INCAP-ICNND (1961) - Table of Composition of Foods Used in Latin America; (3) Ingram and Greenwood-Barton (1962).
[b] Vitamin A value varies with variety.

TABLE 8.7. NUTRITIVE VALUE OF STARCHY FRUITS IN 100g EDIBLE PORTION

			Macroconstituents					Vitamins					Minerals (mg)			
				g					mg							
Crop	Edible part	% refuse	Energy (cal)	Water	Protein	Fat	CHO	A (IU)	B_1	B_2	Niacin	C	Ca	Fe	Mg	P
Sweet corn[a]	Seeds immature	65	116	68	4.0	1.3	22	Trace–640[c]	0.20	0.06	1.7	9	11	1.4	45	125
Plantain[b]	Fruit	—	122	66	1.0	0.3	32	290	0.06	0.04	0.6	20	8	0.8	—	34
Breadfruit[b]	Fruit	—	81	77	1.3	0.5	20	Trace	0.10	0.06	0.7	29	27	1.9	—	33

[a] Howard *et al.* (1964).
[b] INCAP-ICNND. 1961. Table of Composition of Foods Used in Latin America.
[c] Vitamin A value varies with varieties.

TABLE 8.8. WORLD PRODUCTION AND CONSUMPTION OF VEGETABLES AND MELONS

	Production (10^3 MT)	Consumption (kg/person/year)
World	340,342	79
Continent		
Africa	22,217	49
North and Central America	33,182	91
South America	11,405	48
Asia	179,076	71
Europe	63,739	132
Oceania	1,633	73
U.S.S.R.	29,091	110
United States	26,544	121
Developed countries	140,669	122
Developing countries	199,673	63

Source: FAO (1979).

the similarities of physiology, culture, harvest, postharvest responses, and pest and disease management.

BIBLIOGRAPHY

COURSEY, D.G., and HANES, P.H. 1970. Root crop and their potential as food in the tropics. World Crops *22*, 261–265.

FAO. 1980. FAO Production Yearbook, Vol. 33. Food and Agric. Org., Rome, Italy.

HOWARD, F.D., MACGILLIVRAY, J.H., and YAMAGUCHI, M. 1962. Nutrient Composition of Fresh California Grown Vegetables. California Agric. Exp. Stn. Univ. of California, Berkeley.

KAY, D.E. 1973. Root Crops. The Tropical Products Institute, London, England.

KNOTT, J.E., and DEANON, J.R., Jr. 1967. Vegetable Production in Southeast Asia. Univ. of Philippines Press, Manila.

INGRAM, J.S., and GREENWOOD-BARTON, L.H. 1962. The cultivation of yams for food. Tropical Sci. *4*, 82–86.

LEON, J. 1976. Origin, evolution and early dispersal of root and tuber crops. Proc. 4th Int. Symp. Trop. Root Crops, Cali, Colombia. Int. Dev. Res. Center, Ottawa, Canada.

MARTIN, F.W., and RUBERTE, R.M. 1975. Edible Leaves of the Tropics. Antillian College Press, Mayaguez, Puerto Rico.

RAEBURN, J.R., KERKHAM, R.K., HIGGS, J.W.Y. 1950. Report of a Survey of Problems in the Mechanisation of Native Agriculture in Tropical African Colonies. Her Majesty's Stationery Office, London.

TREWARTHA, G.T. 1954. An Introduction to Climate. McGraw Hill Book Co., New York.

WILSIE, C.P. 1962. Crop Adaptation and Distribution. Freeman & Co., San Francisco, California.

Part A: Starchy Roots, Tubers and Fruits

9

White or Irish Potato

Family: Solanaceae (nightshade)
Genus and species: *Solanum tuberosum*

Origin

The cultivated potato originated in the highlands of the Andes in South America (Peru, Colombia, Ecuador, Bolivia) and was first brought to Europe by the Spanish explorers in 1537. The potato had been cultivated by the Incas for over 2000 years prior to the Spanish discovery.

There are over 150 wild species found in Central America, Mexico, and as far north as Colorado in the United States. The natives of these regions used the wild form for food and had not domesticated the potato; they cultivated corn and beans.

The potato was introduced to Europe early in the sixteenth century and to Ireland by Sir Walter Raleigh about 1586. In less than 100 years it became a staple crop in Ireland as well as in Northern Europe. In the early 1700s it was brought to New Hampshire by the Scotch–Irish immigrants. The well-known Irish potato famine, which occurred in 1845 and 1846, was caused by late blight, *Phytophthora infestans*. An estimated million persons died of starvation and another million emigrated, mostly to the United States.

Botany

An herbaceous dicotyledonous annual plant, the potato reproduces asexually by means of tubers (seed potatoes). The potato of commerce is a tetraploid ($2n = 48$); the wild potato is a diploid ($2n = 24$) and there are some hexaploids ($2n = 72$). Tubers form at the end of underground stems called stolons. However, these underground structures are not true stolon; botanically they are rhizomes. The plant flowers and produces small green or purplish-green fruits (a berry) about 1.3–2 cm (½–¾ in.) in diameter which are very poisonous. The seeds are used in breeding and plants from seeds may produce tubers with characteristics unlike the parent.

The only edible part of the plant is the tuber. Morphologically, the tuber is a fleshy stem with buds or eyes in the axil of leaf-scars. The anatomy of the potato tuber is shown in Fig. 9.1.

Culture

Climatic Requirements

Cool season crop. Mean air temperatures in the range of 16°–18°C (60°–65°F) are optimal for high yields. The crop is damaged by freezing temperatures. Tuberization decreases at soil temperatures slightly above 20°C (68°F) and is almost completely inhibited above 29°C (84°F). Different cultivars can vary slightly from these temperatures.

It has been stated that for every 5/9°C or 1°F above the optimum the yield of potatoes decreases by 4%.

Soil temperature. In cool soils the sprouts emerge very slowly; it takes from 30 to 35 days for complete emergence at mean soil temperature of 12°C (52°F). The optimum is about 22°C (72°F); higher temperatures seem to retard emergence. Figure 9.2 shows sprout emergence rates for four night–day soil temperature regimes.

Soil temperatures of 16°C (60°F) night and 18°C (65°F) day gave the highest yield and the highest amount of starch in Russet Burbank variety. In the White Rose variety the highest yield occurred when the night soil temperature was 21°C (70°F) and day soil temperature was 24°C (75°F), but the starch was highest at 16°–18°C (60°–65°F) soil

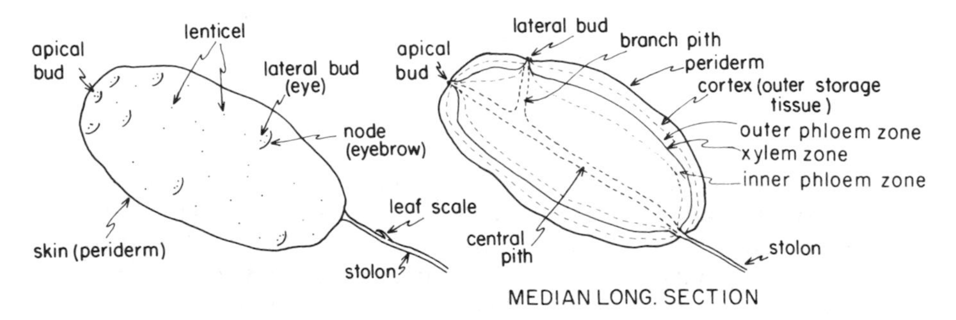

FIG. 9.1. Anatomy of a potato tuber.

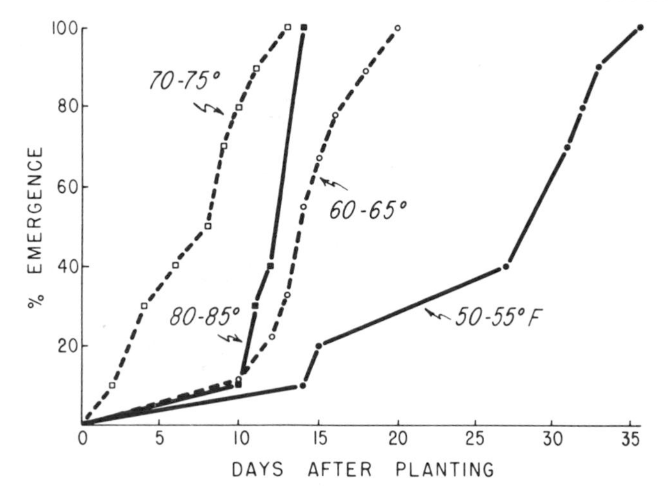

FIG. 9.2. Effect of soil temperature on shoot emergence of White Rose potato. The first temperature indicated is for 12 hr nights and the second temperature is for 12 hr days.
From Yamaguchi et al. (1964).

temperature. High soil temperature seems to increase knobbiness and poor shape in tubers; also several tubers can be found on the same stolon.

Tuber Initiation

Both photoperiod and temperature affect tuber initiation response. Short days induce tuberization. Under long days tuberization occurs if the night temperature is well below 20°C (68°F). Tuberization is optimal at night temperatures of 12°C (54°F). The temperature-sensitive part for tuberization is the tops, and not the stolons.

Low nitrogen level in the plants aids in tuber formation; also, high light intensity seems to enhance the process. Indigenous potatoes of Peru, when planted under the long days of the northern latitudes, will not initiate tubers.

The temperate zone is ideal for growing of the crop because the plants are started in the early spring and the tops established during cool weather. As the weather gets warmer and days longer, carbohydrates are translocated rapidly to the tubers for maximum yields. Night temperature and day-length effects are shown in Fig. 9.3.

Growing season length should be from 90 to 120 frost-free days. Potatoes are grown in areas with much shorter growing seasons in the

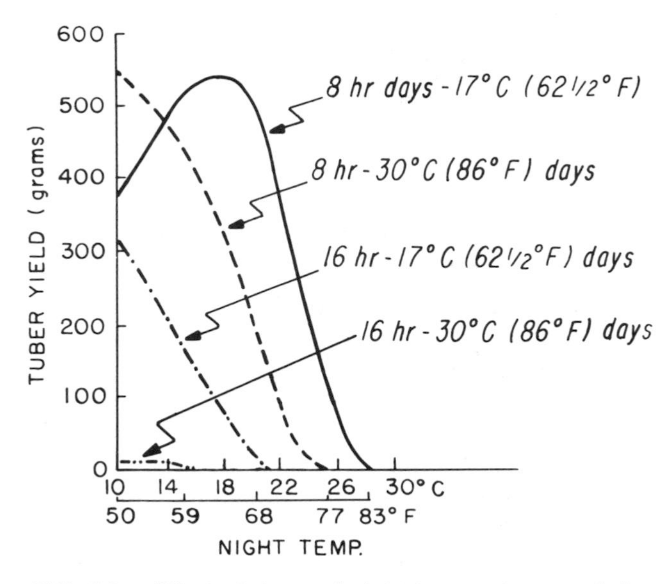

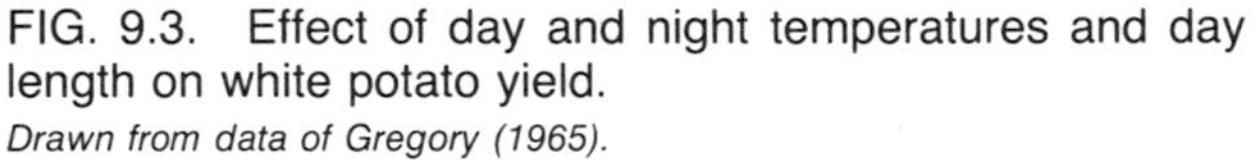
FIG. 9.3. Effect of day and night temperatures and day length on white potato yield.
Drawn from data of Gregory (1965).

northern latitudes where the shortness of the season is compensated by the long days. In northern Europe presprouted seed tubers are often used; this practice shortens the growing period 10 or more days. The growth pattern of plants and tubers are shown in Fig. 9.4 and 9.5.

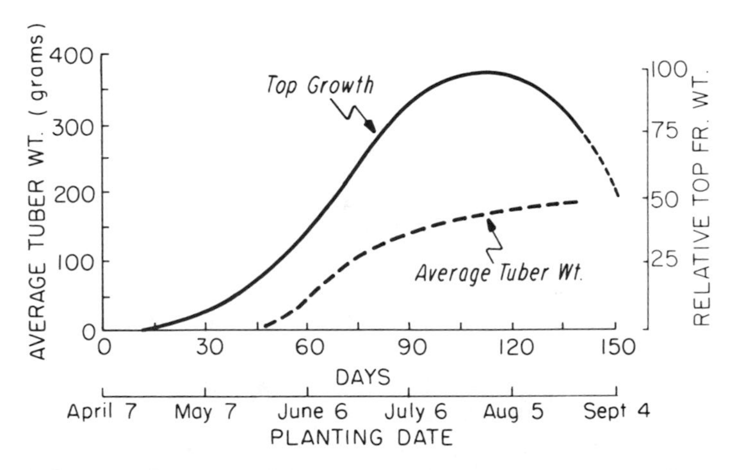

FIG. 9.4. Growth of white potato tops and tubers in Davis, California.

FIG. 9.5. Spatial arrangement of white or Irish potato *(Solanum tuberosum)* tubers in soil. (Instrument at left is a tensiometer for measurement of available soil moisture.)

Soil, Water, and Nutrition

Sandy loam, loam, silt loam, or peat soils produce high quality potatoes. Loose textured, well-drained soil with a pH of 5.0–6.5 is best. Poorly drained, heavy clay soils should be avoided when raising potatoes. In high soil moisture tubers have enlarged lenticles and in heavy soils they are misshapen.

Soil should be at least 60–100 cm (2–3 ft) deep as roots are found below this depth.

A total of 500–750 mm (18–30 in.) of water (rainfall and/or irrigation) during the growing season is required.

Nitrogen (N) [140–220 kg/ha (120–200 lb/acre)], phosphorus (P) [90–100 kg/ha (80–100 lb/acre)], and potassium (K) [110–220 kg/ha (100 to 200 lb/acre)] should be used to fertilize the crop. The amount of N, P, and K will vary according to the fertility of the soil. In heavy rainfall regions (tropics) leaching of N and K and fixing of P could occur. Application at intervals during the initial growth period is recommended. Continued applications of N (also high N) inhibit tuberization.

Planting and spacing. Seed pieces weighing 40–60 g (1½–2 oz) are planted 5–10 cm (2–4 in.) deep in beds 75–90 cm (30–36 in.) wide and 25–30 cm (9–12 in.) between seed pieces. Cut seed pieces having at least one "eye" (node) should be allowed to heal (suberize) at 18°–21°C (65°–70°F) and 85–90% RH for 2–3 days. Seed pieces should be treated with a fungicide. About 2¼ MT of seed tubers are needed per hectare (1 ton per acre).

Certified Seed Potatoes

Because potatoes are asexually propagated and diseases are transmitted to the plant from the seed piece, disease-free seed tubers are required for quality and high yield.

Certified seed potatoes are grown in the cool regions where disease symptoms are expressed in the tops of the plants. The symptoms are often not expressed under warm conditions. Plants showing disease symptoms are removed (rogued) and the tubers and plants eradicated. Seed potatoes are grown in Maine, Idaho, Minnesota, North Dakota, Oregon, and northern California in the United States. Netherlands, Belgium, and Germany ship certified seed potatoes to southern Europe, the Middle East, and Egypt.

Foundation seeds are used to grow *Certified seeds* and the latter are used by the grower for commercial plantings. Both Certified and Foundation seed are inspected and controlled by the states in the United States and by the governments in the various countries where such service exists.

Besides Certified and Foundation seed potatoes, Canada has three additional classes: Elite I, II, and III. Elite I is produced from tubers that have been tested for virus infection through eye indexing and bacterial ring rot. Elite I is used to produce Elite II, and II to produce III. Elite III is used to produce Foundation seed.

Virus-free potato plants can be produced by tissue culture techniques. Using this technique thousands of plants can be produced from a single tuber. This can shorten very much the time between selection and commercial production.

Harvesting and Storage

Harvest

As tubers mature, the vines turn yellow. For early harvest the vines are beaten down or cut a few days before digging to "set" the skin on immature tubers. Tubers should be dug and carefully handled as bruising and subsequent black spot formation can occur.

Curing

A period of 4–5 days at 16°–21°C (60°–70°F) in high humidity will suberize or heal cuts and surface injuries of the tuber. Curing is usually not practiced by growers.

Storage

Storage at 4°–10°C (40°–50°F) at RH of 90% is recommended. Low temperature storage induces conversion of starch to sugar. Reconditioning (conversion of sugar to starch) is accomplished by transferring to room temperature 18°–21°C (65°–70°F) for 7 or more days. Tubers should be stored in the dark to prevent greening (chlorophyll formation) and solanine formation (see Toxins).

Sprout Inhibitors

Maleic hydrazide spray at 1000–6000 ppm solution to the plant 2–3 weeks before harvest has been effective in inhibiting sprouts. Also a 0.5% solution of Chloro IPC (isopropyl *N*-tetrachlorocarbamate) and nonyl or amyl alcohol at 0.05–0.12 mg/liter concentration in the atmosphere of storage rooms have been effective in controlling sprouts.

Rest and Dormancy of Tubers

Freshly harvested tubers will not sprout even when placed under favorable environmental conditions of temperature and moisture. This period is called *rest*, which lasts for a period of about 4–15 weeks, depending on the cultivar, holding temperature after harvest and the maturity at harvest. It is often called "internal dormancy."

Following this period the tubers are *dormant* ("external dormancy"). The dormant tubers, when put under favorable temperature conditions, will sprout.

Breaking of the rest period can be accomplished by storage of the freshly harvested tubers at 21°–27°C (70°–80°F) at high humidity, wounding or cutting of tubers, and by lowering the O_2 concentration surrounding the tuber.

There are chemical methods of breaking rest. One percent ethylene chlorohydrin solution or as a gas in a closed space has been used. This method is very effective but is no longer used because of the dangerous effects of the chemical on humans. Soaking of tubers in a 1% solution of thiocyanate, soaking of tubers in 4½–6 g/liter (6–8 oz/10 gallons) of calcium carbide (acetylene gas) for 4–5 hr, or dipping for 5 min in 5–25 ppm gibberellic acid are effective in breaking the rest.

Nutritive Value

Potatoes are high in carbohydrates, the components of energy. They are also high in minerals and vitamin C but they are low in proteins and provitamin A. Table 9.1 shows the changes in composition during

TABLE 9.1. NUTRITIVE VALUE PER 100 g EDIBLE PORTION OF WHITE ROSE POTATOES GROWN AT DAVIS, CALIFORNIA, AT DIFFERENT STAGES OF GROWTH

								Minerals			Vitamins			
Harvest date	Days after planting	Av. wt. per tuber (g)	Water (g)	Total sugars (g)	Starch (g)	Protein (g)	Energy (kcal)	Ca (mg)	Fe (mg)	P (mg)	C (mg)	B_1 (mg)	B_2 (mg)	Niacin (mg)
June 7	60	50	85.3	0.75	8.12	1.90	43.7	13	3.1	37	38	0.064	0.019	0.57
June 21	75	115	85.0	0.88	8.92	1.73	46.7	10	1.3	32	43	0.072	0.019	0.72
July 5	90	140	84.3	0.98	10.45	1.86	53.7	12	2.0	37	42	0.093	0.016	0.68
July 19[a]	105	155	81.8	0.28	12.61	2.10	60.7	11	1.8	45	48	0.106	0.024	0.62
Aug. 8	120	175	81.6	0.18	11.52	2.09	56.0	12	1.9	45	39	0.101	0.023	0.67
Aug. 17	135	192	80.8	0.25	11.86	2.37	58.7	14	2.0	53	36	0.117	0.025	0.66

Source: Yamaguchi *et al.* (1960).
[a] Usual shipping maturity for Kern County potatoes.

TABLE 9.2. NUTRITIVE VALUE PER 100 g EDIBLE PORTION OF MATURE WHITE ROSE POTATOES GROWN AND STORED AT DAVIS, CALIFORNIA

	Water (g)	Total sugars (g)	Starch (g)	Protein (g)	Energy (kcal)	Minerals			Vitamins			
						Ca (mg)	Fe (mg)	P (mg)	C (mg)	B_1 (mg)	B_2 (mg)	Niacin (mg)
At harvest 8/17/49	81	0.25	11.9	2.4	59	14	2.0	53	36	0.12	0.025	0.66
41°F storage												
3 weeks	81	0.94	10.7	2.4	57	17	1.6	46	26	0.13	0.025	0.70
6 weeks	80	1.23	11.0	2.4	59	13	1.8	49	19	0.13	0.032	0.70
9 weeks	81	0.94	—	2.3	—	11	1.6	46	17	0.13	0.033	0.64
12 weeks	80	0.87	10.5	2.4	56	13	1.7	46	16	0.12	0.034	0.62
18 weeks	81	0.73	9.7	2.7	53	12	1.9	49	11	0.12	0.027	0.60
24 weeks	81	0.63	11.3	2.6	59	11	1.9	54	10	0.11	0.031	0.56
30 weeks	81	0.52	10.1	2.7	54	12	2.6	54	10	0.13	0.027	0.56
50°F storage												
18 weeks	80	0.16	10.2	2.6	53	14	1.7	52	12	0.13	0.027	0.60
24 weeks	80	0.18	12.1	2.6	60	14	2.0	56	10	0.11	0.029	0.58
30 weeks	80	0.20	11.8	2.6	59	14	—	55	8	0.11	0.026	0.60

Source: Yamaguchi *et al.* (1960).

growth of tubers and Table 9.2 shows the changes during storage at 5° and 10°C (41° and 50°F).

Toxins

When potato tubers are exposed to light, an alkaloid, solanine, forms. The amount depends on length of exposure, intensity, and quality of light. Solanine tastes bitter; ingestion of large amounts can cause sickness and, in extreme cases, death. Exposure to light also causes greening, which is chlorophyll, a nontoxic compound. However, greening and solanine usually occur at the same time so eating of green potatoes should be avoided. The highest concentration of solanine occurs in the skin; peeling removes a large percentage but not all of it.

Pests and Diseases

Insects

Colorado potato beetle, potato aphid, leafhopper, wireworm and tuber worm, and spider mites are pests of potatoes.

Diseases

Bacterial ring rot *(Corynebacterium sepedonicum)*, blackleg *(Erwinia atroseptica)*, early blight *(Alternaria solani)*, late blight *(Phytophthora*

TABLE 9.3. WORLD PRODUCTION OF WHITE POTATOES *(Solanum tuberosum)*

	Area (10^3 ha)	Production (10^3 MT)	Yield (MT/ha)
World	18,350	284,471	15.5
Continent			
Africa	528	4,629	8.8
North and Central America	726	19,551	26.9
South America	1,017	10,093	9.9
Asia	3,269	36,928	11.3
Europe	5,796	121,884	21.0
Oceania	45	1,086	24.2
U.S.S.R.	6,970	90,300	13.0
Leading countries			
1. U.S.S.R.	6,970	90,300	13.0
2. Poland	2,441	49,582	20.3
3. United States	518	15,769	30.5
4. China	1,534	14,040	9.2
5. German Democratic Republic	527	12,540	23.8
6. India	790	10,125	12.8
7. Federal Republic of Germany	277	8,747	31.6
8. France	268	7,139	26.6
9. United Kingdom	204	6,485	31.8
10. Netherlands	166	6,277	37.8

Source: FAO (1979).

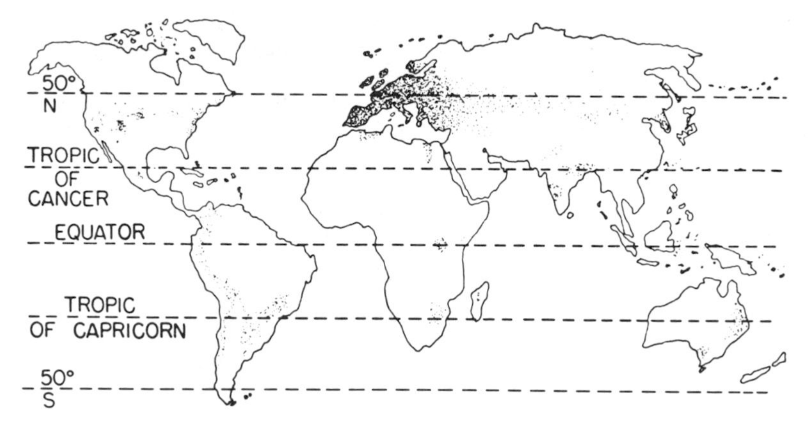

FIG. 9.6. Potato *(S. tuberosum)* growing regions of the world.

infestans), rhyzoctonia *(Corticium vagum)*, scab *(Actinomyces scabies)*, verticillium and fusarium wilts are diseases infecting potatoes. At least 10 viruses transmitted by aphids and insects infect the potato plant.

Economics

World production (Table 9.4 and Fig. 9.6) show that Europe, U.S.S.R., and North America are the principal producers. The per capita consumption in the United States is about 55 kg (120 lb) per year, 50% of which is fresh and the other 50% is as processed products. The trend is toward increased processed products and less for the fresh market.

BIBLIOGRAPHY

ALLEN, E.J., and SCOTT, R.K. 1980. Analysis of growth of the potato crop. J. Agric. Sci. *94*, 583–606.

BURTON, W.C. 1966. The Potato. Veeman and Zonen, Wageningen, Holland.

COX, A.E. 1967. The Potato. Collingridge Ltd., London.

GREGORY, L.E. 1965. Physiology of tuberization. *In* Encyclopedia of Plant Physiology, Vol. XV, Part 1, pp. 1328–1354 Springer-Verlag, Berlin.

HARRIS, P.M. 1978. The Potato Crop. Chapman and Hall, London.

IVINS, J.D., and MILTHORPE, G.L. 1963. The Growth of the Potato. Butterworths, London.

SIMMONDS, N.W. 1976. Potatoes, *In* Evolution of Crop Plants. N.W. Simmonds (Editor). Longmans-Green, London.

SMITH, O. 1968. Potatoes: Production, Storing and Processing. AVI Publishing Co., Inc. Westport, Connecticut.

STEVENSON, F.J. 1951. The potato—Its origin, cytogenetic relationships, production, uses and food value. Econ. Bot. *5*, 153–171.
UGENT, D. 1970. The potato. Science *170*, 1161–1166.
YAMAGUCHI, M., PERDUE, J.W., and MACGILLIVRAY, J.H. 1960. Nutrient composition of White Rose potatoes during growth and after storage. Am. Pot. *37*, 73–76.
YAMAGUCHI, M., TIMM, H., and SPURR, A.R. 1964. Effects of soil temperature on growth and nutrition of potato plants and tuberization, composition and periderm structure of tubers. Proc. Am. Soc. Hort. Sci. *84*, 412–423.

10

Sweet Potato

Family: Convolvulaceae (morning glory)
Genus and species: *Ipomoea batatas*
Other names: Kumara (Oceania),
batata (Spain).

The moist type of sweet potatoes are incorrectly called yams in certain parts of the United States.

Origin

Since prehistoric times the sweet potato has been cultivated in two widely separated parts of the world, Mesoamerica and the South Pacific Islands. Most systematic botanists now agree that, from genetic evidence, the origin of the sweet potato is in tropical America. The closest relative is *I. trifida*, found wild in Mexico. It was believed earlier that sweet potato came from *I. tiliaceae*, a wild species found in the West Indies. *Ipomoea batatas* ($2n = 90$) is a hexaploid.

Botany

The sweet potato is a dicotyledonous perennial plant with long trailing vines with cordate to lobed leaves and morning glory-like flowers. The edible part is the swollen storage roots (Fig. 10.1). From 8 to 15% of the roots in the root system show secondary growth a few weeks after planting. The total number ranges from 4 to 10 tuberous roots per plant. A diagrammatic cross section of a sweet potato root is shown in Fig. 10.2. The structure is a typical root; however, all roots do not have the sweet potato's capability of initiating adventitious shoots. Skin color may be whitish, tan, yellow orange, red, or reddish purple and the flesh color may be white, yellowish orange, orange, salmon orange, or red.

FIG. 10.1. Sweet potato *(Ipomoea batatas)* storage roots of a single plant.

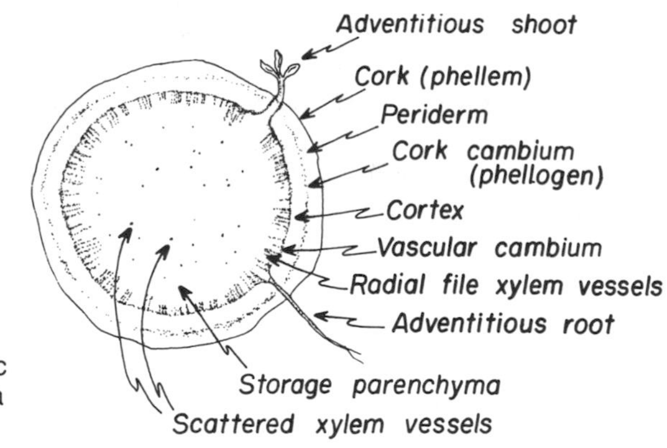

FIG. 10.2. Diagrammatic transverse section of a sweet potato root.

Culture

Climate

Sweet potato is grown in tropical, subtropical, and warmer temperate regions during the frost-free periods. Optimum mean temperature is 24°C (75°F) and it grows best in areas of warm days and warm nights. It can be grown under extremely hot conditions. Plants stop growing at 15°C (59°F) and die due to chilling injury if kept at 10°C (50°F) or below for prolonged periods. Figure 10.3 depicts the regions suitable for sweet potato culture.

Long days promote vine growth. Short days induce root enlargement and flowering. Once root enlargement is induced the process continues under longer day lengths. Yield of storage roots under 11½–12½ hr day length are higher than under very short (8 hr) or very long (18 hr) day lengths.

Propagation

Crop is usually started by transplants, which are obtained by planting "seed" roots in beds of sandy soil. Several small shoots develop from the vascular cambium and emerge through the cortex of the root. The shoots develop adventitious roots and become small plants attached to the seed root (Fig. 10.4). These plants, called transplants, draws, sprouts, or slips, are ready for the field when they have 6 to 10 leaves. The sprouts are developed in beds kept at 24°–27°C (75°–80°F) soil temperature and require about 6 weeks before they are ready for transplanting. Cut sprouts, which are the aboveground portion of stems, are cut at ground level when they are 20–25 cm (8–10 in.) long. These are often used for propagation when roots of the plantlets show disease.

Another method of propagation uses vine cuttings. This method is popular in the tropics because a second crop of sweet potatoes can be planted immediately by using vines from the first crop. It is best not to plant again in the same field but to use a planned rotation system. Cuttings of vines with eight or more nodes about 30–45 cm (12–18 in.) long are used. Vine cuttings with the apex are preferred rather than

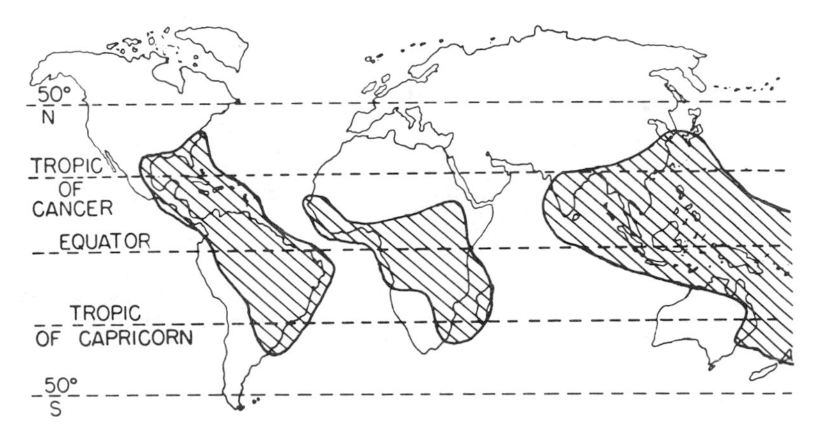

FIG. 10.3. Sweet potato producing regions of the world.

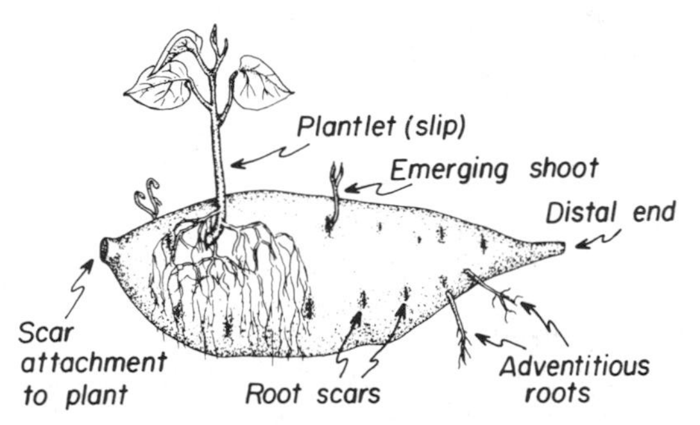

FIG. 10.4. Plantlets (slips or sprouts) growing from a sweet potato root. Plantlets are removed and used for propagation.

from the basal or midportions of the vines. The vines are held for 2 or 3 days under humid conditions for healing at the cut surfaces before planting. When planting about half of the vine is buried in the soil. Vine cuttings and cut sprouts do not transfer soil borne diseases and nematodes as do sprouts with roots. Plant spacing is 30–40 cm (12–15 in.) apart in rows 100–120 cm (3½–4 ft) apart.

Certified Seed Roots

Certified root stocks are available in the southern United States. Many states have Foundation, Registered, and Certified seed stock programs.

Soil

Moderately friable fine sand loam, sand loam, or loamy fine sands, with good drainage, and with pH range of 5.6 to 6.6, is best for growing of sweet potatoes.

Fertilizer

Recommendations for sweet potatoes are

34–67 kg N/ha (30–60 lb N/acre)
56–112 kg P/ha (50–100 lb P/acre)
84–112 kg K/ha (75–100 lb K/acre)

Moisture

About 25 mm (1 in.) of rainfall or applied water per week is necessary during the growing season. A total of 450–600 mm (18–24 in.) is necessary for the crop.

Growth

The crop is an annual in the temperate zone and is usually allowed to grow until the cold weather or frost kills the vine. In the tropics the vines are allowed to grow until the roots reach the desired size. Figure 10.5 shows the growth of the sweet potato vines and roots under temperate conditions.

Harvest and Storage

Harvest

Usually harvesting is from 130 to 150 days from transplanting. Roots are considered ready for harvest when the leaves turn yellow and abscise or the exudate from a cut root dries rapidly and does not change color. However, harvests of sweet potatoes can be made at any time. To avoid chilling injury of the roots, harvests should be made before the soil temperature gets below 10°C (50°F). Yields from 9 to 14 MT/ha (4–6 tons/acre) are considered good.

Curing

Curing heals the wounds on the roots from harvesting. A periderm is formed under the cut area and a corky layer is formed during the curing which prevents microbial invasion and also water loss. Conditions for

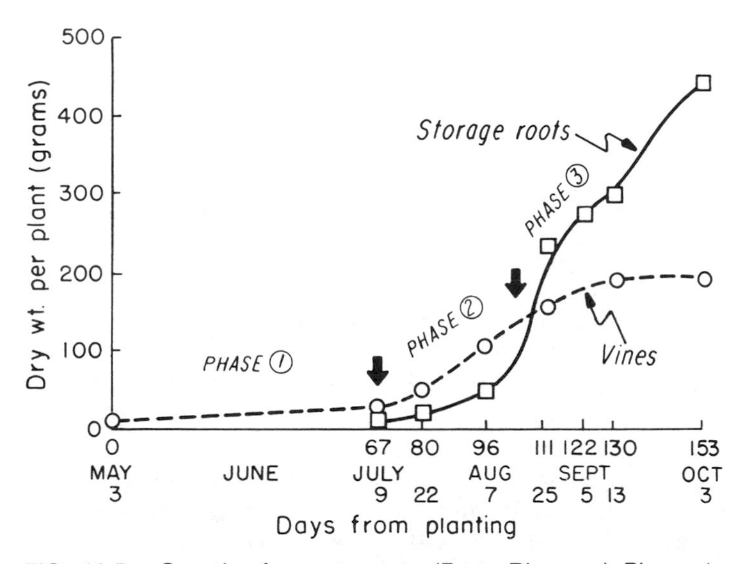

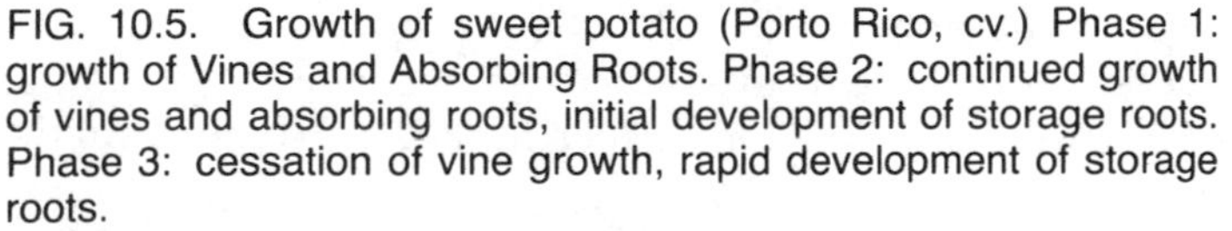

FIG. 10.5. Growth of sweet potato (Porto Rico, cv.) Phase 1: growth of Vines and Absorbing Roots. Phase 2: continued growth of vines and absorbing roots, initial development of storage roots. Phase 3: cessation of vine growth, rapid development of storage roots.

From Scott (1950)

curing are 27°–29°C (80°–85°F) at 85–90% RH for 4 to 7 days. Healing rate is slower below 29°C (85°F) or above 35°C (95°F) and also at RH values below 85%.

Field Curing

In warm climates roots are stacked 60–90 cm (2–3ft) high in piles 1–2 m (3–6 ft) long. They are covered with sweet potato vines 12–15 cm (5–6 in.) thick. Vines act as insulation and also to raise the humidity in the pile. Often paper or plastic sheets are placed over the pile to ensure high humidity and reduce circulation of air through the stacked tubers. Respiration of the roots raises the temperature. Roots are left in the pile 7–10 days for the curing process.

Storage

After curing, temperatures should be lowered to 13°–16°C (55°–60°F) at RH of 85–90%. Lowering temperature increases the storage life of the roots due to a decrease in respiration rate. The 'Porto Rico' cultivar can be stored up to 6 months, while the 'Jersey' roots can be stored for only 2–3 months before they get pithy.

If storage is near 0°C (32°F), damage to roots can occur in a few days; freezing occurs at −1.1°C (30°F). Storage at 10°C (50°F) or lower results in chilling injury, causing internal discoloration and tissue breakdown. Loss of flavor occurs if roots are stored at $<7\%$ 0_2 or $>10\%$ CO_2.

Certain carbohydrate changes occur during curing and storage (Fig. 10.6) and considerable changes occur during cooking (Fig. 10.7).

Nutritive Value

Yellow fleshed sweet potatoes are high in carbohydrates and β-carotene (provitamin A). Roots with orange colored flesh are higher in carotene than the white or pale yellow varieties. The red and purple skin color is from anthocyanins and has no relation to the flesh color.

Roots are high in vitamin C (ascorbic acid), containing as much as in tomato juice (15–30 mg/100 g).

The protein content is "average" with respect to other vegetables, containing 1.5–2.0 g protein per 100 g. The limiting essential amino acids are the sulfur-containing ones, methionine and cystine, and it is low in lysine. The protein quality relative to egg protein is 51% for the sulfur-containing amino acids and 53% for lysine.

Tender leaves and shoots are used as greens in Southeast Asia. These are high in vitamins A and C and much higher in protein than the roots.

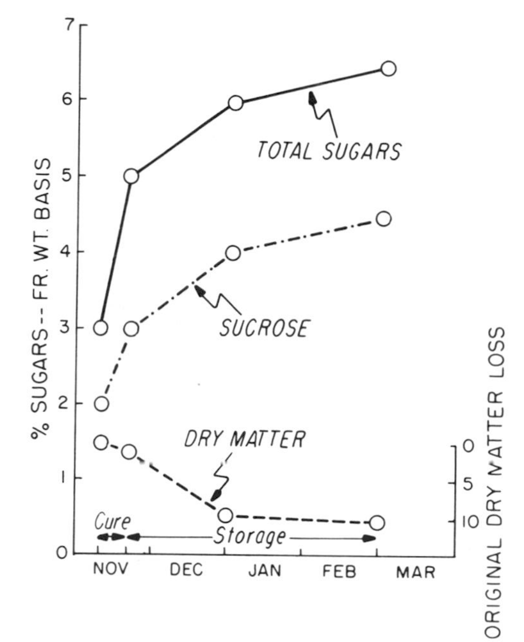

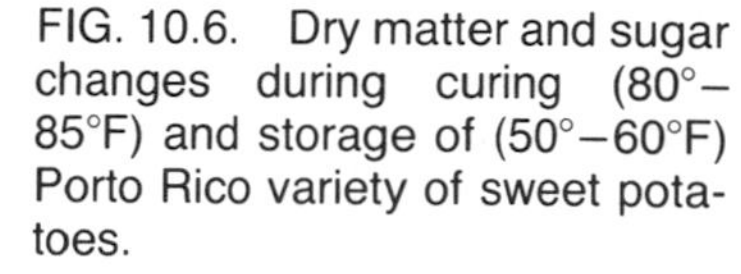
FIG. 10.6. Dry matter and sugar changes during curing (80°–85°F) and storage of (50°–60°F) Porto Rico variety of sweet potatoes.
From Morris and Mann (1955).

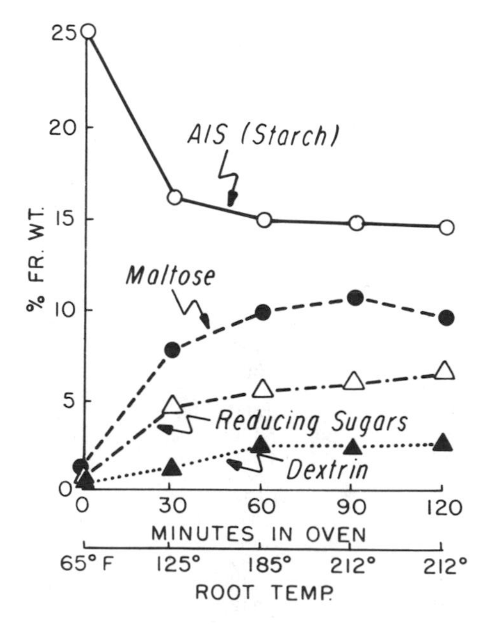

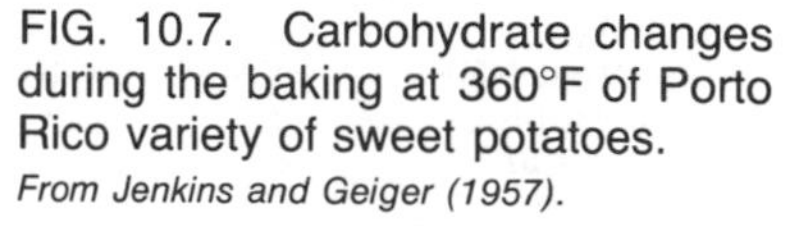
FIG. 10.7. Carbohydrate changes during the baking at 360°F of Porto Rico variety of sweet potatoes.
From Jenkins and Geiger (1957).

Diseases and Pest

Fungi

The fungus diseases are stem rot *(Fusarium oxysporum)* (field), black rot *(Ceratocystis fimbriata)* (field and storage), soft rot *(Rhizopus stolonifer)* (storage), and dry rot *(Diaporthe batatas)* (storage).
For prevention of fungal diseases the following should be employed:

1. Use of clean land and crop rotation every third or fourth year
2. Careful selection of plant roots and clean or fungicide treated beds
3. Store disease free roots, properly cured

Virus

The viral diseases are internal cork, chlorotic leaf spot, yellow dwarf, and russet crack (may be aphid transmitted).

Pests

The pests are vine borer, sweet potato leaf beetle, sweet potato weevil, cotton and peach aphids, white fly, wire worm, white grubs, flea beetle, *Diabrotica* larvae, nematodes, root knot, and others.

Economics

The sweet potato is of great importance in Asia. It is also important in Africa, ranking second in production. It is a secondary source of energy

TABLE 10.1. WORLD PRODUCTION OF SWEET POTATOES *(Ipomoea batatas)*

	Area (10^3 ha)	Production (10^3 MT)	Yield (MT/ha)
World	13,638	113,954	8.4
Continent			
Africa	782	5,053	6.5
North and Central America	178	1,293	7.2
South America	225	2,263	10.1
Asia	12,330	104,617	8.5
Europe	13	135	10.4
Oceania	110	593	5.4
Leading countries			
1. China	10,860	92,600	8.5
2. Vietnam	380	2,400	6.3
3. Indonesia	309	2,350	7.6
4. India	225	1,545	6.9
5. Brazil	136	1,516	11.2
6. Japan	70	1,400	20.0
7. Korea, REP	70	1,387	19.8
8. Philippines	228	1,037	4.6
9. Burundi	99	943	9.5
10. Rwanda	106	842	7.9

Source: FAO (1979).

food in the Americas, Brazil having the highest production. Alcohol can be produced from sweet potatoes by fermentation of the carbohydrates.

Table 10.1 shows the world production and the countries where this crop is important.

BIBLIOGRAPHY

COOLEY, J. S. 1951. Origin of the sweet potato and primitive storage practices. Sci. Monthly *72*, 325–331.

EDMOND, J. B. 1971. Sweet Potatoes: Production, Processing, Marketing. AVI Publishing Co., Inc., Westport, Connecticut.

JENKINS, W. F., and GEIGER, M. 1957. Curing baking time, and temperature affecting carbohydrates in sweet potatoes. Proc. Am. Soc., Hort. Sci. *70*, 419–424.

KIMBER, A. J. 1972. The sweet potato in subsistance agriculture. Papua New Guinea Agric. J. *23*, 80–102.

McDAVID, C. R., and ALAMU, S. 1980. Effect of day length on the growth and development of whole plants and rooted leaves of sweet potato (*Ipomoea batatas*) Trop. Agric. (Trinidad) *57*, 113–119.

MORRIS, L. L., and MANN, L. K. 1955. Wound healing, keeping quality, and compositional changes during curing and storage of sweet potatoes. Hilgardia *24*, 143–183.

ONWUEME, I. C. 1978. The Tropical Tuber Crops: Yams, Cassava, Sweet Potato and Coco Yams. John Wiley & Sons, New York.

PURSEGLOVE, J. W. 1968. Tropical Crops: Dicotyledons. John Wiley & Sons, New York.

SCOTT, L. E. 1950. Potassium uptake by the sweet potato plant. Proc. Am. Soc. Hort. Sci. *56*, 248–252.

STEINBAUER, C. E., and KUSHMIAN, L. J. 1971. Sweet potato culture and diseases. USDA Handb. No. 388. U.S. Dept. Agric., Washington, DC.

THOMPSON, H. C., and KELLY, W. C. 1957. Vegetable Crops, Chapter 22, pp. 405–430 McGraw Hill Book Co., New York.

WILSON, L. A., and LOWE, S. B. 1973. The anatomy of the root system in West Indian sweet potato (*Ipomoea batatas* (L.) Lam.) cultivars. Ann. Bot. *37*, 633–647.

YEN, D. E. 1961. Sweet potato variation and its relation to human migration in the Pacific. Pac. Sci. Congr. Proc. *10*; 93–117.

YEN, D. E. 1976. Sweet potato, In Evolution of Crop Plants, N. W. Simmonds (Editor). Longmans-Green, London.

11

Cassava (Manioc)

Family: Euphorbiaceae (milkweed, castor bean, or rubber)
Genus and species: Manihot esculenta (*Manihot utilissima*)
Other names: Cassava (English-speaking countries), manioc (French-speaking countries), yuca (Spanish-speaking countries of Central and South America), mandioca (Brazil and Paraguay), tapioca plant (India)

Origin

The origin of cassava is somewhere in tropical Brazil. The crop was dispersed to other parts of Latin America thousands of years ago; remains dated at about 800 BC were found near the Colombia–Venezuela border. It was transported to other parts of the world in post-Columbian times. In Africa and tropical Asia it has become established as an important crop supplying carbohydrates.

Botany

Cassava is a dicotyledonous perennial bushy shrub, 1–3 m (4–10 ft) high with large palmate leaves, ranging from three to 11 divisions (Fig. 11.1). Flowers are in loose spreading clusters near the end of the branches, about 1 cm (3/8 in.) in diameter, varying from greenish purple to light greenish yellow. The plant is monoecious; the female or pistillate flowers open first and are receptive to pollination long before the male or staminate flowers open in the same cluster. Hence, cross-pollination usually by insects occurs. Seeds are flat, about 10 × 6 mm (1/2 × 1/4 in.), three seeds to a pod.

Initiation of root enlargement occurs under short days. Under long days root enlargement is delayed resulting in low yields. The storage roots are 4–10 cm (1½–4 in.) in diameter and 20–40 cm (8–15 in.) long; some cultivars have much longer storage roots (Fig. 11.2). The number

FIG. 11.1. Cassava *(Manihot esculenta)* tops and storage roots.

FIG. 11.2. Cassava, manioc *(Manihot esculenta)* roots attached to the propagating stem.

of roots per plant can vary from 5 to 10. They contain from 25 to 40% starch.

Cassava is classified according to the amount of bitter substance the roots contain. *Bitter types* are poisonous. The roots have a cyanogenic glucoside which releases cyanide upon crushing. Most of the poison is in the cortex of the root. Roots of this type must be treated before use as food. *Sweet types* contain little cyanide producing compound; this type requires no treatment before use. There are intermediate types requiring preparation to render the roots nontoxic. It has been reported that soil and local climatic condition influences the amount of cyanogenic glucoside produced by the roots. A sweet cultivar at one location may produce bitter roots grown at another location.

Culture

Climatic Requirements

A tropical or subtropical crop, cassava grows well between 18° and 35°C (65° and 95°F). Growth stops at temperatures below 10°C (50°F). Frost can kill the plant. Cassava is of importance between 30°N and 30°S latitudes. It can be grown up to 2000 m (6500 ft) in elevation and in rainfall regions of 500–5000 mm (20–200 in.). The crop can withstand long periods of drought. During periods of low rainfall, the plant sheds leaves, but quickly grow new leaves after the rains. Figure 11.3 shows the regions of the world where cassava is grown.

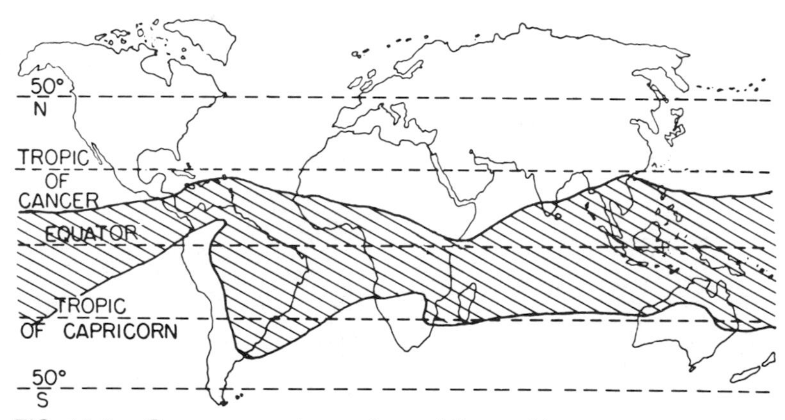

FIG. 11.3. Cassava growing regions of the world.

Soil

Growers prefer sandy or sandy loam soil but all types can be used except waterlogged conditions. Soils with hard pan about 30–40 cm (12–15 in.) below the surface are desirable to prevent deep penetration of roots; shallow roots make digging easy at harvest. Cassava grows well in soils which are very acid (pH 5–5.5) as well as in alkaline (pH 8–9) and low in fertility. It tolerates conditions of high available aluminum and manganese often found in high rainfall tropics, conditions under which most vegetables cannot be successfully grown.

Propagation

Although cassava has viable seeds, propagation is by asexual means. Stem cuttings, 20–30 cm (8–12 in.) long from the previous crop are planted about 10 cm (4 in.) deep. Stem cuttings from an older more mature stem give higher yields than from cuttings obtained from a younger portion of the stem. Spacing is from a minimum of 60–75 cm (2–2½ ft) to 90–140 cm (3–5 ft).

Growth

The crop is allowed to grow from 10 months to 3 years; the longer the period, the greater the yield. After one year of growth, the yield is from 13 to 27 MT/ha (6 to 12 tons per acre). Figure 11.4 shows the growth of cassava over a period of 2 years. In East Africa sometimes the crop is left up to 6 years as a reserve food supply for bad years (crop failures).

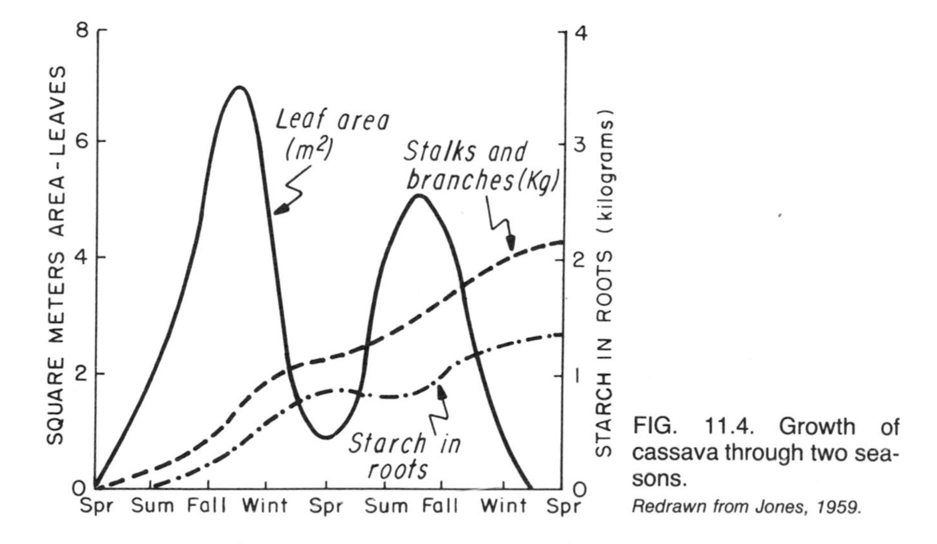

FIG. 11.4. Growth of cassava through two seasons.
Redrawn from Jones, 1959.

In slash and burn agriculture of the humid tropics, cassava is usually the last crop to be grown before the land is abandoned because it can still produce in nutrient depleted soils. High N in the soil and irrigation tend to decrease yield due to excessive top growth.

Storage and Handling

Storage of Roots

Cassava stores fairly well under refrigeration for 1–2 weeks at 5.5°–7°C (42°–45°F) and 85–95% RH; above 20°C (68°F) and high humidity, losses are large. Deterioration of roots starts soon after harvest; roots show internal discoloration of vascular tissues followed by invasion of the injured tissues by microbial organisms.

Field Storage of Harvested Roots

In cool moist seasons the harvested roots are piled on top of 15 cm (6 in.) of straw, dried grass, or sugar cane leaves. The roots are then covered with 15 cm of the same material and then with 15 cm of soil. In dry hot seasons 30–40 cm (12–15 in.) of soil is used and a ventilation system is installed at the top and bottom of the stack for air movement.

Diseases and Pests

Mosaic

Mosaic is a serious viral disease found in Africa and possibly India. It is reported to be transmitted by the white fly, *Bemisia tabaci*. Resistant varieties have been developed.

Brownstreak

Brownstreak is a viral disease prevalent in East Africa causing leaf chlorosis and root discoloration.

Cassava Bacterial Blight

Cassava bacterial blight is caused by the organism *Xanthomonas manihotis (X. cassavae)*. The disease is spread by rain splashes within the field and by cuttings for propagation from field to field. It can also be transmitted by insects; the *Diabrotica* sp. is a suspect. For control, disease free foundation stock is recommended.

Phoma Leaf Spot

Leaf spot disease is found in cooler regions [temperature below 25°C (77°F) at higher elevations]. Disease is caused by *Phyllosticta* spp., which may defoliate leaves and in severe infections there occurs a dieback of the shoots. Leaves have large brown spots, with indefinite margins. Lesions may be found at tips or edges of leaf lobes or along the mid ribs.

Phytophthora Root Rot

The rot is caused by *Phytophthora* spp. under excessive soil moisture condition.

Root-Knot Nematode

Cassava roots can be infected by *Meloidogyne* sp.

Thrips and Spider Mites

Thrips and spider mites are the most important pests which attack the cassava plant. Light infestations of thrips produce yellow dots on leaves and heavy infestations cause leaf deformation and death to the growing shoot. Spider mite infestations occur in the dry season; severe infestations cause leaf drop.

Nutritive Value

Roots are high in starch and vitamin C, but low in vitamin A and protein. In areas of high cassava consumption, concerns have been expressed of chronic cyanide poisoning from residues of the cyanogenic glucoside in cassava products. The "bark" of the root is very high in cyanide as are the young leaves. The inner part of the root has the lowest content.

Economics

The bitter type is used to manufacture starch (tapioca). In Africa, roots are prepared for a popular dish called fufu.

Small amounts of fresh cassava are shipped between countries. It is consumed in the locality where it is grown as the fresh root is quite perishable. Some fresh cassava roots are air freighted to the United States from Venezuela.

Cassava is very important as a starchy staple in tropical Africa. The world production and the leading countries are presented in Table 11.1. The crop has been successfully used for animal feed when combined up to 30% with other types of rations. In Southeast Asia cassava chips are dried and shipped to Europe for animal feed. Brazil is developing the usage of cassava for alcohol production.

TABLE 11.1. WORLD PRODUCTION OF CASSAVA (*Manihot esculenta*)

	Area (10^3 ha)	Production (10^3 MT)	Yield (MT/ha)
World	13,397	117,201	8.7
Continent			
Africa	6,877	45,022	6.5
North and Central America	152	984	6.5
South America	2,610	29,939	11.5
Asia	3,737	41,036	11.0
Oceania	20	220	11.1
Leading Countries			
1. Brazil	2,150	24,935	11.8
2. Indonesia	1,398	13,100	9.4
3. Thailand	1,000	12,500	12.5
4. Zaire	1,800	12,000	6.7
5. Nigeria	1,150	11,500	10.0
6. India	361	6,053	16.7
7. Tanzania	895	4,300	4.8
8. Vietnam	460	3,800	8.3
9. China	217	2,770	12.8
10. Mozambique	450	2,500	5.6

Source: FAO (1979).

BIBLIOGRAPHY

ARAULLO, E.V., NESTEL, B., and CAMPBELL, M. 1974. Cassava, processing and storage. Proc. Interdisciplinary Workshop, Rep. No. IDRC-03ie. Int. Dev. Res. Centre, Ottawa, Ontario, Canada.

BOOTH, R.H. 1975. Cassava storage. Ser. EE16. Centro Internacional de Agricultura Tropical, Cali, Colombia.

COCK, J., MACINTYRE, R., and GRAHAM, M. 1976. Proc. 4th. Int. Symp. Trop. Root Crops, Cali, Colombia. Int. Dev. Res. Centre, Ottawa, Ontario, Canada.

CIAT. Annu. Rep. Centro Int. Agric. Trop., Cali, Colombia.

JONES, W.O. 1959. Manioc in Africa. Stanford University Press, Stanford, California.

KRACHMAL, A. 1969. Propagation of cassava. World Crops, 1193–1195 (July/August).

MORAN, E.F. 1976. Manioc deserves more recognition in tropical farming. World Crops. *28*, 184–188.

NESTEL, B., and MACINTYRE, R. 1973. Chronic cassava toxicity. Proc. Interdisciplinary Workshop, London. Int. Dev. Res. Centre, Ottawa, Ontario, Canada.

ONWUEME, I.C. 1978. The Tropical Tuber Crops: Yams, Cassava, Sweet Potato and Coco Yams. John Wiley & Sons, New York.

PURSEGLOVE, J.W. 1968. Tropical crops: Dicotyledons. John Wiley & Sons, New York.

ROGERS, D.J. 1965. Some botanical and ethnological considerations of *Manihot esculenta*. Econ. Bot. *19*, 369–377.

ROGERS, D.J., and FLEMING, H.S. 1973. A monograph of *Manihot esculenta* with explanation of taximetric methods used. Econ. Bot. *27*, 1–113.

12

Yam

Family: Dioscoreaceae (Yam)
Genus: Dioscorea

Principal species	Common name
D. alata	greater, water or winged yam, Chinese yam, ubi (Southeast Asia)
D. rotunda (D. cayenesis)	yellow or guinea yam (Africa)
D. batatas (*D. opposita* and *D.japonica*)	Chinese yam (China and Japan)
D. esculenta	lesser yam (Chinese yam) (Southeast Asia)
D. bulbifera	aerial or potato yam (Africa and Asia)
D. trifida	cush-cush or yampi (Tropical America)
D. hispida	intoxicating yam[1]

Other names: Namé (Spanish), igname (French)

Origin

Artifacts dated about 50,000 BC from West Africa indicate wild yams were used for food prior to that time. The growing of yams appears to have developed independently in two regions, in West Africa and Southeast Asia about 3000 BC. The present-day *D. bulbifera*, which is grown for the aerial tubers, is indigenous to both regions. The basic chromosome number of the American yams is $n = 9$ while those of Asian and African origins are $n = 10$. Yams show high polyploidy; a race of *D. cayenesis* is reported to have a $2n = 140$.

[1] Tubers are very poisonous containing the alkaloid, dioscorine. Can be used as food after slicing and washing slices in running water for 3–4 days. Used mainly as a source of starch (Asia).

Botany

There are an estimated 300–600 species in the *Dioscorea* genus in the world. It is a monocotyledonous plant with long trailing vines, which twine for many feet through trees and undergrowth. The vines bear racemes of inconspicuous flowers, male and female, borne separately, usually on separate plants (dioecious). The leaves are generally simple cordate (heart-shaped) with several primary veins being arcuate and the secondary veins being reticulate. The vine dies at the end of the rainy season or when temperatures get low. The tubers, which are formed during growth, persist dormant through the dry season and start to grow sending out new shoots when conditions are again favorable. The tuber may be an annually renewed organ or may be perennial becoming larger each year. Depending on the species, from one to several tubers are produced per plant. Some species produce bulbils or aerial tubers at the axes of the leaves.

The storage organ of the yam has been classified as a tuber because it resembles the stem more than it does the root. However, there are some anomalous features of the tuber such as:

1. The lack of remnants of scale leaves or vestigial nodes on the surface
2. No preformed buds or eyes on or near the surface such as in the white potato tuber
3. The growing point has no terminal bud
4. Exhibition of a strong geotropic growth by most species

There is presently strong evidence that the storage organ originates from the hypocotyl tissue, the transition zone between stem and root, as the first meristematic activity in tuber formation occurs in this region.

In tuber growth, the primary meristem is at the distal end. As growth continues, a thin layer of meristematic cells is left beneath the cortex, which, with continued activity, results in the increase in girth. The parenchyma cells just below the epidermis become meristematic and produce cork cells protecting the tuber. The yam tuber is shown in Figs. 12.1 and 12.2.

Culture

Climatic Requirements

The *Dioscorea* are grown principally in the tropical and semitropical regions; plants cannot stand any frost. Only *D. batatas* can be grown in the temperate region. Except for *D. batatas*, plants do not grow well

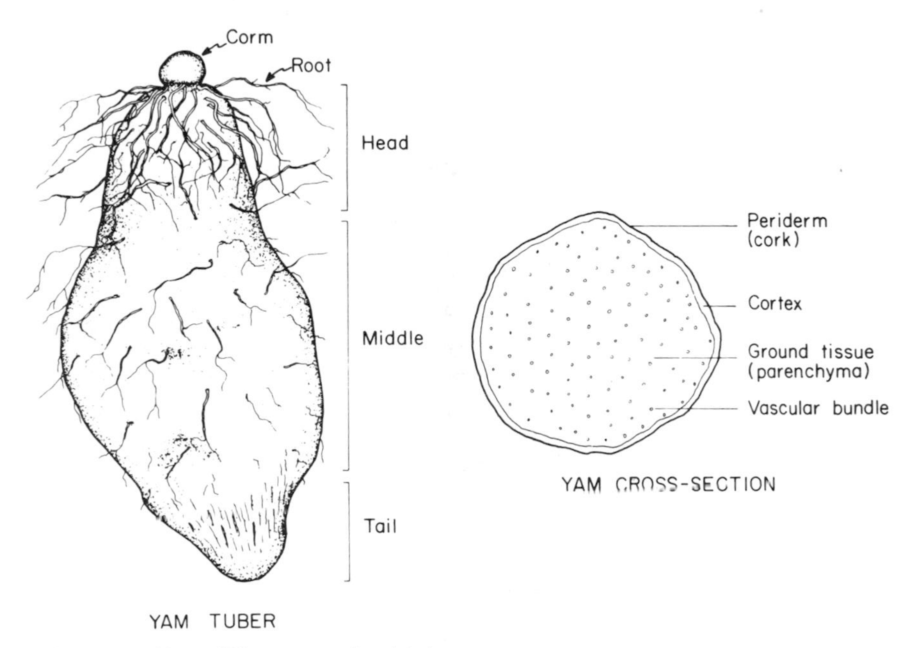

FIG. 12.1. Yam *(Dioscorea alata)* tuber.

below 20°C (68°F). The optimum is 25°–30°C (77°–86°F) mean temperatures. Excessively high temperatures may have deleterious effects. Yams are grown in areas of fairly heavy rainfall from 1200 to 3000 mm (50 to 120 in.) per year. In regions of rainfall of less than 600 mm (24 in.) per year, the yield is poor. Figure 12.3 shows the yam producing regions of the world.

Culture

Yams require loose deep free-draining fertile soil. Good drainage is important as waterlogging causes tuber rot. The ideal soil temperature is about a mean of 30°C (86°F) with a short dry season about the time of planting and 1500 mm (60 in.) rainfall evenly distributed for the remainder of the growing season.

Propagation

Propagation is by asexual means usually by planting of small tubers. About 20% of the previous year's crop is saved for propagating purposes. Sometimes the proximal end (head) from the tuber is cut off and planted. Seed pieces 100–200 g (3–6 oz) are used. In West Africa 3 kg (7 lb) seed tubers are often used. Bulbils (not a true bulb) or aerial tubers can be

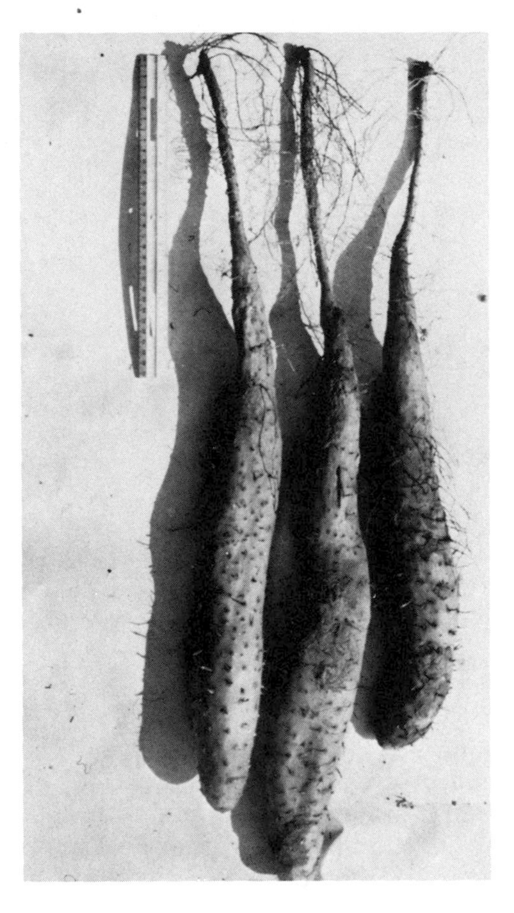

FIG. 12.2. Japanese yam, naga imo *(Dioscorea batatas)* tubers.

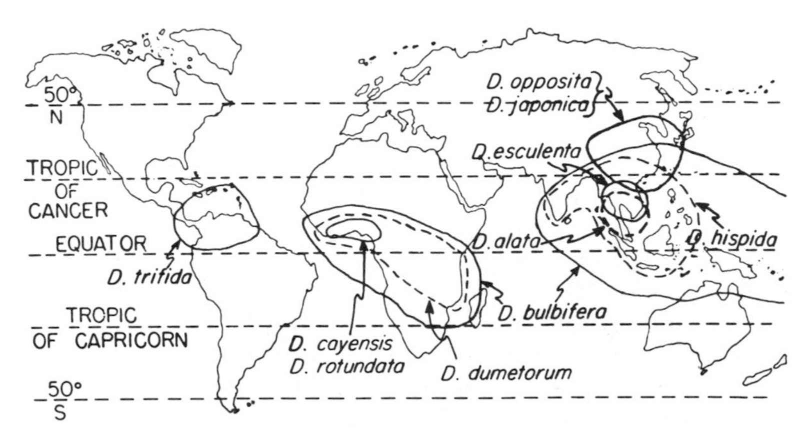

FIG. 12.3. Yam producing regions of the world.
Redrawn from Coursey (1967).

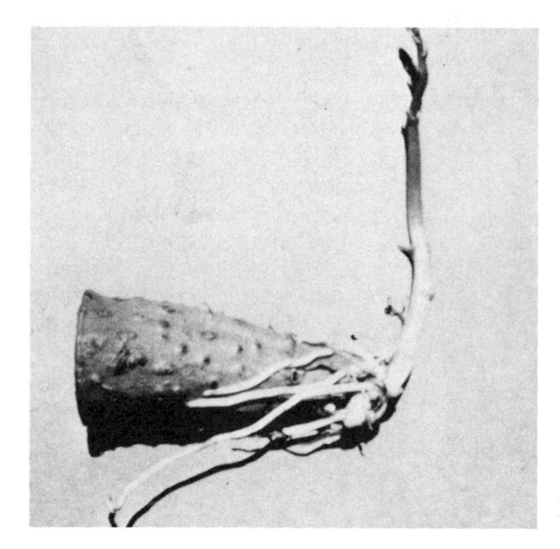

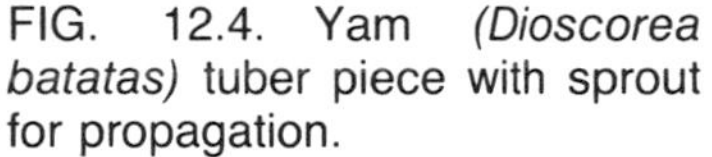

FIG. 12.4. Yam *(Dioscorea batatas)* tuber piece with sprout for propagation.

a. b. c. d.

FIG. 12.5. Stages in the sprouting of *Dioscorea* tuber. (a) Normal tuber piece before planting; (b) formation of sprouting locus; (c) appearance of differentiated shoot buds; (d) enlargement of the shoot bud and development of roots.

From Onwueme (1973). Reproduced by permission of Cambridge Univ. Press, London.

used; it may take 2 or more years before tubers are large enough for harvest. Before the rainy season or with the early rains, the seed tubers are planted about 15 cm (6 in.) in depth in ridges or mounds about 30 cm (1 ft) apart and 1.5 m (5 ft) between rows. Sometimes the tubers are presprouted before planting. Tubers sprout rapidly at 25°–30°C (77°–86°F) and high relative humidity once dormancy is broken. High humidity is necessary for rootlet formation.

Onwueme (1973) has made studies on the sprouting of yam tubers. Sprouts are initiated in a meristematic region about a centimeter from the edge at the proximal end of the tuber (Fig. 12.4). Figure 12.5 shows the stages in the appearance of sprouts on the tuber and Fig. 12.6 shows a diagrammatic representation of changes occurring in the tuber.

Growth

The vines need support; without support from poles or stakes, yields are greatly reduced. Poles vary in length from 2 to 4½ m (6 to 15 ft).

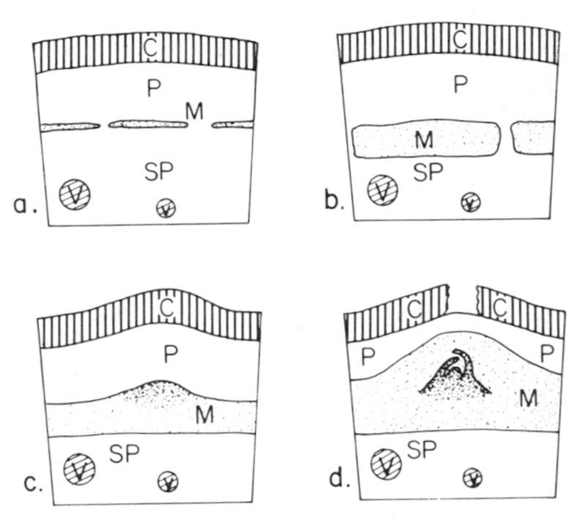

FIG. 12.6. Progressive stages in the formation of the sprout in *D. rotunda*. Newly differentiated bud; C, cork layer; M, layer of meristematic cells; P, parenchyma cells with only small amount of stored starch; SP, storage parenchyma with stored starch and constituting the bulk of the tuber; V, vascular bundles.
From Onwueme (1973). Reproduced by permission of Cambridge Univ. Press, London.

In some regions of the tropics where slash and burn agriculture is practiced, dead trees are left standing so that the yam vines can be supported.

There is little information on the photoperiodic effects on yams. It is believed that vine development is favored by day length of greater than 12 hours and tuber formation under short days. Short days seem to induce tuber initiation and enlargement.

In West Africa it is common practice to mulch the mounds with dried grass, straw, or leaves. This is covered with a layer of stones or soil to prevent the mulch from being blown away during storms. Mulching conserves soil moisture and reduces soil temperature during the dry season. Yams are not usually fertilized in the slash and burn culture as they are usually the first crop planted. However, some manure or NPK fertilizer mixture is recommended for high yields.

The growth of the yam plant over a period of 24 weeks is shown in Fig. 12.7.

Harvest and Storage

Harvest

Yams are harvested at the end of the rainy season or early part of the dry season when vegetative growth is finished for the year. Wooden spades or digging sticks are preferred to iron tools as they are less apt to damage the fragile tubers. Harvesting is a laborious process; injury can lead to infection by rotting organisms and loss during subsequent storage. Sometimes an early harvest is made by carefully removing the soil

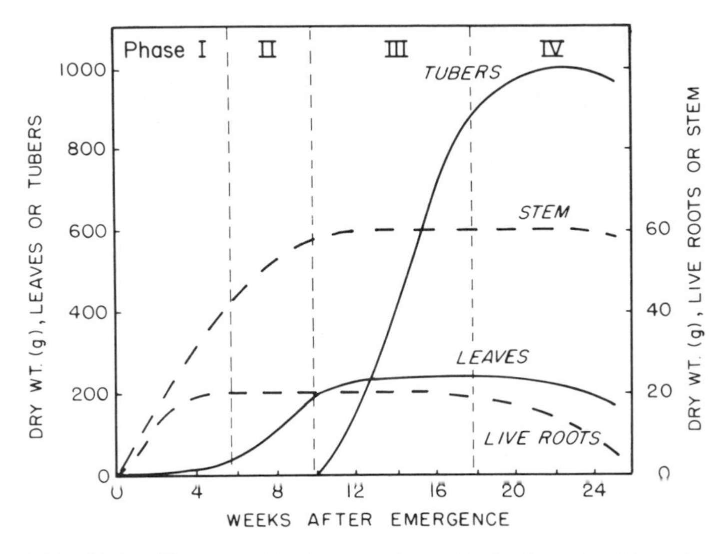

FIG. 12.7. Yam growth phases. Phase I: 0–6 weeks. Sprouting; root and stem growth. Phase II: 6–10 weeks. Continued stem growth; root growth complete, leaf expansion. Phase III: 10–18 weeks. Tuber enlargement. Phase IV: 18–24 weeks. Tuber maturation followed by senescence of tops.
Redrawn from Onwueme (1978).

around the tuber and cutting the lower portion, leaving the upper part of the tuber or the "head" to heal and continue growth. The soil is returned and the plant left to grow to the end of the season. Many varieties produce a single tuber, others form several small tubers in the second growth after early harvest.

Storage

In the tropics after lifting from the soil, the tubers are dried for a few hours, then stored in well-ventilated weatherproof buildings, sheds, or under shade in the open. Two tubers are tied to a fibrous rope at each end hung on horizontal poles in these "yam barns." Under favorable conditions yams can be stored for several months; some varieties are suitable for long storages. In storage sprouting can be a problem but the sprouts can be rubbed off as with the white potato.

Typical of most tropical crops, chilling injury can occur to the yam tubers if stored at or below 10°C (50°F). The recommended temperature range for storage is 12°–16°C (54°–61°F), 15°C (59°F) being optimal.

Pests and Diseases

Yams are very resistant to fungi and very few pests are found to attack the plant. Rusts are sometimes found on leaves. Mosaic is the most common viral disease of yams. The common pests are beetles and rodents, which decrease tuber yields.

Nutritional Value

Tubers are high in starch. The vitamin A value is very low. Vitamin C is from 5 to 10 mg/100 g. The limiting essential amino acids are isoleucine and those containing sulfur.

The shoots of some wild species are boiled and eaten as greens.

Some species of yams contain the alkaloid dioscorine, $C_{13}H_{19}O_2N$. Raw *D. hispida* is poisonous; before use the tuber is boiled so that the alkaloid is leached from the tuber.

Economics

Some yam enters into the world trade but most are used where produced. Yams are a staple crop in Central Africa and Southeast Asia. Yam flour is produced in Africa and used to make *fufu*. Yams contain a steroid sapogenin compound, diosgenin. Both wild and cultivated yams are harvested and the steroid compound extracted for use as a base for the drugs such as cortisone and hormonal drugs.

World production (Table 12.1) shows that most yams are produced in Nigeria.

TABLE 12.1. WORLD PRODUCTION OF YAM *(Dioscorea)*

	Area (10^3 ha)	Production (10^3 MT)	Yield (MT/ha)
World	2,110	20,198	9.6
Continent			
Africa	2,049	19,539	9.4
North and Central America	22	243	11.1
South America	10	48	4.7
Asia	15	168	11.4
Oceania	15	200	13.5
Leading countries			
1. Nigeria	1,350	15,000	11.1
2. Ivory Coast	200	1,700	8.5
3. Ghana	160	800	5.0
4. Togo	100	750	7.5
5. Benin	59	610	10.3

Source: FAO (1975).

BIBLIOGRAPHY

AYENSU, E.S. and COURSEY, D.G. 1972. Guiana yams—The botany, ethanobotany, use and possible future of yams in West Africa. Econ. Bot. *26*, 301–318.

BARKHILL, I.H. 1960. The organography and the evolution of discoreaceae, the family of yams. J. Linn. Soc. London, Bot. *56*, 319:412.

COURSEY, D.G. 1967. Yams. Longmans-Green, London.

COURSEY, D.G. 1976. Yams, *In* Evolution of Crop Plants, N.W. Simmonds (Editor). Longmans-Green, London.

INGRAM, J.S., and GREENWOOD-BARTON, L.H. 1962. The cultivation of yams for food. Trop. Sci. *4*, 82–86.

LIU, T. and HUANG, T. 1962. On the Taiwan species of *Dioscorea*. Bot. Bull. Acad. Sin. *3*(2), 133–149.

MARTIN, F.W. 1969. The species of *Dioscorea* containing sapongenins. Econ. Bot. *23*, 373–379.

MARTIN, F.W. 1972. Yam Production Methods. ARS, USDA Production Res. Rep. 147. U.S. Dept. Agric., Washington, DC.

MARTIN, F.W. 1974. Tropical Yams and Their Potential, Part 1, *Dioscorea esculenta*. USDA Agric. Handb. No. 457. Mayaguez Inst. Trop. Agric., Mayaguez, Puerto Rico.

MARTIN, F.W. 1974. Tropical Yams and Their Potential, Part 2, *Dioscorea bulbifera*. USDA Agric. Handb. No. 466. Mayaguez Inst. Trop. Agric., Mayaguez, Puerto Rico.

MARTIN, F.W. 1976. Tropical Yams and Their Potential, Part 3, *Dioscorea alata*. USDA Agric. Handb. No. 495. Mayaguez Inst. Trop. Agric., Mayaguez, Puerto Rico.

MARTIN, F.W. and SADIK, S. 1977. Tropical Yams and Their Potential, Part 4, *Dioscorea rotunda* and *D. Cayenesis*. USDA Agric. Handb. No. 502. Mayaguez Inst. Trop. Agric., Mayaguez, Puerto Rico.

MARTIN, F.W. and DEGRAS, L. 1978. Tropical Yams and Their Potential, Part 5, *Dioscorea trifida*. USDA Agric. Handb. No. 522. Mayaguez Inst. Trop. Agric., Mayaguez, Puerto Rico.

MARTIN, F.W. and DEGRAS, L. 1978. Tropical Yams and Their Potential, Part 6, Minor cultivated *Dioscorea* species. USDA Agric. Handb. No. 538. Mayaguez Inst. Trop. Agric., Mayaguez, Puerto Rico.

ONWUEME, I.C. 1973. The sprouting process in yam (*Dioscorea* spp.) Tuber pieces, J. Agric. Sci. *81*, 375–379.

ONWUEME, I.C. 1978. The Tropical Tuber Crops: Yams, Cassava, Sweet Potato and Coco Yams. John Wiley & Sons, New York.

PURSEGLOVE, J.W. 1972. Tropical Crops: Monocotyledons. J. Wiley & Sons, New York.

13

Edible Aroids: Taro, Yautia, and Others

Taro and Yautia

Family: Araceae-Arum
Genus and species: *Colocasia esculenta*
Other names: Dasheen, eddo; old cocoyam (Central Africa); tari, also Kolkas (Egypt); gabi (Philippines). Dasheen is sometimes classified as *C. antiquorum*.
Genus and species: *Xanthosoma sagittifolium*
Other names: Tannia, malanga; new cocoyam (Central Africa)

Origin

Colocasia esculenta (taro) occurs wild in India and Southeast Asia; taro cultivation probably originated here and the corms were taken to China and then to Japan. It also spread to the eastern Mediterranean and to Egypt and also to Spain. It was probably taken to the island of Madagascar by the Indonesians as early as 500 AD and then spread across Africa to the Guinea coast. From Southeast Asia the Polynesians took taro to the various islands of the South Pacific. The oriental taro, known as dasheen, is a nonacrid cultivar, which originated in China and was brought to the West Indies by the explorers in post-Columbian times. It is now grown in all parts of the subtropics and the tropics.

Xanthosoma sagittifolium (yautia) was domesticated in tropical America. The people of the West Indies cultivated yautia long before the first white explorers landed there. The crop was introduced to Ghana and elsewhere in West Africa in the early 1840s and since it resembled cocoyam (taro), it was given the name of "new cocoyam." *Xanthosoma violaceum*, a pink fleshed species of New World origin, is grown in the Philippines.

Botany

All aroids are monocotyledonous plants. *Colocasia esculenta* (taro) plants, 90–180 cm (3–6 ft) tall, are grown for their underground storage corms. At the apex of the corm is a whorl of large erect leaves on long slender petioles. The leaves are peltate without a marginal vein and blades are 20–50 cm (8–20 in.) long and about half as wide. The inflorescence is borne on a peduncle; the pale yellow spathe is about 20 cm (8 in.) long and rolled inward at the apex. The spadix is cylindrical. Many cultivars do not flower. Corms are usually cylindrical, 13–20 cm (5–8 in.) high and about 8–13 cm (3–5 in.) in diameter (Fig. 13.1). Adventitious buds are formed to produce daughter corms, called cormels; these are much smaller, ranging in size from 5 to 8 cm (2 to 3 in.) long and 3.5–5 cm (1¼–2 in.) in diameter. Figure 13.2 shows a taro plant with an attached cormel. When cut the petiole exudes a milky watery juice.

Xanthosoma sagittifolium (yautia) plants may grow over 2 m (6 ft) in height. The large stalked leaves are sagittate, with submarginal vein; blades may be more than 1 m (3 ft) long. The corms produced at the base of the plant are 18–25 cm (6–10 in.) long and up to 10 or more cormels attached (Fig 13.3). The flower is similar to the taro. When cut, the exudate from the petiole is thick and milky.

FIG. 13.1. Taro (*Colocasia esculenta*) comels and corm trimmed for marketing.

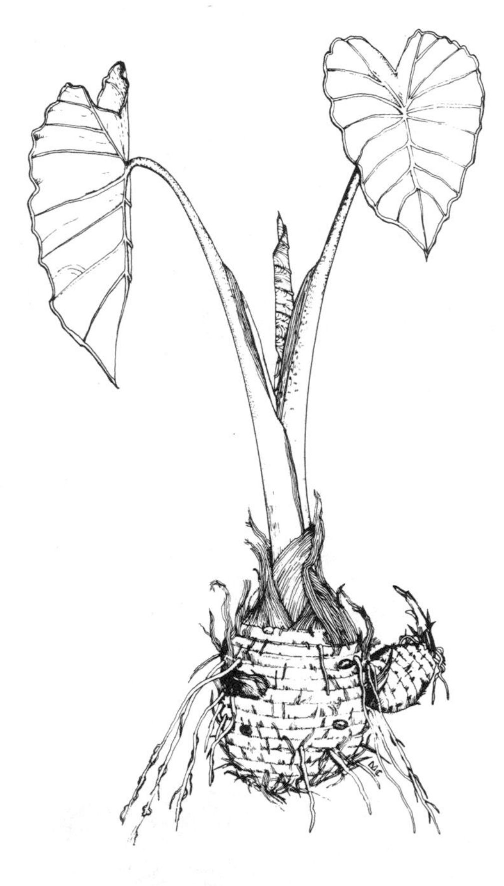

FIG. 13.2. Taro *(Colocasia esculenta)* plant.

Culture

Climate Requirement

Both genera can be grown at any time of the year in the tropics and subtropics from sea level to 1300 m (4500 ft) elevation. Taro can be found as a commercial crop in the warmer areas of the temperate zones. They are warm season crops and cannot stand cold temperatures. Since the natural habitat of taro and yautia are the tropical rain forests, both require large amounts of water during growth. Figure 13.4 shows the Aroid producing regions of the world. Aroids can be grown in saline soils.

Propagation

Propagation of aroids is usually by planting of corms or cormels. In Hawaii, taro is propagated by using the cut stem portion with petiole

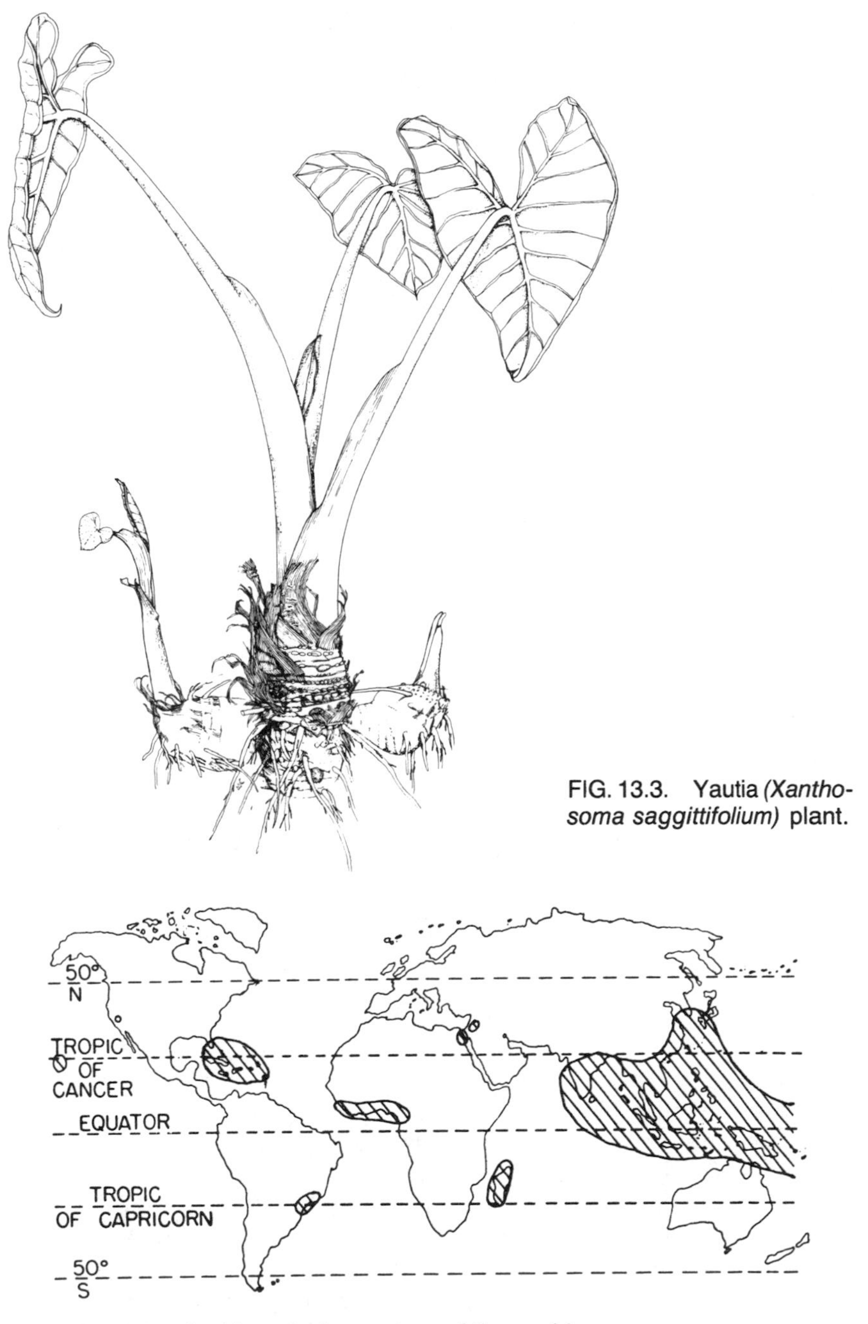

FIG. 13.3. Yautia *(Xanthosoma saggittifolium)* plant.

FIG. 13.4. Aroid producing regions of the world.

attached. These are called *huli* or sets and are obtained at harvest; the rest of the corm is used for food. Aroids bloom occasionally and some produce fertile seeds, which lose viability quickly.

Taro Culture

There are two methods of growing taro. (1) *Lowland or paddy culture method.* The field is first flooded, then plowed and puddled as in low land rice culture. After puddling, it is planted with presprouted corms about 45–60 cm (1½–2 ft) apart and 90–120 cm (3–4 ft) between rows. When plants have one new unfolded leaf, irrigation water is put into the diked field. The water level is then raised to about 10 cm (4 in.), and kept at about this level; water is allowed to flow continuously throughout growth. A hard pan in the soil horizon is recommended for this type of culture. (2) *Upland or dry culture.* The field is prepared and worked to break large clods. After the land has been prepared, furrows 30 cm (12 in.) deep and 75–90 cm (2½–3 ft) apart are dug to plant the presprouted cormels. Spacing is about 38–40 cm (15–16 in.) and the cormels are covered about 5–8 cm (2–3 in.) with soil. Immediately after planting, the field is irrigated. The optimum soil pH is from 6.0 to 7.0; below 5.0 or above 8.0, top growth and corm yields are poor.

A complete fertilizer mix amounting to 45 kg N, 45 kg P, and 34 kg K/ha (40 lb of N, 40 lb of P, and 30 lb of K per acre) is recommended for high yields in the Philippines; higher rates of fertilization are used for taro growing in Hawaii.

Rapid vegetative growth occurs during the first 4–6 months during which time maximum leaf canopy is reached. Corm formation occurs at about 3–5 months after planting. During maturation, the leaves become smaller and smaller, until very few leaves exist; the older leaves turn yellow and die. At this stage the corms are ready for harvest, about 7–11 months from planting.

Yautia Culture

Yautia, unlike taro, is grown mostly as an upland crop and is unable to withstand continuous flooding. However, it can grow in moist or wet conditions. Yautia may be the first crop grown in the humid tropics when the land is cleared. Culture is similar to upland taro. Presprouted corms which were previously cut into pieces each having an "eye" and weighing about 200–900 g (½–2 lb) are used for propagation. Spacing varies from 60 × 60 cm (2 × 2 ft) to 180 × 180 cm (6 × 6 ft), the average being 90 × 90 cm (3 × 3 ft) in Ghana. The soil is mounded around plants to prevent exposure of roots and help in the development of daughter corms (cormels).

When mature, the basal leaves of the yautia start to yellow and the soil covering the cormels starts to crack. Corms are ready for harvest 6–12 months after planting. The main or mother corm grows large and is not used for food but is fed to hogs. The daughter corms are harvested for food.

Harvest and Storage

Harvest

Harvesting of the corms of taro and yautia is usually done by hand. The corms are washed and the roots and fibers removed. The tops, which are cut off below the growing point, are saved for planting in other fields. Taro yields vary from 35 to 45 MT/ha (15–20 tons/acre) in Hawaii. Yautia yields from 4½–11 MT/ha (2–5 tons/acre) in Ghana, in Puerto Rico the nonirrigated fields yield of yautia cultivars range from 8 to 14 MT/ha (3.5 to 6.5 tons/acre); irrigation increases the yield from 10 to 25%. Taro and yautia yields cannot be compared because of different cultural methods and regions of growth.

Storage

Taro can be stored successfully at 10°C (50°F). Taro and yautia are usually stored by placing the dry corms and cormels in the ground and covering them with soil. The top of the soil is dampened occasionally to keep the corms from losing moisture and to keep them cool. The corms and cormels can be stored in a cool place for several months.

Diseases and Pests

Taro plants appear to be subject to few diseases. Leaf blight disease is caused by *Phytophthora colocasiae*. It is worse in flooded taro culture than in upland culture. Root knot nematodes attack feeder roots and corms. Nematodes interfere with the nutrition of the plant and make corms unfit for the market. Soft rot caused by *Phythium* is a soil-borne disease attacking both roots and corms. In the Solomon Islands two serious viral diseases of the aroids occur; elsewhere different and not too serious virus infections are common.

Nutritive Value

Corms are high in starch but otherwise relatively poor nutritionally. Corms can be sliced and dried in the sun. Many aroids contain an acrid substance, which is principally calcium oxalate crystals. The fine needle-like crystals are very sharp, puncturing the delicate skin of the mouth and tongue. Dasheen, a cultivar of taro, contains few calcium oxalate crystals. The acrid element can be removed by thorough cooking, which seems to alter the arrangement of the crystals. Also, there is evidence that cooking removes volatile substances that cause the irritation. Aqueous distillates from acrid cultivars showed irritant effect, which suggest that the irritant substances are volatile and are not destroyed by heat but are lost by volatilization on prolonged cooking.

Some wild Aroids contain cyanogenic glucosides, similar to the cassava; the tubers must be detoxified by cooking and leaching. The starch grains of the *Xanthosoma* are larger than those in *Colocasia.*

Phytic acid is present in leaves of aroids; this compound ties up free calcium to form calcium phytate.

Leaves and petioles are used as greens (Fig. 13.5). The acrid element is also present in the fresh leaves and is removed by boiling in water twice, each time the boiled water is discarded. It is high in provitamin A and even after boiling twice some vitamin C is retained. Blanched leaves of taro are obtained by forcing of the corms in the dark at warm temperatures.

FIG. 13.5. Taro *(Colocasia esculenta)* corms with attached petioles, Papeete, Tahiti.

Uses and Economics

Taro and yautia are usually cooked by boiling and used as a starchy vegetable. Flour from taro can be used for many purposes. The starch is used for making alcohol and for sizing. In Hawaii, it is fermented into *poi* and in Africa into *fufu*. It is reported that yautia makes better *fufu* than taro; hence, its greater popularity. *Xanthosoma* is replacing *Colocasia* almost everywhere in Africa because it produces a better quality product, gives higher yields, and has fewer disease problems. Both taro and yautia can be shipped long distances. The world production in 1975 is shown in Table 13.1. (This was the last year taro production was reported by the FAO-UN.)

TABLE 13.1. WORLD PRODUCTION OF TARO *(Colocasia esculenta)*

	Area (10^3 ha)	Production (10^3 MT)	Yield (MT/ha)
World	816	4,502	5.5
Continent			
Africa	728	3,621	5.0
Asia	55	619	11.2
Oceania	33	262	8.0
Leading countries			
1. Nigeria	320	1,800	5.6
2. Ghana	230	1,400	9.3
3. Japan	35	500	14.3
4. Papua, New Guinea	26	219	8.4
5. Ivory Coast	230	200	8.7

Source: FAO (1975).

MINOR AROIDS

Genus and species: *Alocasia macrorrhiza*
Common names: ape (ahpi) (Hawaii), biga (Philippines), and bira (Indonesia)

The plant is a large succulent perennial herb 90–270 cm (3–9 ft) in height; leaves are similar to yautia, arrow-shaped but pointing upward in line with the petiole. Leaves are saggitate and no submarginal vein is visible. Juice from a fresh cut is watery. Corms are long, thick and woody, from 45–100 cm (18 to over 40 in.) in length and weighing 20 kg (45 lb) or more. The carbohydrate content is about 17% and protein is less than 1%. The corms are acrid and require thorough cooking and leaching. Crops are harvested after 12–18 months but can be harvested after as long as 4 years. Small corms are used for propagation.

Genus and species: *Cyrtosperma chamissonia*

This crop is important in some Pacific Islands. Plants are very similar in appearance to *Alocasia*. It has sagittate leaves without submarginal veins. Crop is allowed to grow 4 years or so and some as long as 10 to 15 years. The rhizomes weigh 40–80 kg (90–175 lb) and contain up to 29% carbohydrates and 1% protein. Some varieties mature in 1–2 years. The culture is in freshwater swamps as with taro.

Cyrtosperma merkussi is grown in the Philippines.

Genus and species: *Amorphophallus campanulatus*

Elephant ear yam. Plant is called *pongapong* and *Tindoc* in the Philippines, *teve* in Samoa, *ol ka chu* in Pakistan, and *kidaran* in Sri Lanka. Corms are short, wider than long, weighing from 1 to 11 kg (2 to 25 lb). The leaf structure is more like a dicotyledenous plant. Young tender leaves and petioles are used in the Philippines. The leaves and corms are very acrid so thorough boiling is required. Propagation is by small corms. Corms are harvested 1–3 years after planting and contain about 24% starch and 1.2% protein.

Genus and species: *Amorphophallus konjak* (*A. rivieri*, var. *Konjak*)

The crop originated from India or Sri Lanka. The corms from this plant are used to make a product called *konnyaku* of Japan. The corms are rich in mannose and mannans; the latter gives *konnyaku* the gel property.

Genus and species: *Xanthosoma brasiliense*

Tanier spinach, Tahitian taro, belembe, and calalou are some names given this crop which is grown principally for the edible leaves. Corms are very small, low in starch content, and used mainly for propogation. Tender raw leaves are sometimes used in salads but the leaves including the petioles are usually cooked for only 10–15 min. Prolonged cooking is not necessary to render the calcium oxalate crystals non-irritating as with other crops in this genus. The protein content of the leaves is 3% and the analysis shows high Ca, P and vitamins A and C. The calcium, however, may not be nutritionally available as it is in the insoluble form.

BIBLIOGRAPHY

BUDDENHAGEN, I.W., MIBRATH, G.M., and HSIEH, S.P. 1970. Virus disease of taro and other aroids. Proc. 2nd Int. Symp. Trop. Root Tuber Crops, Honolulu, Hawaii.

COURSEY, D.G. 1968. The edible aroids. World Crops, *20*, 25–30 (Sept).

HERKLOTS, G.A.C. 1972. Vegetables in South-East Asia. Hafner Press, New York.

HODGE, W.H. 1954. Dasheen—A tropical root crop for the South. USDA Circ. No. 950. U.S. Dept. Agric., Washington, DC.

KARIHARI, S.I. 1971. Cocoyam cultivation in Ghana. World Crops *23*, 118–122 (March/June).

KNOTT, J.E., and DEANON, J.R., JR. 1967. Vegetable Production in Southeast Asia. Chapter 16, pp. 293–300. University of Philippines Press, Manila.

ONWUEME, I.C. 1978. The Tropical Tuber Crops: Yams, Cassava, Sweet Potato and Coco Yams. John Wiley & Sons, New York.

PLUCKNETT, D.L. 1976. Giant swamp taro, a little known Asian Pacific food crop. Proc. 4th Int. Symp. Trop. Root Tuber Crops., Cali, Colombia.

PLUCKNETT, D.L. 1970. Colocasia, Xanthosoma, Alocasia, Cyrtosperma, and Amorphophallus. Proc. 2nd Int. Symp. Trop. Root Tuber Crops, Honolulu, Hawaii.

PLUCKNETT, D.L., and DE LA PENA, R.S. 1971. Taro production in Hawaii. World Crops, *23*, 244–249 (Sept/Oct).

PURSEGLOVE, J.W. 1972. Tropical Crops: Monocotyledons. J. Wiley & Sons, New York.

VOLIN, R.B., and ZETTLER, F.W. 1976. Cocoyam and taro production in Florida. HortScience *11*, 446.

VOLIN, R.B., and ZETTLER F.W. 1976. Seed propagation of cocoyam, *Xanthosoma caracu* Kock and Bouche. HortScience *11*, 459–460.

14

Other Starchy Underground Vegetables

The world production of other starchy root and tuber crops are shown in Table 14.1. The aroid, yautia, is included in Table 14.1. Though the production compared to the major root and tuber crops is small, these minor crops are important in some parts of the world.

EDIBLE CANNA

Family: Cannaceae (canna)
Genus and species: Canna edulis
Other names: Queensland arrowroot, Australian arrowroot, Achira (Peru and Bolivia), *tous-les-mois* (West Indies).

The plant, which grows to about 2.5 m (7 ft) in height, is indigenous

TABLE 14.1. WORLD PRODUCTION OF OTHER ROOTS AND TUBERS

	Area (10^3 ha)	Production (10^3 MT)	Yield (MT/ha)
World	994	4,439	4.5
Continent			
Africa	518	1,446	2.8
North and Central America	30	239	8.0
South America	99	576	5.8
Asia	323	1,785	5.5
Europe		4	
Oceania	23	290	17.2
Leading countries			
1. Indonesia	282	1,417	5.0
2. Cameroon	220	600	2.7
3. Ethiopia	205	500	2.4
4. Papua, New Guinea	15	289	19.0
5. Peru	42	228	5.5

Source: FAO (1975).

to South America. The distribution of this plant is in the Andes below 2500 m (8500 ft) from Venezuela to northern Chile, the Amazon River Basin, Paraguay River Basin, and Central America and West Indies. The edible part is the corm-like rhizome 4–12 cm (1½–4 in.) in diameter and 7–15 cm (5–6 in.) in length. The corms are ready for harvest after 6 months and may be allowed to grow for 18 or more months before harvesting. Young corms (6 months) have about 15–17% starch and 1.25% protein while mature corms (10–12 months old) contain 25–30% starch and 0.9–1.0% protein.

ARROWROOT OR BAMBOO TUBER

Family: Marantaceae (arrowroot or maranta)
Genus and species: Maranta arundinacea

Arrowroot is a perennial monocotyledonous plant 60–90 cm (2–3 ft) tall of South American origin. It is widely distributed from Brazil to the Caribbean. The plant is grown in wet soil and propagation is by small pieces of rhizomes. The rhizomes are harvested when the leaves turn yellow, about 11 months from planting. The edible starchy tuber-like rhizomes, 2½–3 cm (1–1¼ in.) in diameter and 10–20 cm (4–8 in.) long, contain about 30% carbohydrates and about 1.7% protein. It is said that the rhizomes were used to treat wounds from poison arrows; hence, its name.

LEREN

Genus and species: Calathea allouia (Callira)
Other names: Besides the many local names are Guinea arrowroot and sweet corn tuber. *Allouia* is a Carib Indian word.

The plant is native to the Caribbean region, southeast to the Guianas, to Brazil, and south to Venezuela to Peru. The crop has been introduced into India and Southeast Asia.

Mature plants occur in dense growth of rhizomes and pseudostems with elongated leaves. The upright rhizomes, branched continuously from lateral buds, are about 20 cm (8 in.) long. Leren flowers but has not been known to set seeds. Tuberous roots, which are oblong to spherical in shape, arise normally near the end of fibrous roots. They vary from 2 to 4 cm (¾ to 1½ in.) long. The surface of the tuberous roots has a fine bark-like periderm and fine fibrous roots; the flesh is white.

Leren is planted at the beginning of the rainy season in loose loam soil with good drainage. Propagation is by presprouted short sections of rhizomes from the mother plant. Although plants can be grown in the shade, full sunlight gives higher yields. In Puerto Rico, planting is from the middle of September to the middle of December; the best time is mid-November. Harvest occurs from March to July depending on the date of planting. Yield is from 2 to 12 MT/ha depending on the rainfall during the growing period.

The cooked tuberous roots retain their desirable crisp texture. According to reports, they sometimes have a slightly bitter aftertaste. The starch content ranges from 13 to 15% and the protein, low in the cystine, is about 6.6% on a dry weight basis. The roots can be stored for about 3 months at room temperature without much loss in quality.

ARRACACHA

Family: Umbelliferae (parsley)
Genus and species: Arracacia xanthorrhiza
Other names: Arracacha (Colombia), Peruvian carrot or parsnip, mandioquina-salsa (Brazil) (Fig. 14.1).

It is indigenous to the Andean highlands from Venezuela to Bolivia. It is considered to be as important as the white potato in some areas of the Andes. The crop is very popular in southern and central Brazil where

FIG. 14.1. Arracacha *(Arracacia xanthorrhiza)* roots.

sizable acreages are grown commercially. The roots contain from 10 to 16% starch and are used as a starchy vegetable either fried, boiled, or in stews. They have a mild carrot or celery-like flavor. The cultivars can vary in color from white, cream, yellow, orange, to purple.

Production in the equatorial Andes is at elevations from 1700 to 2500 m (5000 to 7500 ft) and in southern Brazil at about 800–1000 m (2500–3000 ft) where the temperature ranges from 15° to 20°C (60° to 70°F) during the growing season. An even distribution of rainfall during growth is essential.

Propagation is not by seeds but by offsets or shoots obtained from the crown of the roots. The shoots, which are allowed to "heal" for 2–3 days, are planted in the early spring in southern Brazil and at the beginning of the rainy season in the Andes about 60 × 90 cm (2 × 3 ft) apart in rich loamy soil about 10–15 cm (4–6 in.) deep.

At maturity there may be as many as 10 lateral slender roots about the size of carrots aggregated around the root stock. The roots are ready for harvest from 9 to 10 months in Brazil and 10 to 14 months in the Andes. If the roots are allowed to grow longer, they turn watery, are strong flavored and fibrous.

OCA

Family: Oxadicaceae (oxalis or wood sorrel)
Genus and species: Oxalis tuberosa
Other names: Oca (Peru, Ecuador, Bolivia, Chile, Argentina), ibia (Colombia), cuiba (Venezuela), papa extranjera (Mexico).

The crop has been cultivated by the Incas in Peru since pre-Columbian times and is still very important in the high elevations of the Andes. The starchy tubers contain crystals of calcium oxalate; some cultivars contain more oxalates than others. It is necessary to cure or ripen the tubers by placing them in the sun for several days; afterward they can be eaten raw or cooked, generally in stews. They are preserved by first soaking in running water for 3 or 4 weeks, then by alternately freezing and drying. The product is called *chuno de oca* or *chaya*. It is reported to be better than white potatoes *(chuno de papa)* prepared in a similar way.

Oca is commercially grown on the South Island of New Zealand, where they are called yams; the tubers are baked or sometimes boiled.

The crop is vegetatively propagated by means of tubers (Fig. 14.2). It is planted at the beginning of the rainy season and when mature the plants are 20–30 cm (8–12 in.) in length. The tubers from different

FIG. 14.2. Oca *(Oxalis tuberosa)* tubers.

cultivars range from white, yellow, orange, to red. Good yields are 5–7 MT/ha (2–3 tons/acre). The tubers contain about 80–86% water, 1.1% protein, and about 13% carbohydrates.

ULLUCA

Family: Basellaceae (basella or Madeira vine)
Genus and species: Ullucus tuberosa
Other names: Ulluco, or olluco (Peru, Bolivia), chigua (Colombia), melloco (Ecuador, Colombia), ruba, timbo (Venezuela), popalisa (Spanish).

Ullucu is very frost resistant; hence, it is a very important crop in the Andes of southern Colombia and Peru at elevations of 3000–4000 m (9000–13,000 ft). The seed tubers are planted with 40–60 cm (16–20 in.) spacing in rows 80–100 cm (30–40 in.) apart in September and October. Mature plants are 20–30 cm (8–12 in.), rarely 50 cm (20 in.) in height. Tuberization occurs in day lengths of 12–12½ hr and the tubers are harvested 140–150 days from planting. Yields vary from 5 to 9 MT/ha (2–4 tons/acre).

External color of tubers vary from white, pale yellow, pale green, pale magenta, or yellow with magenta spots.

Ulluca is used either fresh or dehydrated. The dehydrated product, *chuno de ulluco*, is prepared like *chuno de oca* or *chuno de papa*. Fresh tubers contain 80–85% water, 1.0% protein, and as high as 14% carbohydrates.

ANU

Family: Tropaeolaceae (nasturtium)
Genus and species: Tropaeolum tuberosum
Other names: Isano, mashua (Peru, Bolivia), nabos, cubios (Colombia).

Anu is a native of the highlands of Peru and Bolivia. It is not as widely grown as the white potato, oca, or ullucu but nonetheless it is an important staple crop in this region. An herbaceous climber, anchoring to other plants and surrounding objects, anu is propagated vegetatively by tubers. Plantings are in rows 70–100 cm (30–40 in.) apart and 40–70 cm (16–30 in.) between seed tubers.

Tubers mature in about 200 days. They are 3–6 cm (1⅛–2¼ in.) in diameter and 5–15 cm (2–6 in.) long, and are white to yellowish in color and often mottled or striped with red or purple around the eyes. The tubers are not eaten raw as they have a disagreeable taste; they are usually cooked or dehydrated into *chuno* by freezing and sun drying. Yields vary from 20 to 30 MT/ha (9 to 13 tons/acre).

JERUSALEM ARTICHOKE

Family: Compositae (sunflower or composite)
Genus and species: Helianthus tuberosus
Other names: Sunchoke, girasole.

A native of North America, Jerusalem artichoke can be found growing wild from Kansas north to Minnesota and as far east as Nova Scotia in Canada along the Atlantic Coast. The American Indians were cultivating this plant long before the first colonies were established by Europeans.

The tubers, which are used for propagation, were taken to Europe in the early 1600s where it is an important crop in areas that are too dry or soil too poor for white potatoes. The name Jerusalem artichoke was probably derived from the Italian words *girasole articocco* (sunflower edible).

The plant is a perennial growing 1½–2½ m (4–8 ft) in height. It forms underground storage tubers, shaped somewhat like the white potato but quite knobby in appearance, about 3–6 cm (1¼–2½ in.) in thickness and 7–10 cm (3–4 in.) long. Depending on the variety, the color of the tuber varies from white to yellow and from red to blue.

Vegetatively propagated from whole or cut tubers with two to three prominent buds, Jerusalem artichoke grows best in well-drained sandy loams; however, it can be grown in heavy soils, which give high yields, but from this type of soil harvest of tubers is difficult. Except in very fertile soils, some fertilization with high phosphate is recommended early in the growing season. Plantings should be made early in the spring; in late plantings the yield is reduced. About 4½ month growing season is required. Tuberization occurs with decreasing day lengths in the late summer.

At harvest careful handling is required as the tubers bruise easily and lose moisture rapidly because of a rather thin periderm. Yield of tubers range from 11 to 22 MT/ha (5 to 10 tons/acre). The recommended storage is at 0°–2°C (32°–36°F) and high RH. Under these conditions sound disease-free tubers can be stored for several months. Jerusalem artichoke is best stored in the ground in regions where the ground does not freeze in the winter. The soil should be well drained and not waterlogged. The Jerusalem artichokes should be dug just prior to marketing or use.

Instead of starch, a polymer of glucose, the main storage carbohydrate is inulin, a polymer of fructose. The tubers are used by diabetics since the carbohydrate of this form can be utilized by persons having glucose metabolic dysfunction.

JICAMA OR YAM BEAN

Family: Leguminosae (pea)
Genus and species: Pachyrrhizus erosus

Jicama can be found wild in Mexico and northern Central America. It is a herbaceous climbing vine with large white or blue flowers. The plant produces from one to several turnip-shaped roots. In Mexico, growers remove flower buds to increase yields and to improve the quality of the roots. The young brown-skinned and white-fleshed roots are sweet and watery and often eaten raw as a snack food. The older roots attain sizes up to 30 cm (1 ft) in diameter and contain a high quality starch. The immature pods of this species can be used as a cooked vegetable. Mature pods and seeds are poisonous. Mature seeds and roots contain rotenone, which is used as an insecticide and fish poison.

POTATO BEAN

Genus and species: Pachyrrhizus tuberosus

Native to tropical South America, the potato bean's home is thought to be somewhere near the head waters of the Amazon River. *Pachyrrhizus tuberosus* is similar to *P. erosus*, but the flowers are either violet or white and the immature pods have irritant hairs making them unsuitable for use as vegetable. The tuberous roots are used in the same manner as with jicama.

Genus and species: Pachyrrhizus ahipa

Found in Bolivia and northern Argentina, this is a small nonclimbing plant with violet or white flowers, grown for the edible roots which develop in a short time.

KUDZU

Genus and species: Pueraria lobata (P. thunbergiana, Pachyrrhizus trilobus)

Other names: Fankot (China).

Kudzu is a perennial leguminous vine grown in Japan and China for the starchy enlarged root. It is also grown for the vines which are used for fodder and as a cover crop in many parts of the world. The tropical kudzu *(Pueraria phaseoloides)* is also a perennial vine but is a more vigorous and is used mainly for fodder and as pasture plant.

Kudzu vines grow as long as 8 m (25 ft). In some climates, kudzu set very few seeds so the crop is propagated vegetatively from stem cuttings obtained from just above the enlarged root after harvest. Sections are rooted in nursery beds, and after about 2 months, the rooted stem sections are planted in the field and staked to provide support for the vines.

The enlarged roots can weigh as much as 35 kg (80 lb) and can be almost a meter (3 ft) long and can contain over 25% starch. Kudzu is a very important source of starch in the Orient.

CHINESE ARTICHOKE

Family: Labiatae (mint)

Genus and species: Stachys tuberifera (S. sieboldii, S. affinis)

Other names: Kon loh (China).

Chinese artichoke, a native of the Far East, is a perennial plant grown for the tubers, which are somewhat cylindrical, 5–8 cm (2–3 in.) long and 1½–2 cm (⅝–¾ in.) in diameter. The tuber has several distinct internodal segments that are constricted and appear bead-like. The plant grows from 30 to 45 cm (1 to 1½ ft) in height and has rough nettle-like leaves typical of the mint family. The crop is started in the spring by planting tubers about 7–8 cm (3 in.) deep and 15–20 cm (6–8 in.) apart in rows about 40 cm (10 in.) apart in rich sandy loam. Tubers, which are dug in the fall, contain the tetrasaccharide stachyose, instead of starch. They are boiled or fried and used like white potatoes.

BIBLIOGRAPHY

BOSWELL, V.R. 1959. Growing the Jerusalem Artichoke. USDA Leaflet No. 116. U.S. Dept. Agric., Washington, DC.

CLAUSEN, R.T. 1944. A botanical study of the yam beans *(Pachyrrhizas)*. Mem. 264, Cornell Univ. Agric. Exp. Stn., Ithaca, New York.

FILGUEIRA, F.A.R. 1972. Manual de Olericultura. Editora Agronomica Ceres Ltd. Sao Paulo, Brazil.

GADE, D.W. 1966. Achira, the edible canna, its cultivation and use in the Peruvian Andes. Econ. Bot. *20*, 407–415.

HODGE, W.H. 1951. Three native foods of the high Andes. Econ. Bot. *5*, 185–201.

HODGE, W.H. 1954. The edible arracacha—little known root crop of the Andes. Econ. Bot. *8*, 195–221.

LEON, J. 1964. Plantas Alimenticias Andinas. Bul. Tech. No. 6, Instituto Interamericano de Ciencias Agricolas Zona Andina, Lima, Peru.

KADDY, M.S., JOHNSTON, A., and WILSON, D.B. 1980. Nutritive value of Indian bread root, squawroot and Jerusalem artichoke. Econ. Bot. *34*, 352–357.

MARTIN, F.W., and CABANILLAS, E. 1976. Leren *(Calathea allouia)*, a little known tuberous root crop of the Caribbean. Econ. Bot. *30*, 249–256.

SCHROEDER, C.A. 1968. The jicama, a root crop from Mexico. Am. Soc. Hort. Sci., Trop. Region *11*, 65–71.

TANNER, R.D., HUSSAIN, S.S., HAMILTON, L.A., and WOLF, F.T. 1979. Kudzu *(Pueraria lobata)*: Potential agricultural and industrial resource. Econ. Bot. *33*, 400–412.

15

Sweet Corn

Family: Gramineae (grass)
Genus and species: *Zea mays*
Subspecies:

saccharata—sweet corn. The grains contain a sweetish endosperm at the immature stage. Starch accumulation occurs with increase of maturity.

trunicata—pod corn. Each kernel is enclosed in a pod or husk; the whole ear is also enclosed in husks.

amylacea—flour or soft corn. The kernel has a soft or floury instead of a vitreous endosperm.

indurata—flint corn. This is the field corn which is characterized by a starchy endosperm. Kernels are large, broad and tops rounded. When harvest immature, they are called "roasting ears" and used as a vegetable.

indentata—dent corn. Sides of grain have corneous endosperm and soft white starch. The apex shrinks on drying at the mature stage producing the characteristic "dent."

everta—popcorn. A large portion of the endosperm is horny. The ears are small and also the kernels. When heated, the trapped steam in the kernel causes an explosion, bursting the seed coat and exposing the white fluffy endosperm.

ceretina—waxy corn. Chemical composition of the starch is entirely of the amylopectin type as contrasted to both the amylose and amylopectin type in other subspecies.

Origin

A native of tropical Americas, corn has been the principal food in Mexico, Central America, and many South American countries since pre-Columbian times. No living wild types of corn can be found in these regions, only the domesticated forms. Recent discoveries of corncobs in fire caves in the valley of Tehuacan in southern Mexico have been dated to about 5400–7200 years ago. Fossilized maize pollen grains, dated at about 80,000 years, have been discovered in Mexico. Many remains are thought to resemble the original wild forms of corn. These findings indicate no substantial change occurred in the botanical characteristics except increases in size of ears and productiveness under domestication. The ancestor of the corn is thought to have a short tassel attached at the apex of a small ear, a perfect flower.

Botany

Corn is an herbaceous monocotyledonous annual seed producing plant. The plant is monoecious, bearing its male organ (tassel) and female organs (ears) separately on the same plant. The tassel is borne on the main axis as the terminal inflorescence and the ear as a lateral inflorescence at the axis of the leaf. The plant may bear one or more ears.

Culture

Climatic Requirements

A warm season crop but not suitable for the wet tropics, sweet corn requires from 70 to 110 frost-free days from planting to harvest. The crop is grown from 50°N to 40°S latitudes, from sea level to 3000 m (10,000 ft) in elevation. Soil temperature for planting should be above 13°C (55°F); it will not germinate below 10°C (50°F). The optimum range for seed germination is 21°–27°C (70°–80°F) soil temperature. The optimum mean air temperature for growth is 21°–30°C (70°–86°F); corn can be grown at means as low as 16°C (60°F) to means as high as 35°C (95°F). Where field corn is grown (Fig. 15.1), generally sweet corn can also be grown as fewer days are required from planting to harvest.

Flowering of sweet corn is influenced by photoperiod and temperature. Corn flowers sooner under short days and takes longer under long days (Table 15.1). Many tropical cultivars do not flower in temperate regions until the day lengths decrease to 12–14 hr; often these tropical types remain vegetative and plants may reach heights of 6 m (19 ft) before tasseling. Very short days (8 hr) and low temperature (<22°C/72°F means) can also delay flowering. Tropical types, which do mature normally in temperate regions, are usually extremely early and yield poorly because the plants are small. Cultivars bred for long days and cool temperatures yield less when grown under warm temperatures and short days (Table 15.2).

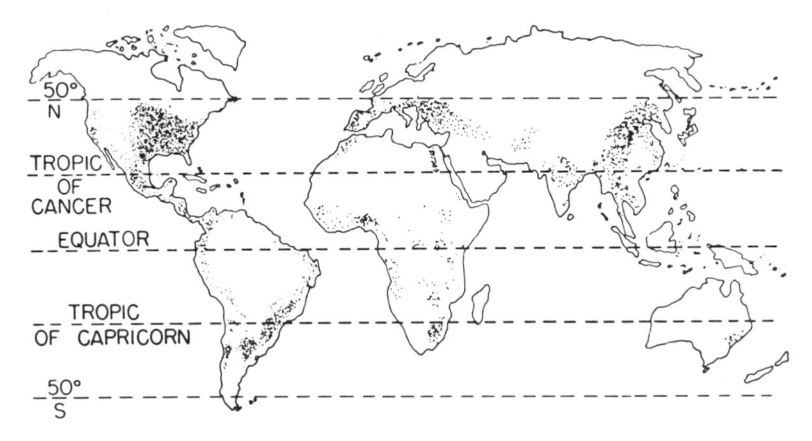

FIG. 15.1. Corn *(Zea mays)* growing regions of the world.

Table 15.3 shows the effect of cool and warm temperatures at early and mid periods of growth on corn development. The temperature may be cool during the seedling stage (fourth leaf stage) but must be warm thereafter for the plant to grow. Coolness during the latter period inhibits plant growth.

TABLE 15.1. EFFECT OF PHOTOPERIOD ON 'MAJOR BELLE' A PUERTO RICAN CULTIVAR

Photoperiod[a] (hr)	Days to tassel initiation	No. leaves visible at initiation	Days to tassel emergence	No. leaves at tasseling
10	26	9.5	66	19.8
13	27	9.7	66	19.2
16	50	17.0	87	25.2

Source: Arnold (1969).
[a] 24°C (75°F) at 32,000 lux (3000 fc).

TABLE 15.2. PHOTOPERIOD AND TEMPERATURE EFFECT ON CORN (HARROW 691 HYBRID) YIELD[a]

Treatment	Total plant dry wt (g/plant)	Grain yield (g/plant)	Average number of kernel	Wt/kernel (g/dry wt)
30°C/10 hr day	174	57	379	0.192
30°C/10 hr day + 10 hr low light	250	55	399	0.171
20°C/10 hr day	220	114	427	0.281
20°C/10 hr day + 10 hr low light	478	194	607	0.338
LSD 5%	27	—	91	0.022

[a] Data from Hunter *et al.* (1977).

TABLE 15.3. EFFECT OF COOL AND WARM TEMPERATURES ON DEVELOPMENT GOLDEN CROSS BANTAM SWEET CORN

Treatment period	Temperature[a]			
Planting to fourth leaf stage	Warm	Cool	Warm	Cool
Fourth leaf to ninth leaf stage	Warm	Warm	Cool	Cool
	Average number of leaves			
Main stalk	17.5	17.5	14.3	14.9
Below first ear	12.6	12.3	9.8	9.6
Above first ear	4.9	5.2	4.5	5.3
Showing at tassel initiation	9.3	9.5	6.5	6.3

Source: Arnold (1969).
[a] Warm: 35°C days/26.7°C nights (95°/80°F); cool: 21°C days/12.8°C nights (70°/55°F); day length: 14 hr at 32,000 lux (3000 fc).

Soil, Nutrition, Moisture

Corn is a shallow rooted crop. It can be grown from sandy loam to clay loam or in peat and muck soils. Optimum soil pH is from 6.0 to 7.0 but it can be grown successfully in the range of 5.0 to 8.0. It is moderately tolerant to salt and alkali.

Corn requires high fertilization, 100–112 kg/ha (90–100 lb/acre) N for heavy soils and late plantings. A heavier application of N, 115 kg/ha (110 lb/acre), is recommended for light soils and for early spring planting. From 45 to 112 kg/ha (40 to 100 lb/acre) P are recommended, and some K, 60 kg/ha (50 lb/acre), for those soils low in this element. If barnyard manure is used, 27.4 MT/ha (10 tons/acre) or 9–11 MT/ha (4–5 tons) of poultry manure is recommended.

Where rainfall is not adequate during the growing season, supplemental watering is necessary. From 300 to 660 mm (12 to 26 in.) of water is required to grow a crop of sweet corn. Flooding is very injurious to corn plants. Water stress at tasseling and silking stages is detrimental to plant development. It causes stalk rot, reduced plant height, ear development, and ultimately yield. Recommendations are that no more than 40% of the available soil water be depleted.

Growth

Several cultivars should be tested for adaptability. Hybrid seeds usually produce more vigorous plants and higher yields. Fungicide treatment of seed is beneficial when planted in cool wet soil. Corn is planted 2.5–4.0 cm (1–1½ in.) deep in rows 90 cm (36 in.) apart. Spacing between plants should be 20–30 cm (8–12 in.). The thinning is to be done when the plants are 10–15 cm (4–6 in.) high.

Pollination is by wind and gravity. "Poor fill" is usually a result of poor pollination. Pollination is poor when the air temperatures are above 36°C (96°F), under hot dry winds, or when the plant is under moisture stress. In small plots, to ensure good pollination, corn should be planted in blocks and not in long single or double rows. Shriveling of kernels at the tip is caused by N deficiency during maturation. Removal of suckers does not increase the yield. (Suckers are tillers or secondary shoots from the axils of the lower leaves.)

Under normal temperatures, about 14–19 days after pollination, the ears are ready for harvest for use as a vegetable; field corn requires about 30 additional days for mature seeds. Figure 15.2 shows the various stages in the growth of corn. Total sugar, starch, and dextrin accumulations in sweet corn is shown in Fig. 15.3; the dextrin accumulation of field (dent) corn is very low.

Harvest and Storage

Harvest

Sweet corn ears remain in marketable condition for about 5 days when the mean temperature is about 16°C (60°F) and 1 or 2 days at 29°C (85°F). At harvestable stage the kernels, when crushed, are milky. To determine harvestable maturity the husks should not be disturbed but

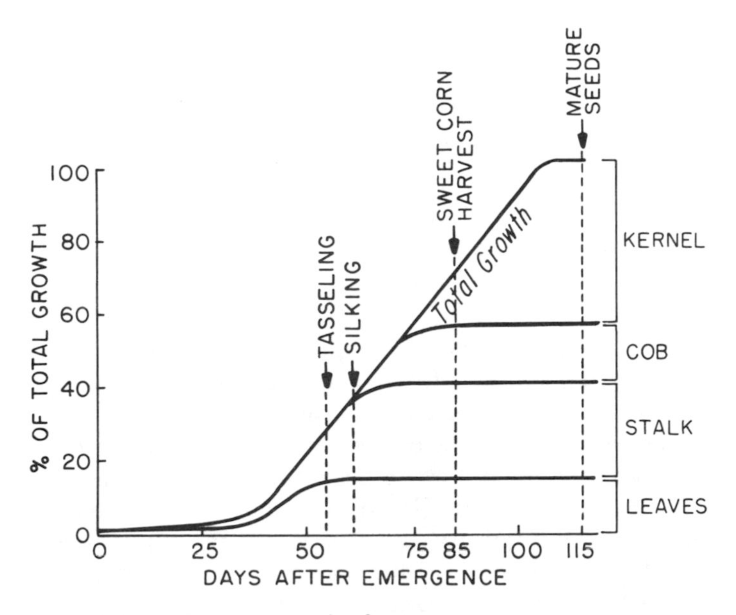

FIG. 15.2. Stages in growth of corn.

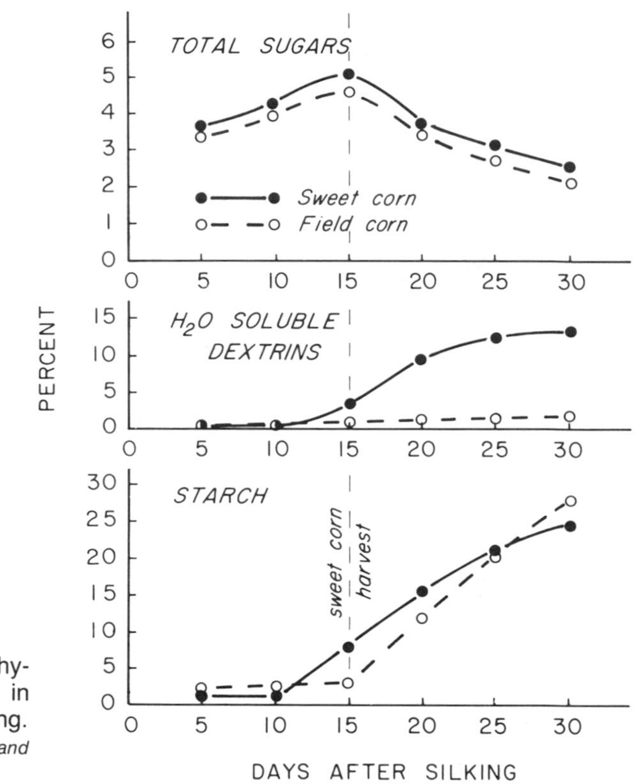

FIG. 15.3. Carbohydrate accumulation in sweet corn after silking. *Data from Culpepper and Magoon (1930).*

the external appearance is used: the silk is somewhat dried, the husks are tight, and the ear feels firm when grasped. Machines are used to harvest large acreages of corn for the freezing and canning industries and also for the fresh market.

Storage

Sweet corn should be harvested early in the morning to take advantage of the cool temperature. Immediately after harvest for best quality corn, the temperature of the ears should be lowered to 10°C (50°F) by immersion into ice water. The ears should be cooled further to near 0°C (32°F) as soon as possible. Low temperature decreases the rate of conversion of sugars into starch. The recommended storage temperature is 0°C (32°F) and high humidity. If sweet corn is vacuum cooled, the ears are wetted down before cooling and packed in ice during transit. Under these conditions satisfactory quality is maintained for about 1 week but if held at 10°C (50°F) the time is reduced to 2 days.

The gene in sweet corn, *sugary 1*, prevents starch synthesis from sucrose translocated into the endosperm. Also, the gene *brittle 1* inhibits the breakdown of sucrose into fructose and a derivative of glucose. These two genes allow sweet corn to remain sweet several days without refrigeration and extend the time for harvest.

Pests and Diseases

Corn ear worm—insecticide for control.

Smut—crop rotation and resistant varieties.

Pink rot or *ear mold*—develops under heavy irrigation and high temperatures (Fig. 15.4).

Sugar cane mosaic virus—stunts plants and produces small ears with kernel blanking. Leaves show mosaic pattern with broken linear stripes between veins.

FIG. 15.4. Sweet corn (*Zea mays* subspp. *saccharata*). (Left), normal; (center) infected with bacterial pink rot; (right) infected with smut and pink rot.

Nutritive Value

Corn is high in energy; yellow corn is high in provitamin A. The yellow pigment is cryptoxanthin, a carotenoid. Corn is low in the essential amino acid lysine. New cultivars with high lysine content are being developed. The *opaque 2* gene increases the proportion of essential amino acids lysine and tryptophan.

Economics

The world sweet corn production has not been compiled. In the United States sweet corn production in 1979 was 2.8 million MT (3.1×10^6 tons) grown on 236,500 ha (584,000 acres). About 20% of the production was for the fresh market and the remainder for processing (canned and frozen). The U.S. per capita consumption has not changed significantly during the past decade. In 1979 the per capita consumption of fresh sweet corn was 3.3 kg (7.3 lb), 6.0 kg (13.3 lb) canned, and 3.3 kg (7.3 lb) frozen (on the cob basis).

Sweet corn is not shipped long distances without proper refrigeration (0°C/32°F). At room temperature there is a rapid conversion of sugars to starch. Quality of sweet corn and sugar content are correlated.

BIBLIOGRAPHY

ARNOLD, C.Y. 1969. Environmentally induced variation of sweet corn characteristics as they relate to the time required for development. J. Am. Soc. Hort. Sci. *94*, 115–118.

CULPEPPER, C.W., and MAGOON, C.A. 1930. Effects of defoliation and root pruning on the chemical composition of sweet corn kernels. J. Agric. Res. *40*, 575–583.

GOODMAN, M.M. 1976. Maize, *In* Evolution of Crop Plants, N.W. Simmonds (Editor). Longmans-Green, London.

HUELSON, W.A. 1954. Sweet Corn. Interscience Publishers, Inc., New York.

HUNTER, R.B., TOLLENAAR, M., and BREUER, C.M. 1977. Effects of photoperiod and temperature on vegetative and reproductive growth of a maize (*Zea mays*) hybrid. Can. J. Plant Sci. *57*, 1127–1133.

MACGILLIVRAY, J.H. 1953. Vegetable Production. Blakiston Co., New York.

MAKUS, D.J., CHOTENA, M., SIMPSON, W.R., and ANDEREGG, J.C. 1980. Water stress and sweetcorn seed production. Current info. ser. No. 521. Univ. Idaho, Coll. Agric. Moscow, Idaho.

MANGELSDORF, P.C. 1974. Corn—its origin, evolution and improvement. The Belknap Press, Harvard Univ., Cambridge, Massachusetts.

MARTIN, F.W., DALOZ, C., and DEL CARMEN VELEZ, C. 1981. Vegetables for the hot humid tropics. Part 8. Vegetable corns, *Zea mays*. Mayaguez Inst. Trop. Agric., Sci. Ed. Admin., US. Dept. Agric., Mayaguez, Puerto Rico.

PIERRE, W.H., ALDRICH, S.D., and MARTIN, W.P. 1966. Corn Production. Iowa State University Press, Ames, Iowa.
PURSEGLOVE, J.W. 1972. Tropical Crops: Monocotyledons. John Wiley & Sons, Inc., New York.
SIMS, W.L., KASMIRE, R.K., and LORENZ, O.A. 1978. Quality sweet corn production. Leaflet No. 1818, Agric. Ext. Ser., Univ. of California, Berkeley.

16

Plantain (Starchy Banana) and Breadfruit

PLANTAIN

Family: Musaceae (banana)
Genus: *Musa*

Principal species	Common name
M. paradisiaca	French plantain
M. corniculata	Horn plantain
M. acuminata	Gros Michel and Cavendish type bananas

Musa paradisiaca and *M. corniculata* are triploids with designated genomes, AAB; *M. acuminata* has triploid genome, AAA.

All edible fruited clones are either *Musa acuminata* or hybrids of *M. acuminata* (AA genome) and a wild species *M. balbisiana* (BB genome). Fruits of the diploid *M. balbisiana* are neither edible nor parthenocarpic. The plantains are naturally produced triploids of *M. acuminata* and *M. balbisiana* (AAB genome). Fruits of the triploid are larger; the plants are more vigorous and give higher yields than the diploids. Plant breeders have produced tetraploids of *M. acuminata*.

Origin

According to Sauer, the banana was probably among the first plants domesticated by man. This occurred in southeast Asia, probably Malaya. At that time vegetative parts were used as food since fruits were inedible. During the evolution of the banana, the first and crucial event to take place was the occurrence of the parthenocarpic fruit. First records of bananas were in India about 500–600 BC; also, there are records of this fruit growing in China about 200 AD. Subsequently, banana spread from its place of origin to Africa and to the Mediterranean region before 650 AD. It was introduced to the New World in post-Columbian times.

Botany

Plants are tree-like huge perennial monocotyledenous herbs with a basal corm. They grow 2–6 m (6–20 ft) in height. The pseudostem, which is trunk-like, is composed of leaf sheaths. The underground stem is a corm with very short internodes. Leaves are formed in spiral succession with the inflorescence, the final organ, emerging through the base of the pseudostem. From three to four buds, called suckers, develop from the corm. As new leaves are produced, older leaves die so that a more or less constant number (about 10–15) remains. The inflorescence bears both male and female flowers; the first five to 15 basal nodes produce female flowers; the intermediate nodes produce hermaphoditic flowers; and the upper or distal nodes produce male flowers. The fruit is a berry; the edible fruit of commerce is parthenocarpic (not pollinated) and therefore seedless. French plantains have seven to 10 "hands" per bunch and numerous "fingers" (individual fruits), while horn plantains have three to five "hands" per bunch and few individual "fingers" (Fig. 16.1).

Culture

Climatic Requirements

Banana and plantain are cultivated in a wide variety of climates varying from wet tropical to dry subtropical; mostly between 30°N and 30°S of the equator. Winter mean temperatures should not fall below 16°C (60°F) and about 1250 mm (50 in.) of rainfall per year are required. The ideal conditions are about 27°C (80°F) mean and at least 100 mm (4 in.) of rainfall per month. Strong winds damage the leaves and reduce crop yields. Figure 16.2 shows the world banana growing regions.

Soil

Plantains require soil with good drainage and aeration. The soil need not be necessarily deep as most of the adventitious fiberous roots are found in the top 30 cm (12 in.) of soil. Because of good drainage requirement, gentle slopes and hill sides are often used for growing of crops. Soils with high organic matter, low salinity and pH range of slightly acid to slightly alkaline are preferred; those containing 0.05% or more NaCl are toxic.

FIG. 16.1. Plantain *(Musa corniculata)* ratoon crop, Palmera, Colombia.

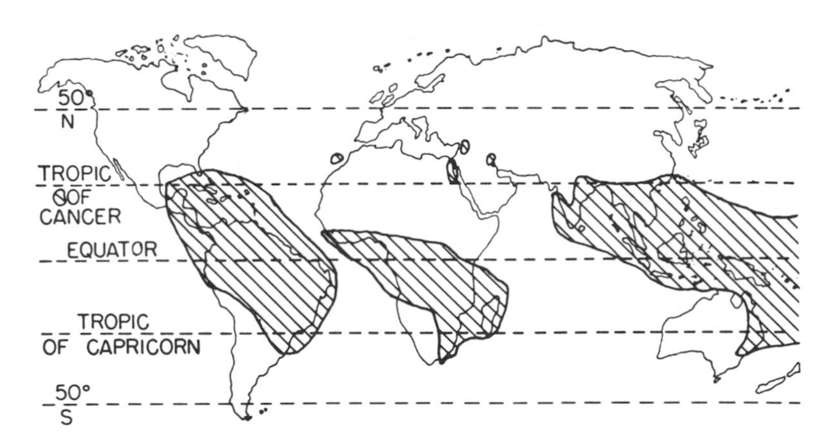

FIG. 16.2. Banana *(Musa)* growing regions of the world.

Propagation

Propagation is exclusively by vegetative means, either by means of suckers or by large corms. Suckers, of which there are several types, develop from the main plants. They are removed and used for new plantings. Spacing varies from 3 to 6 m (10 to 20 ft) between plants. A single sucker is allowed to continue to grow from the main plant for the *ratoon* crop as soon as the main plant has reached the reproductive stage; the rest of the suckers should be removed. A ratoon crop is produced from the plant arising from a sucker (adventitious shoot).

Growth

Plants should be supplied with small amounts of N during growth and with P and K twice a year. If rainfall is less than 100 mm (4 in.) per month, supplemental irrigation is required, especially during the reproductive phase of growth.

The vegetative phase lasts from 8 to 10 months depending on temperature, followed by emergence of the flowering stalk; flowering and fruit development require about 4–6 months. The total life of the pseudostem is 12–22 months, depending on growing conditions and particular clonal characteristics. The growth pattern of the plant is depicted in Fig. 16.3.

In Ghana intercropping is practiced on plantain farms. Aroids (cocoyams) followed by maize, eggplants, peppers, tomatoes, and okra are interplanted.

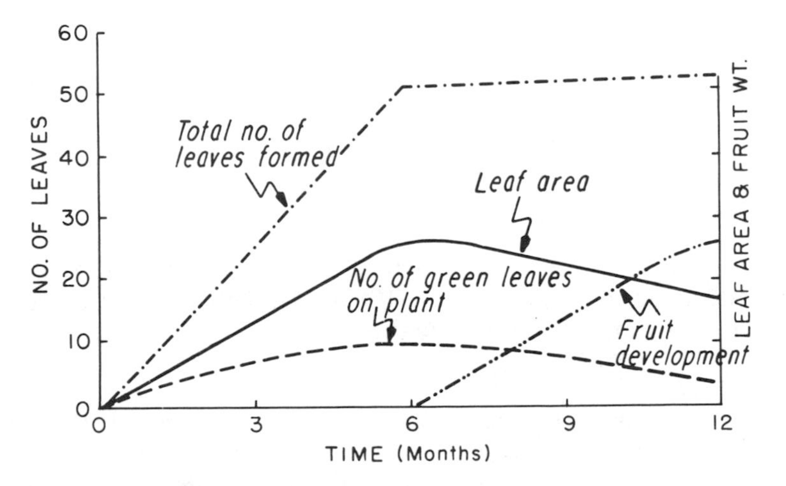

FIG. 16.3. Diagram showing growth pattern of plantain.

Harvest and Storage

Harvest

Bunches are usually harvested when fruits are firm and green, about 80–90 days after the female flowers open. The yield of plantain varies from 7 to 16 MT/ha (3 to 7 tons/acre); yield varies with spacing. As high as 67 MT/ha (30 tons/acre) have been obtained. Bunches are cut into hands in the field and placed into padded boxes to minimize physical damage during transport.

With bananas treatment with ethylene (100 ppm of C_2H_2) will stimulate the entire bunch to ripen at the same time; the color changes from green to yellow with concurrent softening of the pulp and conversion of starch to sugars. Plantains usually require no ethylene treatment as a firm textured and starchy pulp is desired. To delay ripening during transport, plantain can be packed in boxes lined with 40 μm (1.5 mil) polyethylene, which is sealed to create a modified atmosphere; a C_2H_4 absorber may also be enclosed.

Storage

As a chilling-sensitive crop, the fruits should be held at or above 12°–13°C (53–56°F), which is the recommended storage temperature. To prevent ripening, ethylene should be removed from the atmosphere during transport and storage. In controlled atmosphere storage the O_2 concentration should be from 1 to 3%, the CO_2 from 5 to 10% and with little C_2H_4 as possible.

Reduction of pressure (hypobaric) to ½ atmosphere and 16°C (61°F) in the storage room will double the shelf life of bananas. At the same temperature but at ⅓ atmosphere pressure the storage life is doubled again. These effects of hypobaric conditions are due to reduced O_2 tensions and accelerated C_2H_4 removal.

Nutritional Value

Plantains are high in carbohydrates and vitamin A; the vitamin C content ranges from 18 to 28 mg/100 g. (See Table 8.6 for approximate compositions and Table A.2 for the amino acid content.)

Food Uses

Fruits Unlike the ordinary banana eaten raw when ripe; the plantain even when ripe is starchy and unpalatable when eaten without cooking. Boiling, steaming, or baking removes the astringency due to tannins present. The fruits are also used for making of beer, wine, and distilled spirits.

Vegetative Parts In New Caledonia *M. sapientum* var. *oleracea* are specially grown for the fleshy rhizomes, which are eaten as starchy food. The hearts of the stem (pseudostem) of *M. flaviflora* are used in India as vegetables; the inner sheaths of this species are used for food in the East Indies. Plants of *Ensete ventricosum*, which virtually never flowers, are used for food in southern Ethiopia.

Male Flowers The male bud of many varieties, after removal of fibrous bracts, are boiled and eaten as vegetables in southeast Asia. Several changes of water are required to remove the astringency (Fig. 16.4).

Pests and Diseases

Diseases

The plantain group are resistant to Panama disease, caused by *Fusarium oxysporum*, and leaf spot, caused by *Mycosphaerella musicola*, to which most of the commercial banana varieties are susceptible. All

FIG. 16.4. Male banana (*Musa* spp.) blossoms trimmed for use as vegetable.

TABLE 16.1. WORLD PRODUCTION OF PLANTAIN AND BANANA (*Musa* spp.)

	Production (10^3 MT)		
	Plantain		Banana
World	20,584		39,129
Continent			
Africa	13,285		4,502
North and Central America	1,464		7,503
South America	4,336		11,820
Asia	1,495		13,789
Europe	—		432
Oceania	4		1,084
Leading countries			
1. Uganda	3,192	1. Brazil	6,424
2. Colombia	2,236	2. India	4,000
3. Nigeria	2,150	3. Indonesia	2,905
4. Rwanda	2,127	4. Philippines	2,430
5. Zaire	1,420	5. Ecuador	2,391

Source: FAO (1979).

plantains and bananas are susceptible to the bunchy top virus (cabbage top or strangles) transmitted by aphids. Control is by destruction of diseased plants and aphids on them and use of certified disease-free planting material.

Bacterial wilt *(Pseudomonas solanacearum)* is due to poorly drained land.

Economics

The world production of plantains in 1979 was over 20 million MT and bananas was 39 million MT. It has been estimated that half of the production of banana in the world is used as cooked food. Table 16.1 shows that Africa is the leading continent in plantain production and Uganda is the leading country.

BREADFRUIT

Family: Moraceae (mulberry)
Genus and species: Artocarpus altilis
Other names: *Arbre a pain* (Fr.), *Fruta de parr* (Sp.)

The breadfruit is a multiple fruit of a tropical tree 7–20 m (25–65 ft) in height. A native of South Pacific and East Indies, it is an important energy-giving food for the people of this region. The crop grows best in the humid tropics with mean temperatures of 21°–32°C (70°–90°F) with annual rainfall of 1500–2500 mm (60–100 in.). Some cultivars can withstand drought, and others tolerate some salinity in the soil.

Propagation is by suckers or crown cuttings as the fruits are usually seedless. The tree starts to bear in 3–6 years. The fruit, a globular structure, develops from the entire female inflorescence and is 10–30 cm (4–12 in.) in diameter and weighs 1–4 kg (2–10 lb) when mature. However, the fruits are harvested immature when the pulp is still white and mealy. Because of the high starch content (30–40%) they are sometimes ground into flour. After removing the outer layer of the fruit, the flesh, sliced or diced, is either baked, boiled, or fried.

Jackfruit *(A. heterophyllus)*, a native of Malaya and a close relative of the breadfruit, is usually grown for the sweet acid fruits which are pear or oblong in shape and weigh up to 30 kg (60 lb). The fruits at the immature (dark green) stage contain about 24% carbohydrates and are used as a vegetable similar to breadfruit. Seeds are used as vegetables in culinary preparations, and young flower clusters are eaten with syrup and agar in Java.

In the Philippines young fruits of *A. communis* are prepared like the breadfruit while seeds of *A. ovata* and *A. blancoi* are used for food.

Both fresh breadfruits and jackfruits are highly perishable tropical produce which are chilling sensitive. They should be held at temperatures above 12°C (53°F) during postharvest handling and in storage.

BIBLIOGRAPHY

Plantain

BEZUNCH, T., and FELEKE, A. 1966. The production and utilization of the genus *Ensete* in Ethiopia. Econ. Bot. *20*, 65–70.

KARIKARI, S.K. 1922. Plantain growing in Ghana. World Crops, *24*, 22–24.

PURSEGLOVE, J.W. 1972. Bananas, *In* Tropical Crops: Monocotyledons. John Wiley & Sons, New York.

SIMMONDS, N.W. 1966. Bananas. Longmans-Green, London.

SIMMONDS, N.W. 1976. Bananas, *In* Evolution of Crop Plants, N.W. Simmonds (Editor). Longmans-Green, London.

Breadfruit

PURSEGLOVE, J.W. 1968. Breadfruit, *In* Tropical Crops: Dicotyledons. pp. 379–384. John Wiley & Sons, New York.

THOMAS, C.A. 1980. Jackfruit, *Artocarpus heterophyllus* (Moraceae), as source of food and income. Econ. Bot. *34*, 154–159.

Part B: Succulent Roots, Bulbs, Tops, and Fruits

17

Alliums: Onion, Garlic, and Others

Family: Amaryllidaceae (amaryllis) [Sometimes classified in the Liliaceae (lily) family, subfamily Allioideae]
Genus: *Allium*

Principal species	Common name
Allium cepa (common onion group)	Onion
A. cepa (Aggregatum group)	Shallot, multiplier onion, ever-ready onion, potato onion
A. cepa (proliferum group)	Egyptian onion, tree onion, topset onion
A. sativum	Garlic
A. ampeloprasum	Leek, great head garlic, kurrat, pearl onion
A. fistulosum	Welsh onion, Japanese bunching onion
A. schoenoprasum	Chive
A. chinense	Rakkyo, chiao-tou
A. tuberosum	Chinese chive, nira

ONION: *Allium cepa*

Origin

Most botanists believe that the onion originated in the regions around Iran and West Pakistan. The onion is not known in the wild form, although many alliums with onion-like flavor are found growing wild in the temperate regions of the Northern Hemisphere. The history of onions dates back at least to 3200–2800 BC from findings in Egyptian tombs. The culture of onion spread to India about 600 BC and the Greeks and Romans wrote about onions and garlic about 400–300 BC. Onions had spread into northern Europe by the start of the Middle Ages.

Botany

This biennial herbaceous plant is usually grown as an annual except for seed production. It has a superficial root system extending only within 30 cm (12 in.) of the soil surface. A number of adventitious roots about 1.5 mm (1/16 in.) in diameter grow from the stem. There is very little branching of the roots. Continued formation of new adventitious roots (three to four per week) occurs from the stem as the plant grows; concurrently, senescence and death of older roots take place. During early growth the number of active roots increases; as the bulb matures, roots die at a more rapid rate than new roots are produced.

The stem from which the roots arise is very short, the diameter increases with growth, and appears as a shortened inverted cone when mature. Leaves are produced from the apical meristem, and they push through the pseudostem formed by the sheath leaf bases of the older leaves. The green leaf blade, is hollow. Figure 17.1 is a diagrammatic drawing of the onion.

The bulb, for which the crop is grown, is formed under favorable conditions of day length and temperature when the plant has reached a certain stage of growth (see Bulbing below). Inflorescence appears after vernalization the second year. The umbel, which is borne on a scape, has from 50 to 2000 flowers. Pollination is usually by insects. Seeds of onion are very short-lived, usually not more than 2–4 years at room temperature. If kept at cool or cold temperatures and low humidity, seeds remain viable for several years. In the tropics under high temperature and high humidity, onion seeds are viable for less than a year.

Culture

Climatic Requirements

Temperature Onion is a cool season crop, adapted to 13°–24°C (55°–75°F) and tolerant of frost. For high yields, cool temperatures are needed during early stages of growth before bulbing commences. The optimum temperature for seedling growth is 20°–25°C (68°–77°F) (Fig. 17.2); growth starts to decline at temperatures greater than 27°C (81°F). Vernalization is required for flowering.

Photoperiod The day length is important in the growing of onions for bulbs (see Bulbing).

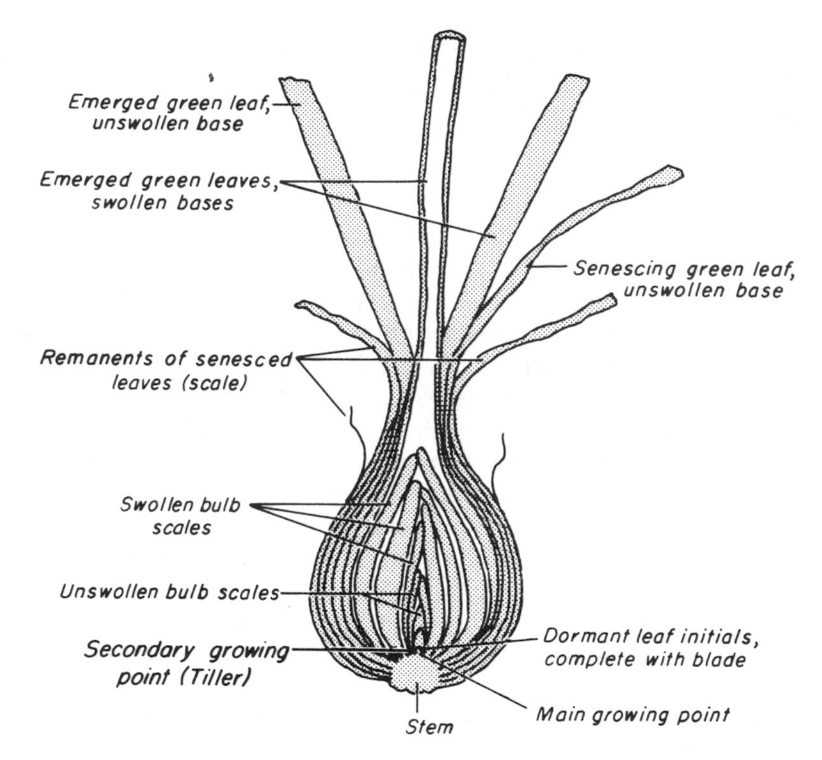

FIG. 17.1. Longitudinal median section of a bulbing onion.

Soil and Fertilization Fertile sandy, silty loam, or peat soils are excellent for onion production. Heavy clay and coarse sand should be avoided. Fertilizers containing nitrogen and phosphorus are usually needed. High rates of commercial fertilizers containing ammonium ion ($NH^{+}{}_{4}$) applied close to the plant should be avoided as ammonia (NH_3) is toxic to onion plants.

Moisture Shallow fibrous roots explore only the upper 30 cm (1 ft) of soil. From 380 to 760 mm (15 to 30 in.) of water are required to raise a crop to maturity in California.

Soil moisture is important in the growth of new adventitious roots, the soil moisture must reach the base of the bulb periodically if the newly formed adventitious roots from the stem are to grow into the soil. Roots do not grow into dry soil.

Propagation

Direct Seeding in Field Onion seeds germinate at 0°–35°C (32°–95°F). It requires about 4½ months at 0°C (32°F) but it takes only 3–4 days at 21°–27°C (70°–80°F).

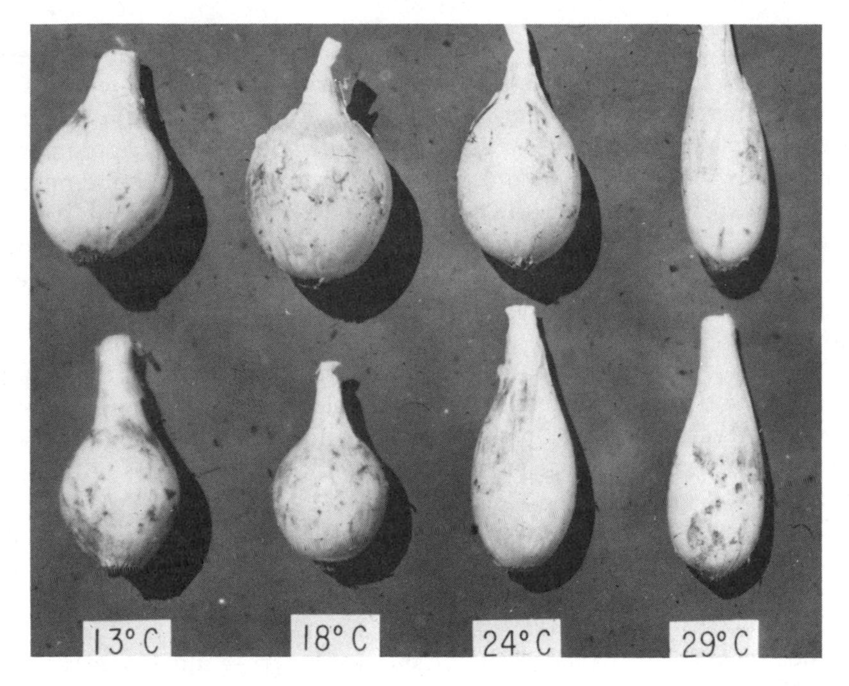

FIG. 17.2. Soil temperature effect on onion *(Allium cepa)* bulb shape.
From Yamaguchi et al. (1975).

Onions are usually grown on raised beds 35–45 cm (14–18 in.) apart for single row and about double the width for two-row beds. Seeding rate is from 2.2–3.4 kg/ha (2–3 lb/acre) for bulb onions. For green bunching onions or for transplanting the seeding rate is about six times the above. The depth of seeding is from 6 to 12 mm (¼ to ½ in.); the soil is kept moist until seedling emergence. For bulb production the spacing between plants is 7–10 cm (3–4 in.).

Onion Sets Onion sets are small bulbs about 1–3 cm (½–1⅛ in.) in diameter. They are generally produced in the fall and planted the following year in late winter or early spring for production of green onions or bulbs.

For production of onion sets a good seed bed is necessary to get a uniform stand of seedlings. A light loam soil is preferred and heavy clays avoided. The small size of bulbs is the result of planting seeds closely together at about 2–4 cm^2 (0.3–0.6 $in.^2$) per seed for high percentage germination seeds. The seedling rate is 70–110 g/m^2 (60–100 lb/acre) planted 6–12 mm (¼–½ in.) deep. Cultural requirements are similar to regular bulb production. The ideal size of sets should be 1.5–20 cm (⅝–¾ in.) in diameter. The sets are harvested after the tops are down and cured on trays and protected from rains. In regions where the winters are cold, the sets should be stored in buildings where the temperatures are not allowed to fall much below freezing. Bulbs greater than 2.5 cm (1 in.) in diameter become vernalized when held at temperatures below 10°C (50°F) for long periods of time; the lower the temperature, the sooner the vernalization. Bulbs less than 2.5 cm (1 in.) in diameter are probably still in the juvenile stage and are less apt to be vernalized under similar conditions. Storage at very cold temperatures, 0° to −1°C (32°–30°F), reduces the number of bolters as compared to storage at 2°–7°C (35–45°F). Onion sets of diameters 1.0–1.5 cm (3/8–5/8 in.) are the least sensitive to cold, and those having diameters of greater than 3 cm (1⅛ in.) are most sensitive to cold. The yield of bulb onions from small sets are often not significantly different or even higher than yields obtained from large sets because of high percentages of bolters from the latter.

Onion Transplants Onion transplants are produced much like onion sets. The seeds are planted very closely, about 80–100 g/m^2 (70–90 lb/acre). About 275,000 transplants are needed per ha when the spacing is 10 cm (4 in.) between transplants in 35 cm (14 in.) rows.

The size of the transplants should be less than 6–7 mm (¼ in.) at the base of the plant if seedlings are to overwinter at temperatures of below 15°C (59°F). However, with nonbolting types for spring plantings, large transplants will give higher bulb yields than small ones. The culture of the transplants is the same as direct seeded onions.

Bulbing

Bulbing is initiated under long-day conditions. Cultivars are classified according to approximate photoperiod necessary to induce bulbing: (a) short day (> 12–13 hr), (b) intermediate (> 13½–14 hr), (c) long day (> 14½–15 hr), (d) very long day (> 16 hr). High temperature seems to shorten the time necessary for the bulbing response.

"Short-day" cultivars are *not* "short-day" plants for flowering. They are cultivars that bulb under relatively shorter days than others, i.e., bulbing occurs when the photoperiod is longer than the minimum day-length characteristic for the cultivar. The plants are actually sensitive to the length of the dark period than the light period.

All cultivars are "long-day" plants with respect to bulbing. They bulb more readily as day length increases.

Temperature Effects Other factors being equal, onions bulb quicker at warm than at cool temperatures. Very high temperatures, maximums of 40°C (104°F) in the tropics, retard bulbing. The minimum photoperiod for bulbing really cannot be specified without also specifying the temperature. When grown under short photoperiods at high temperatures, many cultivars form new leaves indefinitely without bulbing.

Near the critical photoperiod, nitrogen deficiency of the plant has the same effect as increasing the day length, i.e., bulbing commences sooner. High nitrogen has an antagonistic effect, i.e., bulbing is delayed.

For high yields, maximum vegetative growth is desirable before bulbing is initiated. This requires that the temperature not be too high to hasten bulbing or too low for vernalization to occur (see Flowering below). Figure 17.3 shows the growth of onions for bulbs in temperate climates.

Chemical Induction of Bulbing The chemical ethephon (2-Cl-ethylphosphonic acid), sprayed on leaves at a concentration of 1.2 g/liter (1200 ppm), induces bulbing in short-day, intermediate, and long-day types of onions. For continued bulb enlargement under short days, the spraying of the leaves must be repeated.

Flowering

Bolting is induced by temperatures less than 10°C (50°F). Above 21°C (70°F) there is no bolting. Seedlings in the "juvenile phase" of growth [less than about 6 mm (1/4 in.) in diameter] and onion sets [less than 16 mm (5/8 in.) diameter] are not vernalized by cold temperatures. The temperature and length of time required for vernalization can vary

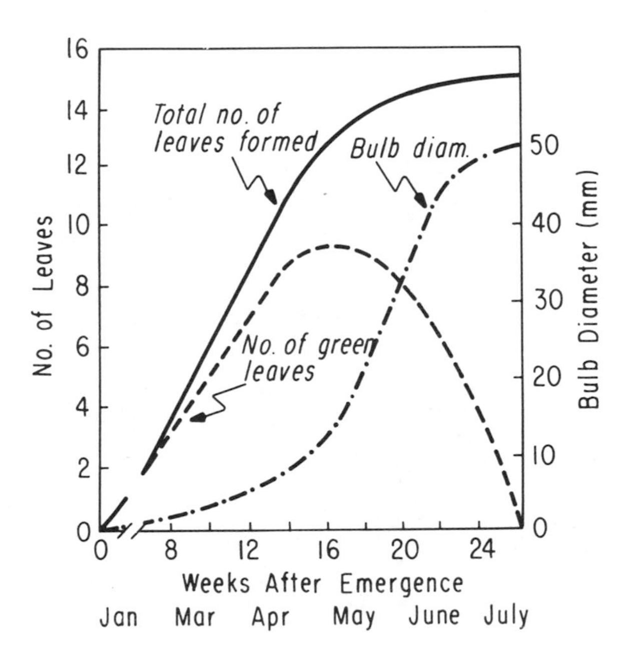

FIG. 17.3. Growth of long-day onions at Davis, California.
From Yamaguchi et al. (1975).

slightly with cultivars. Rapid bulb formation may suppress the emergence of seed stalks that have been initiated.

Ethephon at 2500–5000 ppm (400 liters/ha) reduce bolting of fall and late winter short-day onions.

Table 17.1 summarizes the photoperiod and temperature effects.

Figure 17.4 shows what happens in Davis when short-day, intermediate, and long-day onions are seeded during the various months of the year. *Example I*. Short-day (12 hr) cultivar seedlings emerging April 1, May 1, June 1, July 1, Aug. 1, or Sept. 1. Seedlings are growing under long-day length of 13 hr or more. Under these conditions seedlings receive the long-day stimulus and will bulb almost immediately producing very small bulbs. *Example II*. Short-day (12 hr) cultivar seedlings emerging on Oct. 1. Seedlings will grow into the fall under mild temperature and under less than 12 hr day lengths. The plants will probably have grown sufficiently large (not juvenile stage) before the late fall and winter cold weather to be vernalized. In Dec., Jan., and early Feb. The mean temperatures are less than 10°C (50°F) so the plants will have been vernalized and produce a flower stalk (bolt) when growth activity resumes in the spring (April). *Example III*. Short-day (12 hr) cultivar seedlings emerging Nov. 1, Dec. 1, Jan. 1, or Feb. 1. Seedlings emerging at these dates are still in the juvenile stage during the time when the temperatures are cold, less than 10°C (50°F). Since the plants have not been vernalized, they will continue to grow vegetatively until they are induced to bulb at day-lengths of 12 hr or longer. For the short-day (12 hr) cultivar, this induction occurs toward the middle of March (about March 20) and bulbing commences. The size of the mature bulb will depend on the size of the plant at the beginning of bulbing. Those seedlings that had been planted early will produce larger plants and larger mature bulbs than those that emerged later in the season. *Example IV*. Short-day (12 hr) cultivar seedlings emerging on Mar. 1. Seed-

TABLE 17.1. PHOTOPERIOD AND TEMPERATURE EFFECT ON ONION BULBING AND FLOWERING

	Photoperiod	
Temperature	Short days (11 hr)	Long days (15 hr)
High temperature: 21°C (70°F)	No bulbing: No floral initiation (no emergence of previously formed initials)	Rapid bulbing: No floral initiation (previously formed initials - destroyed)
Low temperature: 10°C (50°F)	No bulbing: Floral initiation (slow bolting)	Bulbing: (Floral initials formed can emerge) No bulbing: Floral initiation (rapid bolting)

Source: Adapted from Brewster (1977).

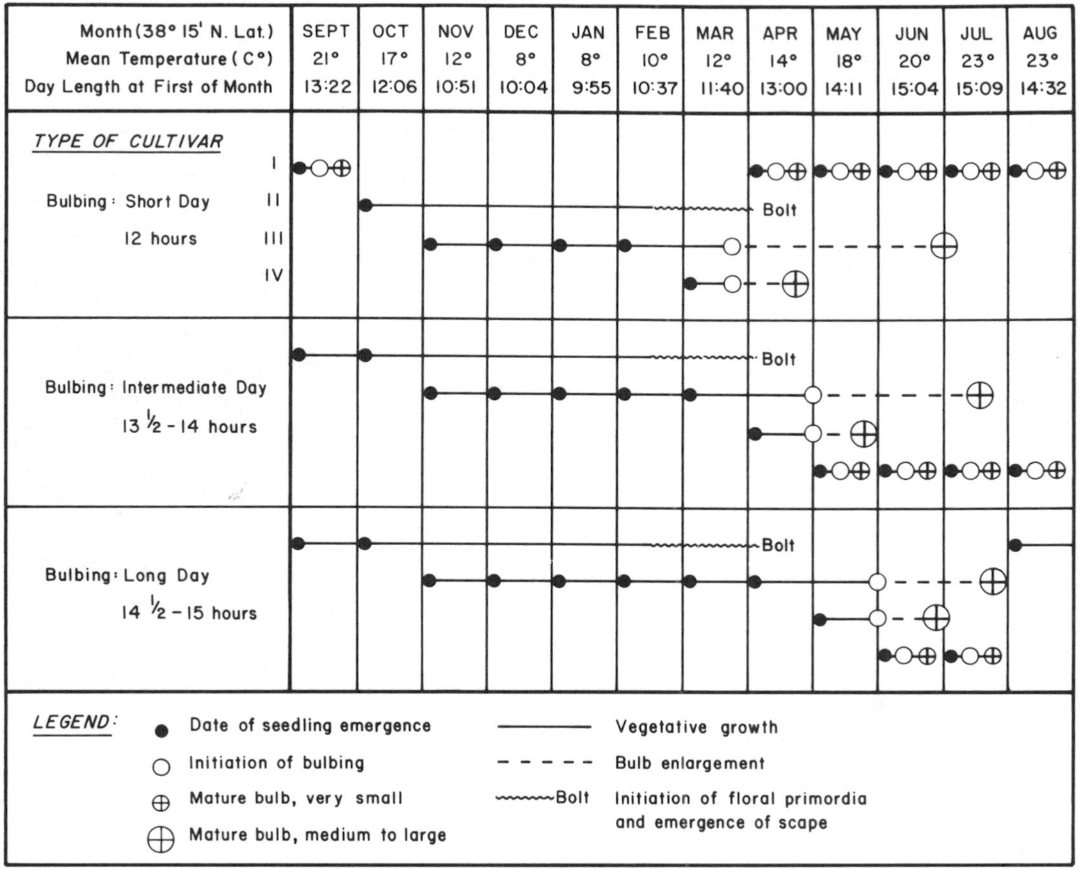

Month (38° 15' N. Lat.)	SEPT	OCT	NOV	DEC	JAN	FEB	MAR	APR	MAY	JUN	JUL	AUG
Mean Temperature (C°)	21°	17°	12°	8°	8°	10°	12°	14°	18°	20°	23°	23°
Day Length at First of Month	13:22	12:06	10:51	10:04	9:55	10:37	11:40	13:00	14:11	15:04	15:09	14:32

TYPE OF CULTIVAR

Bulbing: Short Day 12 hours — I, II, III, IV

Bolt

Bulbing: Intermediate Day 13 ½ - 14 hours

Bolt

Bulbing: Long Day 14 ½ - 15 hours

Bolt

LEGEND:

● Date of seedling emergence

○ Initiation of bulbing

⊕ Mature bulb, very small

⊕ Mature bulb, medium to large

——— Vegetative growth

- - - - Bulb enlargement

～～Bolt Initiation of floral primordia and emergence of scape

FIG. 17.4. Diagram of growth effects due to day length and temperature on onion types at 38°15′N lat. (Davis, California). Examples I, II, III, IV: see text for explanation.

lings will, in a very short growing period, receive the bulbing stimulus about the middle of March (same as in Example III) of 12 hr day length. The plants are small when this critical day length is reached, so the bulbs produced will also be small. *Note*: Except for the initiation of bulbing, the above examples for the short-day types can be used for the intermediate and long-day types of onions.

Harvesting and Storage

Harvesting

When the tops begin to fall down (about half of tops fallen), the bulbs are ready to harvest. In the tropics, this often does not occur as new leaves continue to form; bulbs are harvested with the tops still green and standing.

Curing

Curing is necessary in preventing rot-causing organisms from entering the bulb if the necks are not dry. Curing is accomplished by allowing the harvested onions to air dry either in windrows or open slotted crates in the field for 10–12 days. The dried or partially dried tops are usually cut off. In the field the stacked bulbs are protected from the sun to prevent scalding. This is accomplished by placing the green tops or other suitable materials over the pile but still allowing air circulation through the bulbs. Curing can be also accomplished by forcing heated air at 46°–47°C (115°–118°F) through the pile for a period of 12–24 hr.

Storage

Bulbs store best at low temperatures of 0°–7°C (32°–45°F), or a high temperature of 25°–35°C (77°–95°F) (Fig. 17.5). At these temperatures bulbs can be stored for 3–6 months without sprouting. Poorest storage is at room temperature in the range of 15°–21°C (60°–70°F). If the humidity during storage is kept at 40% or lower RH and temperatures at 3°C (37°F), some cultivars can be stored for almost a year.

Rest and Dormancy

Rest Onion bulbs are in a state of rest for some time following maturation. The length of the rest period varies with cultivars. When at rest the bulb will not sprout when placed under optimal conditions of temperature and moisture for sprouting and growth. Rest period disappears gradually; there is no abrupt change from rest to dormancy.

Dormancy The period during which bulbs stored at suboptimal temperatures for sprouting show no signs of sprout emergence. When re-

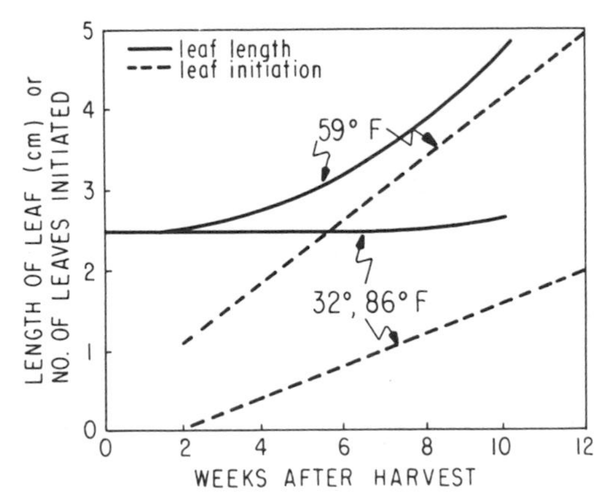

FIG. 17.5. Leaf growth and leaf initiation of onion bulbs in storage. *Redrawn from Abdalla and Mann (1963).*

moved from these suboptimal temperatures, root emergence occurs first followed by leaf emergence. Continued removal of newly formed roots delays sprout emergence.

Sprout Prevention in Storage

Maleic hydrazide (MH-30) is sprayed on the crop before harvest to prevent sprouting in storage. The chemical is sprayed while the tops are still green and about the time when tops begin to fall over. Plants should have at least five photosynthetically active leaves at the time of spraying. From 2.2 to 3.4 kg/ha (2 to 3 lb/acre) of MH-30 control sprouting. γ-Ray radiation has been effective in inhibiting sprouting.

Nutritive Value

Onions are high in energy and about average among vegetables with respect to other nutrients. The bulbs contain no starch; most of the carbohydrates are sucrose, glucose, fructose, and fructosan (polymer of fructose). Green onion tops have high vitamin A content (Table 17.2).

Pests and Diseases

In dry climates red spiders may be a serious problem in growing onions, and onion thrips are one of the main pests attacking the tops. Pea leaf miner larvae feed under the epidermis of onion leaves causing reduction in yields. Cutworms, nocturnal in their feeding habit, can damage or cut off young plants at the base. Nematodes and wireworms are soil-borne pests attacking the roots. Onion maggots infest sets

TABLE 17.2. PROXIMATE COMPOSITION OF SOME ALLIUMS IN 100 g EDIBLE PORTION

			Macroconstituents					Vitamins								
				g					mg				Minerals (mg)			
Crop	Edible part	% refuse	Energy (cal)	Water	Protein	Fat	CHO	A (IU)	B_1	B_2	Niacin	C	Ca	Fe	Mg	P
Chives	Leaves	0	20	92	2.8	0.6	1.1	6400	0.10	0.18	0.7	79	81	1.6	55	51
Garlic	Bulbs	20	39	61	6.4	0.5	2.9	0	0.20	0.11	0.7	15	24	1.7	32	195
Leek	Leaves	70	35	83	1.5	0.3	5.4	95	0.06	0.03	0.4	12	59	2.1	28	35
Onion, dry	Bulbs															
Southport White Globe		5	37	86	1.5	0.6	8.4	0	0.06	0.01	0.1	9	33	0.4	17	43
Sweet Spanish		5	26	90	0.5	0.1	6.5	0	0.02	0.02	0.1	6	27	0.6	16	27
Onion, green	Bulbs	10	21	90	1.3	0.2	4.0	330	0.06	0.05	0.3	32	62	0.5	25	43
Onion, green	Leaves	10	19	92	2.0	0.2	3.4	5000	0.07	0.14	0.2	45	80	1.0	24	30

Source: Howard *et al.* (1962).

and bulbs while the corn seed maggots feed on germinating seeds and seedlings.

Downy mildew is a very serious disease of onion foliage in humid climates. Leaf mold, black mold, fusarium basal rot, smut, and white rot are fungal diseases of leaves and bulb scales. Pink root, a fungal disease, causes the pinking of roots, which eventually die. Yellow dwarf, an aphid-transmitted viral disease, causes chlorotic streaking, stunting, and distorted flattening of leaves. Neck rot, caused by *Botrytis allii*, occurs after harvest in stored bulbs; this organism also infects the seed stalk in onions grown for seed. Soft rot, caused by bacteria, first appears as water-soaked areas in the bulb tissue and then becomes odorous, soft, and watery.

All of these pests and diseases can be problems in the growing of alliums.

Economics

Table 17.3 shows the world onion production, according to continents and the ranking countries of the world. Onions, like white potatoes, are easily shipped between countries. In 1973 the United States imported $11.8 million worth of fresh onion bulbs. Dehydration of onions is a very important industry in the United States, especially in California. It is used mostly for seasoning of prepared foods, canned, frozen, or dehydrated. Other onion products are pickled and French fried onion rings.

Green bunching onions in some countries of the Orient are used much more extensively than onion bulbs. The production of green onions and shallots in 1975 is shown in Table 17.4.

GARLIC: *Allium sativum*

Origin

Allium longicispis is thought to be garlic's wild ancestor; it is endemic to central Asia; hence, it is deduced that garlic also originated in this region. Evidence has been found that garlic was grown and consumed in the age of the building of pyramids of Egypt, about 2780–2100 BC. It has been recorded in a medical treatise in India about 6 BC.

Garlic is grown for its bulbs, which are used as a condiment. The green tops are used in some countries, especially the tropics, as green onions are used.

TABLE 17.3. WORLD PRODUCTION OF BULB ONION AND GARLIC

	Onion, bulb			Garlic	
	Area (10^3 ha)	Production (10^3 MT)		Area (10^3 ha)	Production (10^3 MT)
World	1,500	19,494		387	2,446
Continent					
Africa	121	1,241		8	174
North and Central America	65	1,950		12	118
South America	127	1,702		33	151
Asia	846	8,632		237	1,536
Europe	246	4,293		87	446
Oceania	6	177			
U.S.S.R.	170	1,500		11	21
Leading countries					
1. China	208	2,581	1. China	54	510
2. United States	50	1,746	2. Korea, REP	50	350
3. India	215	1,600	3. Thailand	42	300
4. U.S.S.R.	170	1,500	4. Spain	39	215
5. Japan	30	1,120	5. Egypt	6	161
6. Turkey	70	1,000	6. India	54	150
7. Spain	32	905	7. United States	6	84
8. Brazil	68	691	8. Argentina	17	81
9. Italy	22	540	9. Italy	6	60
10. Egypt	21	536	10. Turkey	12	60

Source: FAO (1979).

TABLE 17.4. WORLD PRODUCTION OF GREEN ONIONS AND SHALLOTS

	Area (10^3 ha)	Production (10^3 MT)	Yield (MT/ha)
World	110	1,542	14.0
Continent			
Africa	12	153	12.6
North and Central America	26	296	11.4
South America	4	35	8.8
Asia	59	947	16.1
Europe	9	111	12.3
Leading countries			
1. Japan	25	610	24.4
2. Mexico	26	295	11.5
3. Turkey	14	85	6.1
4. Korea, REP	7	80	11.7
5. Nigeria	3	80	26.7
6. China	5	72	14.1
7. Iraq	5	70	13.6
8. Tunisia	2	44	22.0
9. Equador	4	35	8.8
10. Greece	3	31	12.1

Source: FAO (1975).

Botany

Garlic is a monocotyledonous plant resembling onion in growth but the leaves are thin blades and the scape (seed stalk) is solid, unlike the tubular form of the onion.

The flowers are apparently sterile and are inconspicuously mixed among the bulbils at the end of the scape. Unlike the onion, the bulb is composed of cloves with several layers of sheath leaves surrounding the cloves on the inside. The cloves are initiated in the axil of the inner foliage leaves and the outer leaves form the sheath of the bulb.

Each clove consists of two mature leaves and a vegetative bud. The outermost leaf is a dry sheath with an aborted blade. The second leaf is very much thickened, accounting for most of the size of the clove and has aborted blade. The vegetative bud is composed of sprout leaf, which is bladeless, and one to two foliage leaf initials, which are still at rest shortly after the bulb is mature. Following rest, the clove remains dormant until conditions are right for sprouting and renewed growth.

Physiology of Bulbing

Bulbing in garlic occurs with lengthening photoperiods in the spring and the process is hastened with increasing temperatures up to 25°C (77°F), similar to the response of onions. However, unlike onions, the bulbing of garlic is influenced by the temperatures of the dormant cloves and during growth of the plant before bulbing. Exposure of either the clove or plant to 0°–10°C (32°–50°F) for 30–60 days hastens subsequent bulbing under long photoperiods. Cloves stored or plants grown above 25°C (77°F) may not bulb. Low temperatures during growth may induce sprouting of cloves already formed.

Too long an exposure to cold temperatures can cause rough bulbs due to clove formation in all axils of the leaves, with no sheath leaves surrounding the cloves as described above under normal conditions.

Garlic usually does not bulb when grown in the warm and short days of the lowland tropics; at high elevations in the tropics where temperatures are cooler, the crop can be grown but the bulbs are small and appear rough.

Culture

Garlic can be grown in wide range of soil types but a well drained soil at least 45–60 cm (1½–2 ft) deep is most satisfactory. Garlic should not be planted in recent onion fields unless it is known that the onion crop was free of diseases.

Propagation is by planting of cloves that have been in storage about 5°C (41°F) for several months. For bulbing to occur the storage of the cloves at the proper temperature is especially important in areas where the crop is grown at temperatures above 10°C (50°F). This is of concern particularly in the subtropics.

Seed clove size is important in yields. Large cloves consistently yield higher than medium size and medium size yield higher than small sized ones. Also, virus-free seed could increase yields up to 50% but often times maturity is delayed in virus-free material.

Before planting the bulb is separated into cloves, the small center ones are discarded as these produce small plants; hence, small bulbs. The cloves are planted just below the surface of the soil 5–8 cm (2–3 in.) apart. Usually the cloves are planted by hand or machine in 1 m (40 in.) beds, two rows per bed, 30 to 36 cm (12 to 14 in.) between rows, and spaced 8–10 cm (3–4 in.) between plants. It takes about 900–1100 kg cloves/ha (800–1000 lb cloves/acre). Cloves should not be planted upside-down in the soil; the plant grows poorly or not at all when planted in this position.

From 70 to 90 kg N/ha (60 to 80 lb N/acre) and 90 kg P/ha (80 lb P/acre) are used to fertilize the crop. Potassium is included if the soil is low in this element.

In the temperate regions garlic is usually planted in the fall or early spring and allowed to grow under the cool condition for the chilling required for bulbing and for the plant to develop to a good size before bulbing commences when the days are long in the spring. Maturity occurs in late July and August when the tops fall down and the blade leaves dry. For high yield, garlic should be planted as early as possible. Since garlic, as with the onion, is a shallow rooted crop, irrigation may be required in regions where rainfall is not adequate.

Harvest and Storage

Harvest

Garlic is ready for harvest when the tops are partially dry and bent to the ground. The bulbs are pulled either by hand or mechanically and windrowed. They are left to cure about 1 week, covered with materials to prevent sunburn but to allow air circulation through the piles. In areas of rainfall or dew, the bulbs should be protected from moisture.

Storage

Garlic bulbs for consumption can be kept for several months at ordinary temperatures but for prolonged storage they should be held at near 0°C (32°F) at humidities below 60%. The cloves sprout quickly at storage at 5°C (41°F) so temperatures near this point should be avoided. As with onions, bulbs can be kept at temperature above 25°C (77°F) but shrinkage occurs at this temperature.

Bulbs to be used for planting stock should be stored between 5° and 10°C (40° and 50°F) but not below 4°C (40°F) or above 18°C (65°F). Too low a temperature can cause rough bulbs and early maturation, whereas high temperatures may delay sprouting and interfere with bulbing and maturation.

Economics

In the less developed countries garlic is used fresh; in the United States, a very large percentage of the garlic production is dehydrated.

Garlic consumption is low in northern European countries, the United States, and Canada. Consumption is high in southern Europe, North Africa, and in some parts of Asia and Latin America. In some countries the usage of garlic almost equals that of onions. For example, in 1979 Korea REP produced 393,000 MT of onion on 15,000 ha, and 350,000 MT of garlic on 50,000 ha (more area was devoted to garlic than to onion). Thailand grew 160,000 MT of onion and 300,000 MT of garlic. The garlic production of world is shown in Table 17.3.

SHALLOT: *Allium cepa*, Aggregatum Group

Shallots are often classified as *Allium ascaloricum*. The plant is similar to the common onion but smaller. Because of its heterozygosity, seedlings are unlike the parent, and the crop is usually propagated by bulbs. A single plant produces a cluster of two to more than 15 distinct small bulbs at the base. In the early spring shallots are used extensively as green onions and the mature bulbs used for flavoring or as small boiled onions. The crop is very popular in the southern United States and in some European countries.

MULTIPLIER ONION OR POTATO ONION: *Allium cepa*, Aggregatum Group

This broad hollow leafed onion forms a cluster of medium-sized oblate bulbs numbering from two to five each about 2–4 cm (3/4–1 1/2 in.) in diameter. Propagation is usually by bulbs as sporadic flowerings produce very few viable seeds. As with shallots, multipler or potato onions are used as green onions early in the season and as the bulbs when mature.

EVER-READY ONION: *Allium cepa*, Aggregatum Group

This perennial plant has narrow hollow leaves and produces 10 or more slender bulbs in a season's growth. It is principally grown by small growers and home gardeners for green onions. The crop seldom flowers so propagation is by bulbs.

EGYPTIAN ONION, TREE ONION, OR TOPSET ONION: *Allium cepa*, Proliferum Group (also *A. cepa* var. *viviparum*)

The plant forms tillers at the base as does the multiplier onion. Early in the season, the plants are used as green onion and when mature as scallions (small bulb onions) and for propagation. The plant produces bulblets or bulbils at the top of the scape (flowers stalk); these often germinate producing miniature onion plants on top. Hence, the name tree onion and topset onion. Because of this peculiar characteristic, it is called *kitsune negi* meaning foxy or mysterious onion in Japan.

LEEK: *Allium ampeloprasum*, Porrum Group *(A. porrum)*

Wild relatives of leek can be found in the eastern Mediterranean region and into western and southern Russia. The plant does not bulb, has a flat (not hollow) long green leaf with white leaf bases. Whole plants are harvested and used much like green onions, particularly in soups because of the mild and delightful flavor.

Good germination of leek seeds occurs at soil temperatures of 11°–23°C (52°–73°F) but above 27°C (81°F) germination is drastically reduced. Depth of seeding is best at 0.5 cm (¼ in.); germination is reduced significantly when planted 2.0 cm (¾ in.) deep.

Leek is particularly adapted to cold weather and is more resistant than the onion. However, like the onion, it is vernalized by cold temperatures and subsequently bolts (produces a seed stalk) making it unsalable. Leek grown continuously at 15°C (59°F) forms flower stalks regardless of photoperiod; even at continuous 21°C (70°F), some bolting can occur. Under most temperature conditions leek does not develop into a conspicuous bulb. However, under long-day conditions (19–24 hr day lengths) of the extreme north, leek does form a distinct bulb. Bulbing is greater at temperature between 15° and 18°C (59° and 64°F) than at either 12°C (54°F) or 21°C (70°F).

KURRAT: *Allium ampeloprasum*, Kurrat Group

This plant is leek-like in stature but is much smaller than leek. In Egypt and eastern Mediterranean countries this crop is very important and is used mainly fresh and for seasoning. Kurrat is propagated from seed.

GREATHEAD GARLIC OR ELEPHANT GARLIC: *Allium ampeloprasum*, Ampeloprasum Group

The leaves and flowers of greathead garlic resemble those of leek but the plant produces large garlic-like bulb with many cloves. The flavor somewhat resembles onion and garlic (see Table 17.5 for flavor components). As most cultivars are sterile hexaploids, the crop is usually propagated with bulbs. The few types that do flower are sterile or produce very few seeds. The flowers are leek-like umbels but do not produce small bulblets as does garlic *(Allium sativum)*. Plants that do not flower produce a single large clove, whereas those that flower form a cluster of several cloves around the central flower stalk very much like garlic.

PEARL ONION: *Allium ampeloprasum*, Ampelosprasum Group

The pearl onion produces a small white bulb used mainly for pickling. The bulb is a single storage leaf similar to the garlic clove and quite different from that of the onion bulb.

JAPANESE BUNCHING ONION (NEGI) OR WELSH ONION: *Allium fistulosum*

This allium has been the main garden onion used as greens for many centuries in China and Japan (Fig. 17.6). Unlike the common onion, the plant does not form a well-developed bulb but has the hollow tube-like leaves of *Allium cepa*. The cultigen is not known in the wild state. In China, Taiwan, and Japan there are many cultivars of *Allium fistulosum* separated mainly according to growing season, blanching characteristics, and degree of tillering. Propagation is generally by seed but there are top-set cultivars that are smaller in stature and are propagated by bulbs. *Allium fistulosum* is quite adaptable to wide range of climates; it can be grown in extreme cold regions of northern China and Japan as well as in the warm regions of southern China and Southeast Asia. Blanching of the sheaths is achieved by mounding soil around the lower leaf base to a height of 25 cm (10 in.) or more.

CHIVES: *Allium schoenoprasum*

Chives can be found wild in North America mainly in the northern United States and southern Canada extending to both coasts and also throughout northern Europe and Asia. The small plants grow in clumps and the clumps grow larger by tillering. No well formed bulbs are produced. In the spring the plant produces purple to white flowers. The small hollow green leaves are harvested for garnish and flavoring.

RAKKYO, CHIA-TOU: *Allium chinense (A. bakerii)*

The names *rakkyo* and *chia-tou* are Japanese and Chinese, respectively, and the crop is important in the Orient. *Rakkyo* has hollow leaves,

FIG. 17.6. Japanese bunching onion, negi *(Allium fistulosum)*. Onion sheaths blanched by covering with soil, Tokyo, Japan.

grows in clumps and forms many well developed bulbs some as large as 2.5 cm (1 in.) in diameter. The purple flowers are borne on a solid scape but do not set seeds as most cultivars are tetraploids; hence, propagation is by bulbs. This *Allium* species is unusual in that the leaves are hollow and the scape is solid; usually both the leaves and the scape are alike.

CHINESE CHIVE, KAU TSOI, NIRA: *Allium tuberosum*

The plant spreads by means of well-developed rhizomes and grows into clumps. Propagation is by seed or division of clumps. There are four to five leaves in a bulb. The green leaves, which are harvested, are about 5 mm (¼ in.) in width and flat. Flowers are borne on a solid scape about 35–40 cm (14–16 in.) long; these are harvested at the bud stage and

FIG. 17.7. Chinese chives *(Allium tuberosum)*, flower stalks bunched, Taipei, Taiwan.

marketed. (Fig. 17.7). In the late summer and early fall, the leaves are often blanched by covering with clay tile pipes or in tunnels covered with opaque materials supported with semicircular bamboo stakes; the etiolated leaves are cut, bunched, and marketed.

TABLE 17.5. VOLATILE FLAVOR COMPOUNDS FROM DIFFERENT ALLIUMS

	Radical of sulfur compounds[a]					
Crop	Me_2	Me–Pr	Me–Al	Pr_2	Pr–Al	Al_2
Onion, common	+	+	±	+++++	+	–
Welsh onion	+	++	±	++++	+	–
Chive	+	++	±	++++	±	–
Leek	–	++++	±	+++	±	–
Garlic	±	±	++	–	–	++++
Greathead garlic	±	±	++	–	+	++++
Rakkyo	+++++	+	±	–	–	–
Chinese chive	+++++	–	+	–	–	–

Source: Saghir *et al.* (1964).
[a] Me = methyl, Pr = propyl, Al = allyl.

ALLIUM FLAVOR AND LACHRYMATOR

The flavor of allium is due to the enzyme, alliinase, acting on certain sulfur compounds when the tissues are broken or crushed. The volatile flavor compounds in onions are mainly propyl disulfide and methyl propyl disulfide. The tear- or lachrymator-causing compound is thiopropanyl sulfoxide. The volatile flavor components and the lachrymator are liberated by the same enzyme. Table 17.5 summarizes the type of volatiles in different alliums.

BIBLIOGRAPHY

Onion

ABDALLA, A.A. 1969. Effect of temperature and photoperiod on bulbing of the common onion (*Allium cepa* L.) under arid tropical conditions of the Sudan. Exp. Agric. *3*, 137–142.

ABDALLA, A.A., and MANN, L.K. 1969. Bulb development in the onion (*Allium cepa* L.) and the effect of storage temperature on bulb rest. Hilgardia *365*(5), 85–112.

BREWSTER, J.L. 1977. The physiology of the onion. Hort. Abstr. *47*, 17–23, 102–112.

CORGAN, J.N. and IZQUIERDO, J. 1979. Bolting control by ethephon in fall-planted short-day onion. J. Am. Soc. Hort. Sci. *104*, 387–388.

JONES, H.A., and MANN, L.K. 1963. Onion and Their Allies. Leonard Hill Publishers, New York.

HOLDSWORTH, M., and HEATH, O.V.S. 1950. Studies on the physiology of the onion plant. J. Exp. Bot. *1*, 353–375.

MAGRUDER, R., and ALLARD, H.A. 1937. Bulb formation in some American and European varieties of onions as affected by length of day. J. Agric. Res. *54*, 719–752.

MAGRUDER, R., WEBSTER, R.E., JONES, H.A., RANDALL, T.E., SNYDER, G.B., BROWN, H.D., HAWTHORN, L.R., and WILSON, A.L. 1941. Description of types of principal American onions. USDA Misc. Publ. No. 435. U.S. Dept. Agric., Washington, DC.

MAGRUDER, R., WEBSTER, R.E., JONES, H.A., RANDALL, T.E., SNYDER, G.B., BROWN, H.D., and HAWTHORN, L.R. 1941. Storage quality of principal American varieties of onions. USDA Circ. No. 618. U.S. Dept. Agric., Washington, DC.

PALTA, J.P., LEVITT, J., and STADELMANN, E.J. 1976. Alternate method of onion storage without application of growth regulator. J. Environ. Sci. Health Part A *11*, 663–671.

SAGHIR, A.R., MANN, L.K., BERNHARD, R.A., and JACOBSEN, J.V. 1964. Determination of aliphatic, mono- and disulfides in *Allium* by gas chromatography and their distribution in common food species. Proc. Am. Soc. Hort. Sci. *84*, 386–398.

SCULLYU, M.H., PARKER, N.W., and BORTHWICK, H.A. 1945. Interaction of nitrogen nutrition and photoperiod as expressed in bulbing and flower stock development of onions. Bot. Gaz. (Chicago) *107*, 415–419.

VOSS, R.E. 1979. Onion production in California. Div. of Agric. Sci., Univ. of California, Berkeley.

YAMAGUCHI, M., PRATT, H.K., and MORRIS, L.L. 1957. Effect of storage temperatures on keeping quality and composition of onion bulbs and on subsequent darkening of dehydrated flakes. Proc. Am. Soc. Hort. Sci. *69*, 421–426.
YAMAGUCHI, M., PAULSON, K.N., KINSELLA, M.N., and BERNHARD, R.A. 1975. Effects of soil temperature on growth and quality of onion bulbs (*Allium cepa* L.) used for dehydration. J. Am. Soc. Hort. Sci. *100*, 415–419.

Garlic

JONES, H.A., and MANN, L.K. 1963. Onions and Their Allies. Leonard Hill Publishers, New York.
MANN, L.K. 1952. Anatomy of the garlic bulb and factors affecting bulb development. Hilgardia *21*, 195–251.
MANN, L.K., and MINGES, P.A. 1958. Growth and bulbing of garlic (*Allium sativum* L.) in response to storage temperature of planting stocks, daylength, and planting date. Hilgardia *27*(15), 385–419.
SIMS, W.L., LITTLE, T.M., and VOSS, R.E. 1976. Growing garlic in California. Leaflet 1948, Agric. Ext. Ser., Univ. of California, Berkeley.

Alliums

DRAGLAND, S. 1972. Effect of temperature and day-length on growth, bulb formation and bolting in leeks (*Allium porrum* L.). Rep., No. 46, Dept. of Vegetable Crops Agric. Univ. of Norway Sci. Rep. *51*, NR21. Oslo, Norway.
HERKLOTS, G.A.C. 1972. Vegetables in South-East Asia. Hafner Press, New York.
HOWARD, F.D., MACGILLIVRAY, J.H., and YAMAGUCHI, M. 1962. Bull. No. 788. California Exp. Stn., Univ. of California, Berkeley.
JONES, H.A., and MANN, L.K. 1963. Onion and Their Allies. Leonard Hill Publishers, New York.
MANN, L.K., and STEARN, W.T. 1960. Rakkyo or Chiao Tou (*Allium chinense* G. Don Syn *A. bakeri*, Regel), a little known vegetable crop. Econ. Bot. *14*, 69–83.
McCOLLUM, G.D. 1976. Onion and allies, *In* Evolution of Crop Plants, N.W. Simmonds (Editor) Longmans-Green, London.
PURSEGLOVE, J.W. 1972. Alliaceae, *In* Tropical Crops: Monocotyledons. John Wiley & Sons, New York.
SINNADURI, S. 1963. Shallot farming in Ghana. Econ. Bot. *27*, 438–441.

18

Composites

Family: Compositae (sunflower)
Genera: *Lactuca, Cichorium, Cynara, Articum, Chrysanthemum, Petasites, Trapapogon*

Principal species	Common name
Lactuca sativa	Lettuce
L. sativa var. *augustana* (asparagina)	Celtuce
L. indica	Indian lettuce
Cichorium intybus	Chicory, witloof chicory
C. endiva	Endive, escarole
Cynara scolymus	Globe artichoke
C. cardunculus	Cardoon, cardoni
Articum lappa	Edible burdock, gobo
Chrysanthemum coronarium	Garland chrysanthemum, shungiku
Petasites japonica	Butterbur, fuki
Tragopogon porrifolius	Salsify, oyster plant

Lettuce is the most important of the crops listed; *Lactuca* and *Cichorium* genera are used mainly for salads. Chicory roots are sometimes used as a coffee substitute. The rest are used as cooked vegetables.

LETTUCE: *Lactuca sativa*

Origin

Somewhere east of the Mediterranean Sea, encompassing Asia Minor, Transcaucasia, Iran, and Turkistan, is considered to be the region where lettuce originated. Lettuce was first used for its medicinal properties and as early as 4500 BC for food. By the first century AD lettuce was in general usage by the Greeks and Romans. The head types were known since the 1500s and the leafy types much earlier. The cultivated lettuce was probably derived from the wild lettuce, *L. serriola*.

Celtuce or asparagus lettuce is *Lactuca sativa* var. *augustana*; grown for the succulent thick stem, used raw or cooked; the tender leaves are also eaten. It is very popular in China and Taiwan. Indian lettuce, *Lactuca indica*, is a native to China. It is a perennial plant grown for the succulent leaves, which are used raw or cooked.

Botany

Lettuce ($n = 9$) is an annual plant with a milky juice (latex). There are five types: (1) crisphead, (2) butter head, (3) cos or romaine, (4) loose leaf or bunching, and (5) stem lettuce (celtuce). The basic chromosome number in the genus *Lactuca* is 8 and 9. There are wild types in the New World with a basic number of $n = 17$ chromosomes, probably a natural amphidiploid of crosses between $2n = 16$ and $2n = 18$ chromosomes species.

Culture

Lettuce is a cool season crop. The climatic requirement for crisphead types is more critical than other types. Cultivars within the crisphead types differ in their climatic adaptation. Lettuce is usually grown in areas in which the mean temperatures are 10°–20°C (50°–60°F). Heading is prevented and seed stalks form in the temperature ranges of 21°–27°C (70°–80°F). Cool nights are essential for quality lettuce production; high temperatures tend to produce strong flavors (bitterness). Adequate moisture and cool temperatures are necessary at the time of heading. Low moisture and high temperature may cause a disorder called tip burn in which the tips of the inner leaves in the head show necrosis.

A well-drained fertile soil about pH 6 is most desirable for growing lettuce. It has a fair tolerance of salt. Seeds are sown 1/2–1 cm (1/4–1/2 in.) in depth on 1 m (40 in.) beds, two rows per bed, and 36 cm (14 in.) between rows. Seeds germinate from 7° to 24°C (45° to 75°F). The plants are thinned to 30–40 cm (12–16 in.) between plants. Leaf lettuce may be closer spaced. In many regions of the world lettuce is transplanted from seedling beds.

Head lettuce removes about 53 kg N, 8 kg P, 130 kg K, and 22 kg Ca/ha (47 lb N, 7 lb P, 117 lb K, and 20 lb Ca/acre). Fertilization would depend on the nutrients available in the soil and should be applied early, long before the rapid growth which occurs just before harvest. It is recommended that one-third to one-half of the N be applied at or before seeding and the rest about thinning. For optimal growth lettuce requires a constant and abundant supply of moisture throughout its growth.

Harvest and Storage

It requires 60–80 days from planting to harvest in the summer and fall and 90–145 days during the winter and spring depending on temperature conditions. Over half of the fresh weight of the head is attained in the two weeks before harvest (Fig. 18.1).

Lettuce is a highly perishable crop as it has a very high water content. Ideally, lettuce should be precooled to about 1°C (34°F) and held at this temperature at high humidity (95–97%) for storage or transport. It will

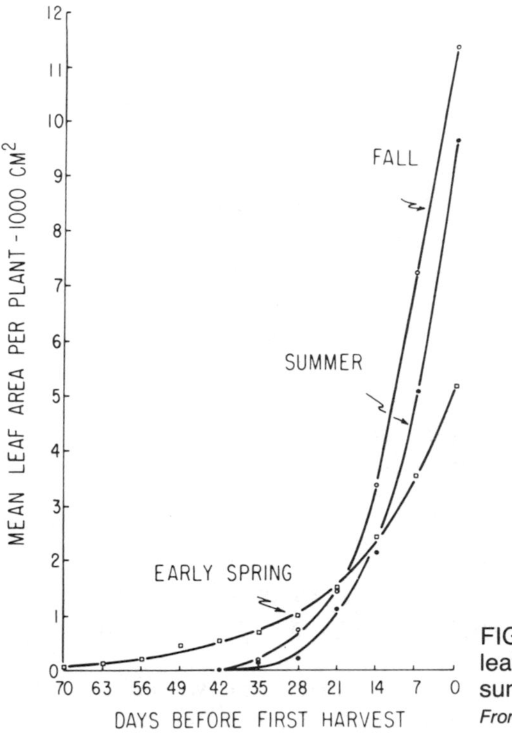

FIG. 18.1. Growth expressed as leaf area per plant for spring, summer, and fall lettuce.
From Zink and Yamaguchi (1962).

remain in good quality for 10–14 days under the above conditions. Deterioration increases with increasing temperatures.

Nutritive Value

Leaf lettuce is more nutritious than the head type lettuce, mainly because of its high vitamin A and vitamin C values. Lettuce is also a good source of Ca and P. The nutritive value of lettuce and other composites are given in Table 18.1.

Economics

Lettuce is the major salad vegetable in North America, in most countries in Europe, in Australia, New Zealand, and in many countries of South America. The production is increasing in the Middle East, Africa, and Japan. Celtuce is an important vegetable in Taiwan. The U.S. production of lettuce in 1980 was 2.8 million MT (3.1 million tons). European production in 1975 was 1.1 million MT (1.2 million tons).

Pests and Diseases

Aphids transmit diseases of lettuce; therefore, they should be controlled. Cut worm, army worm, cabbage looper, corn ear worm, leafhooper, and spider mites are some of the more common insects that attack lettuce.

Mosaic, which is seed-borne, is the most serious disease of lettuce. Spotted wilt, aster yellow, big vein, downy and powdery mildew, sclerotina, anthracnose, bottom rot, and botrytis are other diseases of lettuce. Tip burn, russet spotting, and rib discoloration are physiological disorders, the latter is probably caused by the storage atmosphere conditions.

ENDIVE, ESCAROLE: *Cichorum endiva;* CHICORY, WITLOOF CHICORY, FRENCH ENDIVE, CHICON: *C. intybus*

Endive has been used by Egyptians for many centuries. It had spread to northern Europe by 1200 AD. Chicory is probably of Mediterranean origin and has been used as a salad and cooked vegetable for centuries throughout Europe.

Endive is a loose-headed plant with curled and serrated narrow leaves. The broad-leaf type is called escarole. The older outer leaves are dark green and strong flavored (bitter) while the light-green younger

TABLE 18.1. PROXIMATE COMPOSITION OF SOME COMPOSITES IN 100 g EDIBLE PORTION

			Macroconstituents					Vitamins					Minerals (mg)			
				g					mg							
Crop	Edible part	% refuse	Energy (cal)	Water	Protein	Fat	CHO	A (IU)	B_1	B_2	Niacin	C	Ca	Fe	Mg	P
Artichoke, globe		60	20	83	2.7	0.2	2.3	160	0.08	0.06	0.8	11	53	1.5	48	78
Large bud, 200 g	Bracts	60	20	82	2.6	0.1	2.4	220	0.09	0.07	0.7	12	57	2.1	39	70
Large bud, 200 g	Receptacle	20	20	84	2.8	0.2	2.0	100	0.08	0.04	0.8	10	44	1.4	50	80
Small bud, 100 g	Bracts	60	21	83	2.7	0.2	2.6	220	0.07	0.09	0.8	11	37	1.5	39	75
Small bud, 100 g	Receptacle	30	22	81	2.5	0.3	2.0	100	0.07	0.04	0.7	10	47	1.7	60	83
Burdock root		10	40	72	1.1	0.1	6.7	0	0.01	0.03	0.3	3	41	0.8	—	51
Cardoon	Petiole	55	10	94	0.7	0.1	1.8	120	0.02	0.03	0.3	2	70	0.7	42	23
Celtuce	Leaves	20	12	94	1.1	0.4	1.3	3500	0.09	0.12	0.5	33	59	0.8	38	34
Celtuce	Stalks	70	12	95	0.6	0.2	2.4	70	0.02	0.02	0.6	6	18	0.3	17	43
Chicory	Leaves	20	13	92	1.7	0.3	1.1	4000	0.06	0.10	0.5	24	100	0.9	30	47
Chicory	Roots	0	23	80	1.4	0.2	4.6	0	0.04	0.03	0.4	5	41	0.8	22	61
Endive	Leaves	20	11	95	1.3	0.2	1.2	2500	0.07	0.08	0.4	8	42	2.0	20	30
Escarole	Leaves	15	12	94	1.2	0.2	1.5	1600	0.09	0.07	0.4	5	50	0.7	14	21
Lettuce	Leaves															
Butterhead		20	11	96	1.2	0.2	1.2	1200	0.07	0.07	0.4	9	40	1.1	16	31
Cos (romaine)		25	16	94	1.6	0.2	2.1	2600	0.10	0.10	0.5	24	36	1.1	6	45
Crisphead (Great Lakes)		15	11	95	0.8	0.1	2.3	300	0.07	0.03	0.3	5	13	1.5	7	25
Salsify (vegetable oyster)	Roots	25	34	77	3.3	0.2	5.2	0	0.08	0.22	0.5	8	60	0.7	23	75

Source: Howard *et al.* (1962).

leaves are mild flavored due to partial protection from light. Endive flowers after vernalization; it has been reported that flowering is hastened by long bright days. Gibberellin treatment also induces flowering. There are annual types. The culture of endive is similar to lettuce. From 90 to 95 days are required from seeding to harvest.

Chicory, a perennial composite, is an important salad vegetable in Europe. Over half of the production is forced; the blanched cluster of leaves (head) is called witloof chicory, French endive, and chicon (Fig. 18.2). Compared to the green strong-flavored unblanched heads grown in southern Europe, witloof chicory has a delicate flavor and is used in salads in France, Belgium, and Holland. The green leafed chicory is cultured similar to lettuce and endive. The roots, dried, roasted, and ground, are used as coffee substitute or in blends with coffee.

In forcing special cultivar hybrids are grown. The roots are dug in the fall after a year's growth and stored between 3° and 6°C (37° and 43°F) for vernalization. Vernalization produces quality heads. However, prolonged low temperatures cause the core to elongate excessively resulting in poor quality heads. Cultivars vary in chilling requirements; early cultivars require less chilling than late ones.

Chicory is forced in pits, cold frames, or houses. Pits are dug 18–20 cm (6–8 in.) deep. Roots, trimmed (growing point intact), about 1.3 cm (1/2

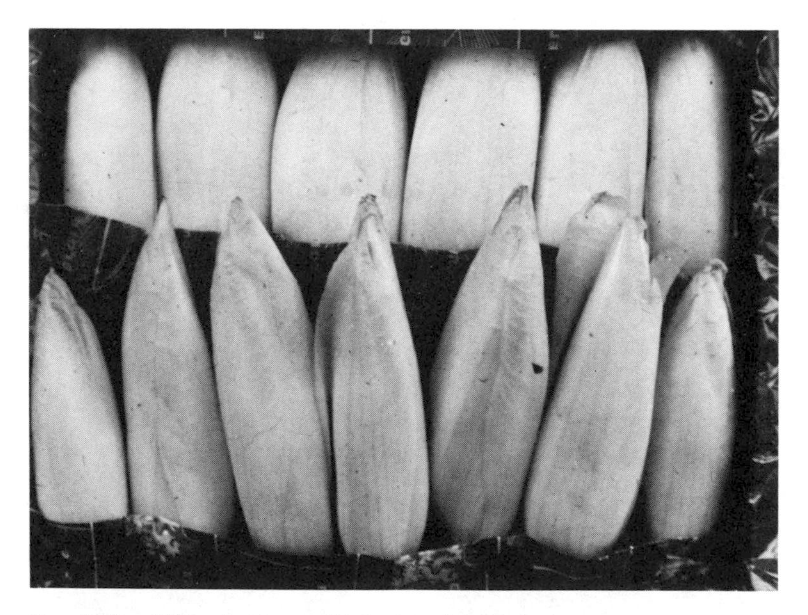

Fig. 18.2. Witloof chicory *(Cichorium intybus)*.

in.) above the shoulder to the depth of the pit, are planted closely packed in upright positions and covered with moist finely aggregated soil, peat, sand, or combinations of the three. A layer of straw mulch is often used. The pit or bed is covered with opaque material allowing at least 40 cm (16 in.) space over the bed for shoot growth. Provisions for heating the soil or air space may be included in the design. Roots can be grown in the range of 5°–21°C (41°–70°F); 18°C (64°F) is optimum. Proper moisture and humidity (high) are maintained throughout growth at 18°C; 20–22 days are required from planting to harvest.

More recently, air-conditioned rooms, forcing without soil cover, and use of black plastic housing have been used for witloof chicory productions.

GLOBE ARTICHOKE: *Cynara scolymus*; CARDOON OR CARDONI: *Cynara cardunculus*

Artichoke is an important crop in the southern European countries. There is evidence that artichoke was derived from wild cardoon, indigenous to the Mediterranean region. The fleshy petioles of the cardoon (Fig. 18.3) are sometimes blanched to make them more succulent. The Romans ate the receptacle of the cardoon flowers. The artichoke petals and receptacle of the flower buds are eaten; also, the very immature flower buds are prepared as "hearts." Both artichoke and cardoon are perennial, propagated usually by suckers which emerge from the crown. Propagation can be accomplished by seed but plants from seeds are quite variable and usually quite different and inferior in quality from the mother plant. It is a cool season crop grown in areas of mean temperatures of 13°–18°C (55°–60°F). In hot and dry climates, the flower buds are tough and the petals tend to spread and the production period is shortened.

SALSIFY, VEGETABLE OYSTER, OYSTER PLANT: *Tragopogon porrifolius*

Salsify is a biennial native to Europe, North Africa, and Asia. It is a hardy plant growing to over 1 m (4 ft) in height with long linear grass-like leaves and light purple flowers. It is grown for the long tap root, which grows as long as 30 cm (1 ft) and 5 cm (2 in.) in diameter, and tastes like oysters. Seeds are planted in the early spring in rows 60–90 cm (2–3 ft) apart and thinned to stands of 8–15 cm (3–6 in.) between

Fig. 18.3. Cardoon *(Cynara cardunculus)*.

plants. The roots are very winter-hardy and freezing does not seem to injure them. Harvest is in the winter or spring.

GARLAND CHRYSANTHEMUM, SHUNGIKU (JAPANESE), KOR TONGHO (CHINESE): *Chrysanthemum coronarium*

Garland chrysanthemum (Fig. 18.4) is a very popular green in Japan, China, and Taiwan. It is grown mainly in the fall, winter and early spring. Small quantities are grown even during the summer months in Japan. There are three types, classified according to leaf size: (1) a cultivar with finely parted narrow dark green leaves, (2) intermediate size leaves and (3) broad somewhat clefted pale green leaves. The broad leaf type is adapted to warm regions whereas the intermediate and

Fig.18.4. Garland chrysanthemum *(Chrysanthemum coronarium).*

narrow leaf types have a wide range of climate adaptability being both cold and heat tolerant.

This annual plant grows to almost a meter (3¼ ft) in height and produces a characteristic yellow composite flower. The young seedlings are harvested when about 20 cm (8 in.) high, about 25–40 days after sowing. Only the tender shoots are harvested from older plants.

EDIBLE BURDOCK, GOBO: *Arcticum lappa*

A native of Siberia and northeastern China, edible burdock (Fig. 18.5) is mainly cultivated in Japan for the long tapered edible tap root, which grows to over a meter (3½ ft) in length and as large as 4 cm (1½ in.) in diameter. Wild burdock grows in the temperature regions of the northern United States; the roots, though edible, are very small, fibery, and strong flavored.

The dicotyledonous biennial crop, grown as an annual, is started from seed which requires light to germinate. A deep, well-drained light sandy soil is recommended; waterlogging is to be avoided as this causes root

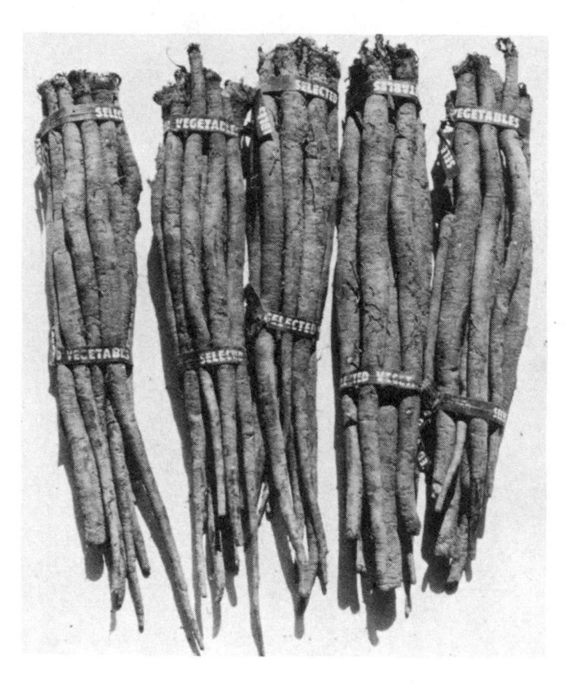

Fig.18.5. Burdock, gobo *(Articum lappa)*.

rot. The optimum temperature for growth is 20°–25°C (68°–77°F); the tops die when the temperature drops to 3°C (37°F) for several days. However, the roots are cold resistant and can survive temperatures well below freezing. The plant is high temperature tolerant; hence, it can be grown in the subtropics but will not flower unless subjected to a period of cold temperatures. Seeds can germinate at minimum temperatures of 10°C (50°F); optimum is 20°–25°C (68°–77°F); light enhances germination and also scarification.

Planting is usually in the early to midspring in deep well-drained sandy loam soils; heavier soils can be used. Harvesting of the succulent roots occur in July to end of December. For fall plantings the seedling are allowed to overwinter and growth resumed in the early spring for harvests in May and June. Fall planting date is critical as the larger overwinter roots will bolt. Bolting occurs in plants with roots larger than 3 mm (1/8 in.) in diameter after exposure to low temperature and followed by day lengths greater than 12½ hr.

Burdock is a good source of B vitamins and also of fibers. Originally burdock was used for medicinal purposes, particularly to purify the blood and relief from pain of arthritis; it is reported to help those suffering from gout and skin diseases.

BUTTER BUR, FUKI (JAPANESE), FENG TOU TSAI (CHINESE): *Petasites japonica*

A perennial composite, butter bur is grown for its succulent petioles mainly in Japan and in China and Korea to a lesser extent. The plant is mentioned in Japanese literature of the tenth century.

Butter bur is a commercial crop in Japan; there are three cultivars of importance, 'Aiichi Early Fuki,' 'Akita Fuki,' and 'Mizu Fuki.' 'Aiichi Early' is the most common one grown. The petioles are graded according to size; they range from 20 to 80 cm (8 to 30 in.) in length. Regular harvest is in April through May; by forcing under plastic houses, harvests can be made in mid-February through April. The temperature range for growing is 10°–23°C (50°–73°F).

The crop is propagated by planting disease free rhizomes 10–30 cm (4–12 in.) in length, spaced 30–50 cm (12–20 in.) apart. Plantings are usually maintained commercially for 5–6 years, although the plants can continue to grow for 10–20 years.

Before use, the petioles are first parboiled to remove astringency and to facilitate the peeling of the epidermal layer.

BIBLIOGRAPHY

BEATTIE, W.R. 1946. Production of salsify or vegetable oyster. USDA Leaflet 135. U.S. Dept. Agric. Washington, DC.

HERKLOTS, G.A.C. 1972. Vegetables in Southeast Asia. Hafner Press, New York.

HOWARD, F.D., MACGILLIVRAY, J.H., and YAMAGUCHI, M. 1962. Bull. No. 788. California Agric. Exp. Sta., Univ. of California, Berkeley.

KUMAZAWA, S. 1956. Vegetable Crops. Yokondo, Tokyo, Japan. (In Japanese)

RYDER, E.J., and WHITAKER, T.W. 1976. Lettuce, *In* Evolution of Crop Plants, N.W. Simmonds (Editor). Longmans-Green, London.

RYDER, E.J. 1979. Leafy Salad Vegetables. AVI Publishing Co., Inc. Westport Connecticut.

SIMS, W.L., RUBATZKY, V.E., SCIARONI, R.H., and LANGE, W.H. 1977. Growing artichokes in California. Leaflet No. 2675, Div. of Agric. Sci., Univ. of California, Berkeley.

WHITAKER, T.W., RYDER, E.J., RUBATZKY, V.E., and VAIL, P.E. 1974. Lettuce production in the United States. USDA Handb. No. 221. U.S. Dept. Agric., Washington, DC.

ZINK, F.W., and YAMAGUCHI, M. 1962. Studies on the growth rate and nutrient absorption of head lettuce. Hilgardia *32*, 471-500.

19

Crucifers

Family: Cruciferae (mustard)

The important cultivated vegetables are listed below with their genome (sum of genes or characters) group relationships.

Principal genera/species	Common name	Genome
Brassica nigra	Black mustard	(genome: b) (n = 8)
B. oleracea	Cole crops	(genome: c) (n = 9)
Capitata group	Cabbage	
Alba subgroup	White cabbage	
Rubra subgroup	Red cabbage	
Sabauda subgroup	Savoy cabbage	
Botrytis group	Cauliflower	
Italica group	Broccoli, calabrese	
Gemmifera group	Brussels sprouts	
Acephala group	Kale (common and Scotch), collard, marrow-stem kale, borecale	
Gongylodes group	Kohlrabi	
B. campestris	Turnip group	(genome: a) (n = 10)
Rapifera group (also *B. rapa* and *B. septiceps*)	Turnip	
Chinensis group	Chinese mustard, celery mustard, *pak-choi*	
Pekinensis group	Chinese cabbage, celery cabbage, *pe-tsai*	
Perviridis group	Mustard (tender green)	

Ruvo group	Broccoli raab, rapa, Italian turnip	
B. carinata	Abyssinian (Ethiopian) mustard	(genome: bc) (n = 17)
B. juncea	Leaf, brown, or Indian mustard	(genome: ab) (n = 18)
B. hirta (Sinapis alba)	White and yellow flowered mustards	
B. alboglabra	Chinese broccoli, Chinese kale	(genome: ?) (n = 9)
B. napus	Swede group	(genome: ac) (n = 19)
Napobrassica group	Rutabaga, swede, Swedish turnip	
Pabulavia group (also *B. fimbriata*)	Siberian kale, Hanover salad, winter rape	

Within the genus *Brassica*, *B. oleracea*, the cole crops, and *B. campestris*, to which the turnip, Chinese cabbage, and mustards belong, are the most important.

Raphanus sativus	radishes
Longipinatus group	winter radish
Eruca sativa	roquette, rochet salad
Rorippa nasturtium-aquaticum (Nasturtium officinale)	watercress
Armoracia rusticana	horseradish
Crambe maritima	seakale
Lepidium meyenii	maca
Lepidium sativum	garden cress

COLE CROPS: *Brassica oleracea*

The many cultivated forms of *Brassica oleracea* are illustrated in Fig. 19.1.

Origin

The word cole is from Middle English *col* derived from Anglo Saxon, *cal*, *cawl*, *cawel* or from Old Norse *Kal*; both probably originated from

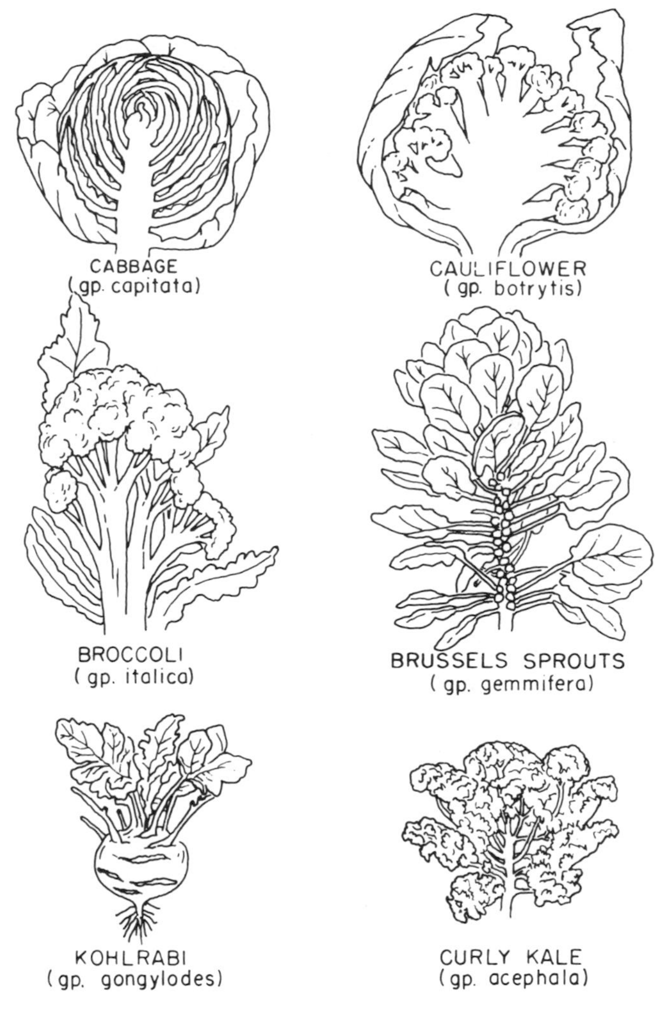

FIG. 19.1. Appearance of cole crops *(Brassica oleracea)*; gp. = group.

the Latin, *caulis*, meaning stem of a plant particularly of cabbage stem or stalk. The ancient Greeks referred to the cole crops as *Kaulion* meaning stem. The wild types of *Brassica oleracea* var. *sylvestris* are found growing along the coast of the Mediterranean Sea. Also, wild forms occur along the Atlantic coast of Europe. It is assumed that the cabbages and kales originated in Western Europe and the cauliflower and broccoli came from the Mediterranean region.

Cabbages and kales were the first to be domesticated. Prior to cultivation and use as food, they were used mainly for medicinal purposes for such ailments as gout, diarrhea, deafness, and headache. Cabbage juice was prescribed as a remedy for poisonous mushrooms.

The first description of cauliflower appeared in 1544. Brussels sprouts, named after the city in Belgium, did not appear until about the beginning of the nineteenth century. Sprouting broccoli of today originated in the Mediterranean region but it was not a popular vegetable in the United States until the Italian immigrants brought seed to California. The production of broccoli started in California in the early 1920s.

Botany

Cole crops are dicotyledonous biennial herbaceous plants, grown as an annual or biennial depending on the part harvested. At the seedling stage, the different crops are hard to distinguish. After some growth, the crops begin to develop into their recognizable characteristics. Except for certain strains of cabbage, broccoli, and tropical-type cauliflower overwintering is necessary for flowering. The yellow petaled flowers are insect pollinated. The seeds are borne in a fruit called *silique*.

Culture

All *B. oleracea* are cool season crops. They are of quality when they can grow into maturity in cool temperatures. In the Northern Hemisphere they are winter crops in the southern temperate regions and a summer crop in the northern temperate regions, and spring and fall crops in the intermediate regions, where temperatures generally conform to the requirements for the particular crop.

Cabbage

Optimum range for cabbage is 15°–20°C (59°–68°F); above 25°C (77°F) growth is arrested. Minimum temperature is 0°C (32°F) but cold hardened plants can withstand temperatures as low as –10°C (14°F) for short periods of time. Young plants with stems less than 6 mm (¼ in.) diameter can tolerate colder and hotter temperatures than older plants. Plants out of the juvenile stage will flower when temperatures are less than 10°C (50°F) for 5–6 weeks; the lower the temperature, the shorter the time required. It is not photoperiod sensitive.

Cauliflower

For the early type cauliflower (snowball) the maximum temperature for curd formation is 20°–25°C (68°–77°F); the optimum is about 17°C (63°F). There is little difference in the quality of the curds at 14°–20°C (57°–68°F); but above 20°C (68°F), the quality is poor. In the tropics, the biennial types under high temperatures remains vegetative. However, through selection, there are cultivars that produce normal curds above 20°C (68°F).

No cold temperature is necessary for head formation for the early type cauliflower (snowball type). The late type cauliflower (winter) requires a period of cold temperature before heading. Winter types must not be sown too late; often they will not form heads because the plants were too small during cold weather to become vernalized. There are intermediate types that form heads below 10°C (50°F).

The curds in the early type (snowball) cauliflower are not true flowers but undifferentiated shoot apices. Only the winter or late cauliflower types have curds which are true floral primorida.

The East Indian cauliflowers are in a group by themselves. They are tolerant of high temperatures and high humidity. Cauliflower was introduced to India from England about 1822. During the century after its introduction, the crop had undergone selections by growers for adaption to local climate and for earliness; also recombinations resulting from natural crossing of different types occurred between different European and snowball types which resulted in the Indian cauliflowers. Cultivars such as 'Early Patna,' 'Early Market,' and 'Early Benares' produce good curds above 20°C (68°F). These cultivars can produce large curds at maximum temperatures of 30°C (86°F) and minimum of 23°C (73°F). Most snowball types curd formation occurs around 17°C (62°F). Apparently, the tropical or Indian types do not require as low a temperature as the snowball or winter types; mean temperatures of about 16°C (60°F) in the winter are low enough for seed production.

Types of Defects in Cauliflower Curd

Blindness. No curd formation due to injury caused usually by low temperature slightly above 0°C (32°F). This occurs when plants have just passed the seven-leaf stage. Also, blindness can occur as a result of freezing injury during initial stage of curd formation. Besides cold temperatures, there are other unexplainable causes of blindness.

Button. Caused by premature generative stage. This is caused by plants that are quite large (with thick stems) at time of transplanting to the field. Such plants go quickly into the generative phase producing a smaller than normal head. Poor environmental conditions, which arrest vegetative growth, often cause buttoning.

Ricey. A disorder of the head in which the curds acquire a velvety appearance somewhat like a pot of boiled rice. This is caused by the development of small white flower buds. This defect has been attributed to high temperatures during curd development. However, some cultivars are more prone to riceyness than others. This defect increases with rapid growth and heavy N side-dressing.

Leaves in curd. Small leaves in the curd occur when the plant reponds to warm temperature after the curd forms. The cause is due to reversion to vegetative growth.

Green curds. Greening is due to excessive exposure to sunlight and chlorophyll formation.

Head shape. Low temperature promote flat heads while high temperatures promote conical shaped heads.

Broccoli

The climatic conditions for broccoli are less exacting than cauliflower. For quality crop, monthly mean averages of 16°C (60°F) or slightly lower are necessary. The crop is injured by freezing temperatures after the inflorescences have formed.

Early and medium strains of broccoli do not require a cold period for flowering, whereas the late or overwintering strains do require cold temperatures (vernalization) before flowering. Hybrid broccoli produce more uniform heads; with some a single head is as large as a medium size cauliflower.

Brussels Sprouts

Late winter cultivars can stand minimum temperatures as low as −10°C (14°F). This crop is suited for regions of moderately severe winters.

Kale

Kale is the hardiest of all cole crops. It can stand temperatures of −15° to −10°C (5° to 14°F). It is also tolerant of high summer temperatures.

Kohlrabi

Kohlrabi is very sensitive to cold temperatures. A week at 10°C (50°F) will cause the plant to bolt. The recommended temperature range is

18°–25°C (65°–77°F), with the optimum at 22°C (72°F).

The temperature effects on the different cole crops are shown in Table 19.1 for the harvested part and for flowering.

TABLE 19.1. TEMPERATURE REQUIREMENTS OF COLE CROPS

	Harvested part		
Crop	Organ	Vernalization requirement[a]	Flower initiation vernalization requirement[a]
Cabbage	Head (leaves)	–	+
With tropical (t) gene	Head	–	–
Cauliflower			
Snowball (early)	Curd (head) (undifferentiated shoot apices)	–	+ (floral primordia)
Winter (late)	Curd (head) (differentiated floral primordia)	+	+
Tropical (India)	Curd (head)	–	+[b]
Broccoli, sprouting			
Early and medium strains	Flower buds	–	–
Late strain (overwintering)	Flower buds	+	+
Brussels sprouts	Axillary buds	–	+
Kohlrabi	Enlarged stem	–	+[c]
Kale and collard	Leaves	–	+

[a]Vernalization: 4–8 weeks at 4.5°–10°C (40°–50°F), less time at lower temperatures.
[b]Indian cauliflowers do not require as low a temperature as Snowball or Winter types for initiation of floral primordia; mean temperatures of 16°F (60°F) are apparently low enough to produce seeds.
[c]Partially germinated kohlrabi seeds can be vernalized at −1.1° to 1.1°C (30° to 34°F).

Culture

Soil

Cole crops grow well in all soil types but for early spring crops, they do best on rich sandy loam, loam, or silt loam. The soils should be slightly acid to slightly alkaline (pH 6.0–8.0). If too acid, the soil should be limed.

Propagation

Cole crops may be either direct seeded in the field or plants started in a seedbed for transplanting; the latter method is practiced in cold climates and when hybrid seeds are used. Crops such as kohlrabi are not transplanted. Before transplanting cole crops to the field, the small plants are hardened by gradual exposure to lower temperatures for a period of about 10 days. Plants can be hardened also by withholding of water. Both methods together are often used in the hardening process.

Cole crops are usually grown on beds and spacing varies according to the size of the plant at maturity. Large headed cabbages are spaced 90 × 90 cm (3 × 3 ft) while smaller cultivars are planted about 45 × 90 cm (1½ × 3 ft). Water is applied when transplanted, or irrigated immediately afterwards.

Fertilization

The crops remove much NPK from the soil; hence, fairly heavy fertilization is required depending on the availability of these elements in the soil. In the Netherlands, about 220 kg N/ha (200 lb/acre) is recommended for an early crop of cabbage. About 100–112 kg N/ha (90–100 lb/acre) are used in other regions. N deficiency not only reduces yields but also delays maturity and keeping quality; it also gives the crop a strong taste. From 22 to 112 kg P/ha (20 to 100 lb/acre) and from 70 to 220 kg K/ha (60 to 200 lb/acre) are recommended for growing of cabbage. Boron deficiency causes "hollow stem" in broccoli and molybdenum deficiency causes "whiptale" of cauliflower.

Harvest and Storage

Harvest

Cole crops reach harvestable maturity according to climatic conditions, nutrients, and moisture available during growth. It is difficult to predict the time required to reach harvestable stage. Ranges are shown in Table 6.1.

Storage

As with most cool season vegetables, cole crops should be stored at 0°C (32°F) at high humidity, 95–98% RH. Under these conditions, cabbage can be stored for 4–6 months; Brussels sprouts, kohlrabi, and cauliflower for about a month; and broccoli and kale for about 2 weeks.

Nutritive Value

All cole crops are very high in vitamin C. Savoy cabbage, kale, and broccoli are high in provitamin A (β-carotene). These vegetables are

TABLE 19.2. APPROXIMATE COMPOSITION OF COLE CROPS *(Brassica oleracea)* IN 100g EDIBLE PORTION

Vegetable	% refuse	Macroconstituents						Vitamins					Minerals (mg)					
		Energy (cal)	g						mg									
			Water	Protein	Fat	Total sugar	Other CHO	A (IU)	B_1	B_2	Niacin	C	Ca	Fe	Mg	P	K	Na
Cabbage																		
White	15	21	92	1.2	0.1	3.8	0.4	200	0.05	0.03	0.3	60	38	0.4	22	34	220	20
Red	15	19	92	1.4	0.1	3.3	0.2	40	0.05	0.03	0.3	57	51	0.7	17	42	190	17
Savoy	15	20	91	2.0	0.1	2.9	—	1,000	0.07	0.03	0.3	31	35	0.4	28	42	230	28
Cauliflower																		
Snowball	50	22	91	2.2	0.1	2.3	0.9	40	0.00	0.02	0.6	71	30	0.5	12	45	230	20
Winter	60	22	90	2.2	0.3	2.5	0.4	0	0.08	0.06	0.6	72	35	0.6	19	60	340	20
Broccoli, sprouting	20	23	90	3.6	0.3	1.6	0.4	3,800	0.11	0.10	0.6	110	78	1.0	39	74	360	40
Brussels sprouts	5	26	88	3.5	0.2	2.2	0.5	950	0.13	0.04	0.6	85	39	0.9	23	60	390	30
Kale	25	27	85	3.3	0.7	2.0	0.4	5,800	0.11	0.13	1.0	120	135	1.7	34	56	400	40
Kale, Scotch (braschette)	40	26	87	2.8	0.6	2.3	0.2	3,100	0.07	0.06	1.3	130	205	3.0	88	62	450	70
Kohlrabi	40	23	91	1.7	0.1	1.5	0.1	30	0.05	0.02	0.4	62	24	0.4	19	46	350	20

Source: Howard *et al* (1962).

much higher in protein than the starchy vegetables. Table 19.2 shows the proximate composition of cole crops; the essential amino acids are given in Appendix Table A. 2.

Brassicas contain 3-methylcysteine sulfoxide, a compound known as "kale anemia factor." Cattle and sheep are often poisoned in the winter feeding of rape, cabbage, Brussels sprouts and rutabaga roots and tops. This same compound has been shown to lower blood cholesterol in experimental animals.

Economics

Cabbage and cauliflower are very important crops in the temperate regions of the world (Table 19.3).

Sauerkraut is the main processed product from cabbages. Some cabbage is dried; blanching is necessary. Considerable broccoli, Brussels sprouts, and cauliflower are frozen in the United States. Some cauliflower and cabbage are canned and pickled.

TABLE 19.3. WORLD PRODUCTION OF CABBAGE AND CAULIFLOWER

	Cabbage			Cauliflower	
	Area (10^3 ha)	Production (10^3 MT)		Area (10^3 ha)	Production (10^3 MT)
World	1,600	33,560		331	4,343
Continent					
Africa	25	621		6	139
North and Central America	93	1,825		22	257
South America	23	575		4	72
Asia	810	13,090		172	1,690
Europe	332	26,023		120	2,062
Oceania	4	11		4	112
U.S.S.R.	380	8,700		2	10
Leading Countries					
1. U.S.S.R.	380	8,700	1. China	60	730
2. China	412	5,715	2. India	88	640
3. Japan	110	4,170	3. Italy	28	595
4. Poland	62	1,902	4. France	36	425
5. United States	70	1,508	5. U.K.	15	222
6. Korea, REP	83	1,034	6. United States	17	189
7. U.K.	41	984	7. Spain	8	185
8. Romania	24	830	8. Poland	9	141
9. Yugoslavia	48	700	9. Germany, D.R.	5	128
10. Italy	29	603	10. Japan	5	98

Source: FAO (1979).

TURNIP, CHINESE CABBAGES, MUSTARD: *Brassica campestris*

Origin

Brassica campestris is thought to have originated in the Mediterranean region and also the Near East, Afghanistan, Iran, and West Pakistan. Turnip was derived from annuals grown for the oilseed.

There are many types of Chinese cabbages and mustards, all probably indigenous to China and eastern Asia. The scientific names and the common names are very much confused. The classification according to Smith and Welch is followed. These crops have been under cultivation in China at least from fifth century AD. Therefore, it is not surprising that natural crossings have taken place in the field, and by man's selections of the desired traits, distinct types within this species have arisen.

Botany

Plants are biennial, herbaceous dicotyledons. Turnips are grown as an annual for the fleshy roots (swollen hypocotyl), which can vary in shape from flat, long to globe shaped. The flesh color may be white or yellow, but the skin color of the roots could range from shades of white, yellow, red, purple, or black. The plant is vernalized in the winter. Flower color is correlated with the flesh color of the roots; white-flesh rooted plants have bright yellow petals and the yellow-fleshed ones have pale orange-yellow petals.

Chinese cabbages and mustards are also grown as an annual. The Chinese cabbage belongs to the pekinensis group, and resembles lettuce in forming an elongated compact head; the sessiled leaves are slightly wrinkled with broad light colored midrib. Chinese mustard belongs to the chinensis group; it somewhat resembles celery or chard; the leaves are oblong or oval with shiny dark green blades, and thick white petioles (Fig. 19.2). The leaves do not form a head but are rosette. Another green mustard is gai choy (*B. juncea* var. *rugosa*) (Fig. 19.3). Chinese broccoli kailan is *B. alboglabra* (Fig. 19.4).

Culture

The growing of *B. campestris* is similar to *B. oleracea*. It is a cool season crop and can be grown in the same regions, and by the same methods, used for cole crops.

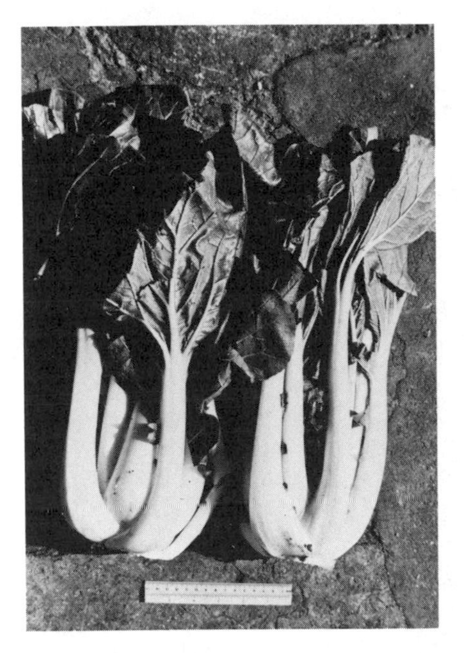

FIG. 19.2. Pak choy (*Brassica campestris*, chinensis group).

FIG. 19.3. Chinese green mustard, gai choy (*Brassica juncea* var. *rugosa*).

FIG. 19.4. Chinese broccoli, kailan, *(Brassica alboglabra)*.

Chinese cabbage prefers cool conditions, average temperatures of 13°–21°C (55°–70°F). Temperatures above 24°C (75°F) may cause some burning of the tips and prolonged temperatures of 13°C (55°F) can cause premature bolting. Chinese cabbage is not only sensitive to low temperatures but is also sensitive to photoperiods for flowering. Long photoperiods (16 hr days for a month) induce flowering in some cultivars. Short days and warm temperatures keep the plant in the vegetative phase. In the tropics and subtropics Chinese cabbages are grown in the cool season and at high elevations.

Turnip, Chinese cabbages, and mustards are seeded directly in the field about 12 mm (½ in.) in depth. In the Philippines, transplants are made from the seedbed to the field. A deep soil at pH of 5.5–7.0 is recommended. Highly acid soils should be limed. Plants are thinned and spaced according to the size of the mature crop. The thinned plants are often used for transplants in "skips" or transplanted into other fields; the larger plants are often used as greens. Fertilization with 60–112 kg N/ha (50–100 lb/acre), 60–85 kg P/ha (50–75 lb/acre), and 60–85 kg K/ha (50–75 lb/acre) are recommended, about half applied at seeding and the other half shortly after thinning.

Harvest and Storage

Depending on cultivar and season, from 35 to 70 days from seeding to harvest are required for Chinese cabbages. In the Orient, harvest is often made before the plant is full grown.

Storage recommendations are similar to cole crops: 0°C (32°F) and 95% RH.

Nutritive Value

Like the cole crops, *B. campestris* crops are high in vitamin C and, except for turnip roots, they are also high in provitamin A. Turnip greens are also eaten; they are more nutritious than the roots (Table 19.4).

WINTER RADISH, LOBOK, DAIKON: *Raphanus sativus* (Longipinnatus Group)

Origin

From the inscription on the walls of the pyramids, radish was an important food crop in Egypt. It dates back to about 2000 BC. The spread to China occurred about 500 BC and much later to Japan (700 AD). The Eastern Mediterranean region is thought to be the origin of radish. Wild species, *R. raphanistum* in Eastern Europe (Volga, Mediterranean, and Caspian Sea regions), *R. maritimus*, found in the same region as *R. raphanistrum* and along the coasts of Great Britain, Holland, Belgium, and France, and *R. rostratus*, found from Greece eastward to the Caspian Sea, all have the same basic $n = 9$ chromosomes.

Botany

Radish is a dicotyledonous annual or biennial herbaceous plant usually grown as an annual for the enlarged succulent tap root. Vernalization is usually required, and long photoperiods seem to hasten flowering. Radishes are normally self-incompatible and are insect pollinated. *Raphanus sativus* var. *radiculata* roots are more or less spherical and grow to be about 2–2½ cm (¾–1 in.) in diameter. The outer color may be white or red. *Raphanus sativus* (Longipinnatus group) called *daikon*, Chinese, or Oriental winter radish produce long

TABLE 19.4. PROXIMATE COMPOSITION OF SOME CRUCIFERAE IN 100g FRESH EDIBLE PORTION

Vegetable	% refuse	Macroconstituents						Vitamins					Minerals (mg)					
			g						mg									
		Energy (cal)	Water	Protein	Fat	Total sugar	Other CHO	A (IU)	B_1	B_2	Niacin	C	Ca	Fe	Mg	P	K	Na
Brassica																		
B. campestris																		
Cabbage, Chinese																		
Pak-choi	5	13	95	1.5	0.2	1.0	0.2	3,000	0.04	0.07	0.5	45	105	0.8	27	37	180	100
Pe-tsai	10	11	91	1.2	0.2	1.3	0.1	1,200	0.04	0.05	0.4	27	92	0.5	14	31	230	70
Mustard greens	5	15	91	2.7	0.2	0.8	0.1	5,300	0.08	0.11	0.8	70	140	2.0	48	45	340	50
Turnip																		
Greens	10	14	91	1.5	0.3	1.3	0.1	3,400	0.07	0.10	0.6	60	190	1.1	31	42	250	40
Roots (mature)	55	18	92	0.9	0.1	3.8	0.2	0	0.04	0.03	0.4	21	30	0.3	11	27	170	40
Tops and roots (immature) (rappini)	20	18	92	1.8	0.2	1.7	0.2	2,700	0.05	0.07	0.5	70	125	1.5	45	45	250	40
B. napus, rutabaga	40	25	90	1.2	0.2	5.0	0.5	Trace	0.09	0.04	0.7	33	31	0.4	19	41	220	20
Raphanus sativa, radish																		
Chinese winter	50	13	94	0.6	0.1	2.5	0.2	0	0.02	0.02	0.2	22	27	0.4	22	24	190	30
Icicle	35	15	94	1.1	0.1	2.5	0.1	0	0.03	0.02	0.3	29	27	0.8	9	28	280	16
Scarlet globe	40	14	94	0.7	0.1	2.7	0.1	0	0.02	0.03	0.4	21	20	0.8	11	27	190	30
Crambe maritima, sea kale																		
Blanched	0	13	94	2.0	0.3	0.7	0.1	100	0.04	0.04	0.3	26	35	0.5	—	34	—	—
Green	0	22	90	3.5	0.3	1.2	0.1	4,600	0.16	0.10	0.5	87	110	0.9	64	63	360	30
Rorippa nasturtium-aquatica, Watercress	0	11	95	2.3	0.1	0.2	0.2	4,700	0.09	0.12	0.2	43	120	0.2	13	60	270	32
Amorcia rusticana, Horseradish	0	55	77	3.1	1.7	1.8	5.2	0	0.06	0.03	0.5	95	150	2.4	81	41	420	16

Source: Howard *et al.* (1962).

roots 5–10 cm (2–4 in.) in diameter and 15–50 cm (6–20 in.) long (Fig. 19.5). A red-fleshed cultivar is grown in China.

Culture

A cool season crop, radish for salads matures in 3–5 weeks after planting. The soil should be light and well drained. The winter radish is grown similarly but grown in deep soil and matures 50–90 days from planting depending on the cultivar.

Winter radish cultivars have been selected for production in summer, autumn, and winter seasons. Also selection for late spring and summer

FIG. 19.5. Oriental winter radish, daikon (*Raphanus sativus*, Longipinnatus group).

cultivars have lead to decreased sensitivity to day length. Pithiness has been attributed to rapid growth rate in the late spring.

Economics

Radishes are a very important crop in Asia. In Japan, daikon ranked first in 1974 among the vegetables grown but in Taiwan it fell from first in 1964 to fourth in 1978. Radishes are eaten raw in salads. The Chinese winter types are also used mainly as a cooked vegetable and in pickles. The leaves are often eaten.

Raphanus caudatum, known as rat-tailed radish, is grown mainly in India; the immature seed pods, called silique (fruit), which grow to lengths of 30 cm (12 in.) or more, are eaten raw or pickled.

Diseases and Pests

Club root, black leg, rhizoctonia root rot, and phytophthora are diseases of the roots of cole crops. Wire stem, pythium, powdery and downy mildews, black spot, black rot, yellows, white rust, and verticillium wilt are more serious of the fungal and bacterial diseases. Black leg and black rot are seed transmitted. Cauliflower mosaic, cabbage black ringspot, and turnip yellow mosaic are viral diseases.

The more common pests of the crucifers include cabbage looper, aphids, wireworm, cut worm, cabbage maggots, harlequin bug, diamondback moth, white butterflies, flea beattles, leaf miners, and nematodes.

ROCKET SALAD, ROQUETTE, GARGEER, ROKA: *Eruca sativa*

Origin

Eruca sativa originated in Southern Europe and Western Asia. It is reported to have been grown by the ancient Romans before the birth of Christ. Presently, the crop is important in southern Europe, Egypt, and Sudan where the leaves are harvested as greens and for salad. It is grown as a salad crop (rucula) in southern Brazil. In India it is grown for the pungent seeds and for the oil in the seed.

Botany

Herbaceous annual, plants rarely exceed 75 cm (30 in.) in height. Leaves and flowers resemble the radish or turnip. The plant is self-sterile and cross-pollination is required for seed production.

Culture

A cool season crop, it is generally planted in the early spring in the temperate climates in rich moist soil. In the subtropics it is planted in the late fall. For salad and greens, the seeds are broadcasted on large beds and the tender leaves harvested in about 60 days. Growing in warm temperatures causes the leaves to be bitter and pungent. For oil and seed production, the growing temperature is not important since the high pungency in the seed is not important.

WATERCRESS: *Rorippa nasturtium-aquaticum*

Origin

Watercress grows wild in streams in the cool regions of the world. It has also been found growing wild above 1500 m (5000 ft) in the Himalayas of Nepal. History records that watercress was used as a medicinal plant about the time of Christ and the English pharmacopoeia of the nineteenth century lists its use for cure of scurvy. Until about 200 years ago, it was not cultivated but obtained from the wilds. Presently, it is grown throughout the world for the tender shoots, eaten raw in salads or as a cooked vegetable in the Far East.

Botany

An herbaceous perennial plant, watercress grows in areas of abundant water supply, in streams, springs, or artesian wells. There are two species, a diploid and a tetraploid; also a triploid, probably a cross of the two, which is sterile, exists. Propagation is either vegetative or by seed. Roots emerge readily from nodes of the stem under damp conditions.

Culture

A cool season crop, it can be grown all year round in the temperate zone and in the winter in semitropics. Adequate flowing water is needed, preferably slightly alkaline with adequate nitrate content. Water from limestone areas are considered good.

Plants can be started from seed in beds kept moist and shaded or from cuttings from established plantings. Cuttings about 15 cm (6 in.) long are planted 15 cm apart between plants about 10 cm (4 in.) deep. The beds are sloped 5 cm/30 m (2 in./100 ft) to allow water to flow. Water depth is 3–5 cm (1–2 in.). Some phosphatic fertilizers may be applied and also some NO_3^- if the NO_3^- content of the water is low.

The top 15 cm (6 in.) of the shoot is harvested. The bunched shoots should be precooled with ice and water and shipped refrigerated.

MINOR CRUCIFERS

HORSERADISH: *Armoracia rusticana*

The condiment crop, grown mainly for the pungent root, is a native of southeastern Europe. The pungent sulfur compound is allyl isothiocyanate. Horseradish is a perennial but is grown as an annual commercially in the northern temperature regions. The crop is propagated from root cuttings or from rhizomes. Plantings are made in the early spring after the last frost and allowed to grow until the first killing frost in the fall. At harvest the larger roots are dug for marketing and the small side roots are saved for next season's propagating material in cool moist sand, generally in cellars.

SEA KALE: *Crambe maritima*

Sea kale is a native along the sea shores of western Europe; it is also found near the Black Sea. The crop is propagated from seeds. Usually the petioles are blanched and the harvested leaves are cooked like asparagus.

MACA: *Lepidium meyenii*

A very minor root crop of the Peruvian Andes, maca is grown at 3500–4000 m (10,000–13,000 ft) elevations for food and medicinal purposes. Although it grows wild in the desolate cold and barren regions of the high altitudes and foraged by grazing animals, it is cultivated in the central highlands of Peru.

The frost-resistant crop is seed propagated. The plant has a rosette of 12–20 leaves, outer leaves die, and there is a continuous formation of new leaves from the center of the rosette. Colors of the enlarged root resembling a turnip, range from cream-yellow, part purple, purple, to almost black. The purple and black color is due to various amounts of anthocyanin pigments in outer layer of cells.

Maca is used both fresh and dried. The fresh roots are roasted in hot ashes. Roots are sun dried and can be stored in this state for several years; however, it is reported that the taste deteriorates after the second year.

It is believed by the natives of this region that the consumption of macas increase the fertility rate of both humans and animals.

GARDEN CRESS, PEPPER GRASS: *Lepidium sativum*

Of European origin garden cress is an annual grown for its leaves used in salads and for garnish. Seedlings with the hypocotyl and green cotyledons are cut and used in Britain for salads and sandwiches as alfalfa sprouts are used in the United States. Garden cress has a pungent taste.

BIBLIOGRAPHY

BANGA, O. 1974. Radish, *In* Evolution of Crop Plants, N.W. Simmonds (Editor). Longmans-Green, London.

COURTER, J.W., and RHODES, A.M. 1969. Historical notes on horseradish. Econ. Bot. *23*, 156–164.

HEMINGWAY, J.S. 1974. Mustards, *In* Evolution of Crop Plants, N.W. Simmonds (Editor). Longmans-Green, London.

HERKLOTS, G.A.C. 1972. Vegetables in South-East Asia. Hafner Press, New York.

HOWARD, H.W. 1974. Watercress, *In* Evolution of Crop Plants, N.W. Simmonds (Editor). Longmans-Green, London.

HOWARD, F.D., MACGILLIVRAY, J.H., and YAMAGUCHI, M. 1962. Bull. No. 788, California Agric. Exp. Stn., Univ. of California, Berkeley.

KNOTT, J.E., and DEANON, J.R., Jr. 1967. Vegetable Production in Southeast Asia. Univ. of Philippines Press, Manila.

LEON, J. 1964. The "Maca" *(Lepidium meyenii)*, a little known food plant of Peru. Econ. Bot. *18*, 122–127.

MACGILLIVRAY, J.M. 1953. Vegetable Production. McGraw-Hill Book Co., New York.

MCNAUGHTON, I.H. 1974. Turnips and relatives, *In* Evolution of Crop Plants, N.W. Simmonds (Editor). Longmans-Green, London.

MCNAUGHTON, I.H. 1974. Swedes and rapes, *In* Evolution of Crop Plants, N.W. Simmonds (Editor). Longmans-Green, London.

NIEUWHOLF, M. 1969. Cole Crops. Leonard Hill Publishers, London.
SHEER, G.M. 1968. Commercial growing of water cress. USDA Farmers Bull. No. 2233. U.S. Dept. Agric., Washington, DC.
SIMMONDS, N.W. Editor. 1974 Evolution of Crop Plants. Longmans-Green, London.
SWARUP, V., and SCHATTERJEE, S.S. 1972. Origin and genetic improvement of Indian cauliflower. Econ. Bot. *26*, 381–393.
THOMPSON, K.F. 1974. Cabbages, kales, etc., *In* Evolution of Crop Plants, N.W. Simmonds (Editor). Longmans-Green, London.

20

Umbellifers: Carrot, Celery, and Condiment Herbs

Family: Umbelliferae (parsley)

Principal genera/species	Common name	Origin
Foeniculum vulgare	Anise, sweet; also Fennel, sweet Florence, finocchio	S. Europe
Daucus carota	Carrot	S.E. Europe and Mideast (Afghanistan)
Apium graveolens (Rapaceum group)	Celeriac; also Celery root	S. Europe, N. Africa, W. Asia
Apium graveolens (Dulce group)	Celery	S. Europe, N. Africa, W. Asia
Anthriscus cerefolium	Chervil; also Chervil, salad	S.E. Russia, W. Asia
Chaerophyllum bulbosum	Chervil, turnip-rooted	Cent. Europe, Caucasus
Petroselinum crispum (Tuberosum group)	Parsley, turnip-	S. and Cent. Europe
Pastinaca sativa	Parsnip	Cent. and S. Europe
Sium sisarum	Skirret	Siberia, N. Iran
Arracacia xanthorrhiza	Arracacha, apio	N. and Cent. Andes
Cryptotaenia japonica	Japanese honewort	probably Japan
Oenanthe javanica	Water dropwort	East Asia
Coriandrum sativium	Coriander	E. Mediterranean
	Condiment Herbs	
Angelica archangelica	Angelica	Europe and Asia
Pimpinella anisum	Anise; also sweet Alice	Greece and Egypt

Foeniculum vulgare	Anise, sweet	S. Europe
Carum carvi	Caraway	Asia Minor
Apium graveolens (Dulce group)	Celery; also smallage	S. Europe and N. Africa
Anthriscus cerefolium	Chervil; also Chervil, salad	S.E. Russia, W. Asia
Myrrhis odorata	Cicely, sweet; also Myrrh	Europe
Coriandrum sativum	Coriander	E. Mediterranean
Cuminum cyminum	Cumin	Mediterranean
Anethum graveolens	Dill	Europe
Levisticum officinale	Lovage	S. Europe
Petroselinum crispum	Parsley	S. and Cent. Europe

These plants are called "umbellifers" or "umbels". This is a very large group consisting of about 250 genera and from 1500 to 2000 species. Many are more or less oily and highly aromatic. They are widely distributed throughout the world from the sub-Arctic to the subtropics; they are rare in tropics except at elevations of above 500 m (1500 ft).

The names umbellifers or umbels is derived from the umbrella-shaped inflorescence. The pedicels of each flower radiate from a common point at the summit of the pedicel (the stalk of an inflorescence). In this family, the umbels are usually compound, each primary ray is terminated by a secondary umbel (umbellet).

The carrot and celery are the two most important vegetables in this family. The arracacha has been covered in the minor root crop section. Florence fennel, an important umbellifer in the Mediterranean region is shown in Fig. 20.1.

CARROT: *Daucus carota*

Origin

Carrot is the most important crop of this group. It has worldwide distribution. Carrots were first used for medical purposes and gradually used as food. Written records in Europe indicate that carrots were cultivated prior to the tenth century. The species *D. carota* is a native of western Asia (Afghanistan). Wild types of the genus are found in Southwest Asia, southern Europe, north Africa and North America. The crop was introduced into China about the thirteenth or fourteenth century and into Japan about the seventeenth century. The first settlers to

FIG. 20.1. Florence fennel, sweet anise *(Foeniculum vulgare)*.

Virginia brought seeds, and soon carrots were grown by the American Indians.

Botany

Carrots are a dicotyledonous herbaceous crop grown for the enlarged taproot. The wild form is an annual but the cultivated crop, which is believed to have been derived from the wild type, is biennial. The shape of the root can vary from a small tap-like root 2–6 cm (1–2½ in.) in diameter and 6–90 cm (2½–36 in.) in length. Colors of the flesh may be white, yellow, orange, red, purple, or black (very dark purple). The anatomy of a carrot root is shown in Fig. 20.2.

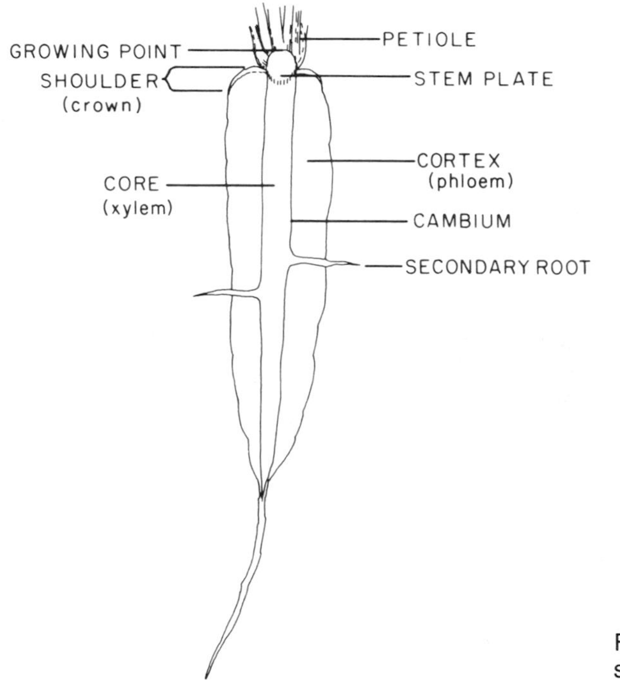

FIG. 20.2. Longitudinal section of a carrot root.

Culture

A cool season crop carrots grow best at mean temperatures of 16°–18°C (60°–65°F). Top growth is reduced at mean air temperature of 28°C (82°F) and the roots become very strong flavored.

At low temperatures the color is poor; roots become long and tapered below 16°C (60°F). At 4°–10°C (40°–50°F) there is little root enlargement and very little top growth (see Seedstalk Formation).

There are many areas meeting temperature requirements for carrot production. Carrots are grown in the sub-Arctic regions of Sweden, Finland, and Alaska as far north as 65° N latitude, where there are less than 60 frost-free days per year.

Carrots are grown in Costa Rica at 1000–3000 m (3000–9000 ft) altitudes, and also in the mountains of Guatemala; in the Philippines at 2000 m (6,000 ft) elevations, carrots can be grown all year. Cool night temperatures are necessary for carrot production in the tropics.

Day lengths between 9 and 14 hr has no effect on root color; under a day length of 7 hr, the color of the roots is much lighter.

Seedstalk Formation

When seedlings with roots 6 mm (1/4 in.) diameter or larger (not at juvenile stage) are subjected to cold temperatures, below 10°C (50°F), for a period of time, flower induction takes place. Experiments at the National Vegetable Research Station in England indicated that vernalization for 8 weeks at 8°–9°C (47°F) is required for flowering of carrot, with the largest roots flowering more readily. The red-colored anthocyanin types are reported to bolt under long-day conditions.

The Japanese variety 'Kintoki' [a red-fleshed (lycopene) carrot] bolts under either low temperature or long days.

Soil, Moisture, and Nutrition

Carrots grow well on many soil types, preferably on well-drained, neutral soils (pH 5.5–7.0). Sandy loam or organic soils (peat or muck) free of clods and rocks are preferred for smooth, straight roots. Soil compaction effects root growth and length; very loose or highly compacted soils are detrimental to growth of roots. In very compacted soils the taproot grow conical in shape instead of long tapered roots in friable soils.

Carrot seed is scattered thinly in a row 8–10 cm (3–4 in.) wide which permits production without thinning. It has been found that for maximum yields of carrots, the population is 100–150 plants/m^2 planted in row spacing of 1:6 (rectangular spacing) rather than a 1:1 (square spacing).

Adequate moisture during growing season is important. A total of 45–90 cm (1½–3 acre-ft) of water (including rainfall) is required for good quality and high yields. Low moisture grown carrots have a very strong and pungent flavor. In high soil moisture or waterlogged soils the color of the root is poor.

Carrots require 70–120 kg/ha (60–100 lb/acre) of N, 30–35 kg/ha (25–30 lb/acre) of P and 0–55 kg/ha (0–50 lb/acre) of K for a good crop, depending on the nutrients available in the soil.

Harvest and Storage

Harvest

Carrots for the fresh market are harvested when majority of the roots are 2½–3½ cm (1–1½ in.) in diameter at the crown. They can be harvested 65–85 days and longer after planting, depending on the temperature. Processing carrots are left in the ground longer, as both the dry matter and color increase with maturity.

Storage

Carrots store best at 0°–1°C (32°–34°F) with high humidity. The roots store better with the tops removed as moisture and nutrients are translocated to the tops from the roots during storage.

Diseases and Pests

Carrots are relatively free from destructive insects but occasionally injury occurs from vegetable weevil, carrot beetle, parsley caterpillar, wingless May beetle, and cut and army worms.

Carrot scab (bacterial) can be avoided by seed treatment and use of disease-free land. Early and late blights are caused by fungi and are controlled by copper-containing sprays. Mosaic, a viral disease, is aphid transmitted; and aster yellow is leafhopper transmitted; soft rot (bacterial) is caused by excessively high soil moisture.

Nematodes cause galls and distorted roots. For carrot growing nematode-free land must be used or the soil should be fumigated with methyl bromide or other recommended chemicals. Wireworms are also controlled with the same type chemicals.

Nutritive Value

Yellow and orange fleshed cultivars are particularly high in provitamin A (70–80% β-carotene and 20–30% α-carotene); the cortical region contains more carotenes than the core. Also, the cortex is sweeter than the core. White flesh cultivars contain little provitamin A pigments. The red water-soluble anthocyanin pigment and the red water-insoluble lycopene pigment in the roots of some cultivars do not contribute to the provitamin A value. The carotene content of the orange and yellow fleshed cultivars increases with growth. Roots contain much iron but are low in ascorbic acid (vitamin C) (Table 20.1). The essential amino acids are given in Table A.2. A bitter coumarin compound is formed when carrots are stored with produce that gives off ethylene.

Economics

Among the succulent vegetables, carrots rank third in world production. This is because carrots not only are popular and inexpensive, but they are easily grown, have a long period of harvest, ship well, and have a relatively long storage life at low temperatures. Table 20.2 shows

TABLE 20.1. PROXIMATE COMPOSITION OF SOME UMBELLIFERS IN 100 g OF EDIBLE PORTION[1]

			Macroconstituents					Vitamins								
				g					mg				Minerals (mg)			
Crop	Edible part	% refuse	Energy (cal)	Water	Protein	Fat	CHO	A (IU)	B_1	B_2	Niacin	C	Ca	Fe	Mg	P
Carrots	Roots															
Chantenay		20	31	89	0.8	0.2	6.6	13000	0.05	0.04	0.3	6	30	1.4	17	36
Danvers		25	32	87	1.0	0.2	7.2	15000	0.05	0.04	0.4	6	33	1.1	—	40
Imperator		20	33	86	1.0	0.2	7.7	14000	0.05	0.04	0.5	7	39	1.3	17	43
Nantes		15	32	88	0.8	0.2	6.8	10000	0.05	0.03	0.4	5	31	0.9	—	32
Celeriac	Roots	55	20	88	1.5	0.3	3.5	0	0.05	0.06	0.7	8	43	0.7	20	115
Celery																
Self-blanching	Petiole	5	7	96	0.7	0.1	1.2	90	0.03	0.02	0.3	7	25	0.3	10	27
Green	Petiole	5	8	95	0.9	0.1	1.2	120	0.03	0.04	0.3	10	70	0.5	14	34
Florence fennel (sweet anise)	Petiole	60	15	93	1.1	0.1	2.6	100	0.04	0.02	0.4	9	44	0.8	23	38
Parsley	Leaves	5	16	90	2.2	0.3	1.3	5200	0.08	0.11	0.7	90	125	2.0	79	40
Parsnip	Root	15	53	81	1.2	0.3	11.6	0	0.09	0.05	0.7	17	40	0.7	29	69

Source: Howard *et al.* (1962).

TABLE 20.2. WORLD PRODUCTION OF CARROTS

	Area (10^3 ha)	Production (10^3 MT)	Yield (MT/ha)
World	468	10,325	22.0
Continent			
Africa	28	336	12.1
North and Central America	43	1,282	29.5
South America	25	419	16.7
Asia	131	2,474	18.9
Europe	127	3,618	28.4
Oceania	5	136	29.6
U.S.S.R.	109	2,060	18.9
Leading countries			
1. U.S.S.R.	109	2,060	18.9
2. China	96	1,730	18.0
3. United States	30	946	31.7
4. U.K.	15	672	43.6
5. Poland	28	619	22.3
6. Japan	24	590	24.2
7. France	20	521	25.8
8. Italy	9	319	37.6
9. Germany, D.R.	8	281	35.6
10. Canada	7	239	33.8

Source: FAO (1979).

that carrots are grown practically everywhere in the world; production is low in the tropical regions. In most production areas carrots are used as fresh vegetables. In the developed countries, they are canned, frozen, and dehydrated.

CELERY: *Apium graveolens*

Origin

Celery is an important crop in the temperate regions of the world, especially in Europe and North America. Wild celery can be found in damp or marshy areas from Sweden in the north and south to Algeria, Egypt, and to the Caucasus in western Asia. Celery was first used for medicinal purposes and later as food. Until very recently, celery was usually blanched to make the stalks tender and to diminish the bitterness.

Celeriac or celery root is of the same species as celery but belongs to the Rapaceum group. It is not as important a crop as celery. The enlarged roots are used mainly in soups.

Botany

The plant is normally a biennial but can complete the life cycle in a year if the plant has been subjected to low temperatures during the development. In the vegetative phase (the first year) the plant is mainly leaves above ground as with the carrot; the stem is very short and does not elongate, and the stalk grows to about 60 cm (2 ft) in height. The stem elongates after vernalization and the plant grows several feet (from 1 to 2 m) in height, the terminals end in compound umbellate flowers typical of the family. There are two types grown commercially, the green and the golden, self-blanching types.

Culture

Climate

Celery has rather exacting climatic requirements. It is grown in areas where the monthly mean temperature is 16°–21°C (60°–70°F). With exposure of plants at the five true leaf stage or larger to temperatures of 5°–10°C (40°–50°F) for 10 days or more, bolting is apt to occur; lowering the temperature shortens the time required for vernalization. There are cultivar differences as to proneness to bolting; some may bolt after prolonged temperatures of 13°C (55°F). Hence, celery plants for transplanting should be grown above 16°C (60°F) minimums.

Soil, Fertilizer, Moisture

Loams, clay loams, and peat or muck soils which are well drained can be used for growing celery. For succulent, "string-free" quality celery, large amounts of nutrients are applied to grow a crop. One ha of crop removes about 330 kg N, 80 kg P, and 70 kg K (300 lb of N, 70 lb P, and 600 lb K/acre). Depending on the available nutrients in the soil 220–450 kg N, 120 kg P, 180 kg K/ha (200–400 lb N, 100 lb P, and 160 lb K/acre) are applied to the crop. About half of the N and all of the P and K are applied before or at the time of planting and the remaining is side-dressed in the middle of the growth period.

Celery is a shallow rooted crop with most of the roots within the first 60 cm (2 ft) of soil. Some roots can be found to about 100–120 cm (3½–4 ft) in fertile deep soils.

For succulent and tender stalks, high soil moisture is necessary. Water must be regularly supplied at frequent intervals and more than adequate moisture is needed in the last month before harvest as the most rapid growth occurs at this time. About 75 cm (30 in.) of water are required to grow a crop of celery to maturity.

Culture

Celery is either direct seeded in beds 60–100 cm (24–40 in.) apart in the field or transplanted when plants are about 10–15 cm (4–6 in.) in height (about 2–2½ months after seeding). In the early spring plants for transplanting should be protected against temperatures below 16°C (60°F). Celery seeds are very tiny; they should be planted no more than 1/2 cm (1/4 in.) in depth and kept moist throughout the germination period. When plants are at the four to six true leaf stage, they are thinned to about 15–20 cm (6–8 in.) between plants.

In some countries blanched celery is preferred. The process of blanching is started several weeks before harvest by wrapping paper around the stalks, placing boards, or mounding soil to exclude light. Also, plants are often spaced very closely so that the plants shade each other.

Harvest and Storage

Harvesting occurs when plants have attained marketable size. The outer petioles become pithy with increase in maturity so the crop should be harvested before this occurs. The plants are usually topped so that the stalk is mostly petiole tissue. The outer leaves are removed and plant with the upright petioles is marketed. Celery is stored at near 0°C (32°F) and at very high humidity. At these conditions it can be held for about a month maintaining good quality.

Pests and Diseases

Pests attacking celery include aphids, spider mites, leafhoppers, cut and army worms, cabbage lopper, white flies, leaf miners, and nematodes.

Early blight caused by the fungus *Cereopori apii* and late blight caused by *Septoria apicola* can be controlled by use of fungicides, some of which contain copper. Celery mosaic, transmitted by aphids, and aster yellows, transmitted by leafhoppers, are viral diseases. Pink rot, a fungal disease, attacks the petioles while the bacterial disease called phoma rot causes decay of the roots.

Physiological disorders of celery include blackheart, the dying of the growing point, caused by calcium deficiency; cracked stem by boron deficiency; and brown checking by low boron in conjunction with excess potassium. Boron deficiency affects the elasticity of the colenchyma (strings) causing them to break while brown checking is the necrosis of the inner surface of the petiole giving a brown and checked appearance. Yellowing of the leaf blades is caused by low magnesium in the soil. The yellowing can become severe in soils with high available calcium as it interferes with magnesium absorption.

MISCELLANEOUS UMBELLIFERS USED AS VEGETABLES

PARSNIP: *Pastinaca sativa*

Parsnip is grown for the enlarged tapered taproot, 5–8 cm (2–3 in.) in diameter and 20–30 cm (8–12 in.) long; tops grow from 40–60 cm (16–24 in.) tall. It is a native of the eastern Mediterranean region and has been used for medicinal purposes as well as for food by the Romans and Greeks in ancient times. The crop, a biennial, is grown like the carrot; the roots are more resistant to cold temperatures than the carrot. Roots become sweeter when exposed to cold soil temperatures. Flesh is creamy in color. From 95 to 120 days are required from seedling to harvestable roots.

PARSLEY AND TURNIP-ROOTED PARSLEY: *Petroselinum crispum (Petroselinum hortense)*

There are two foliage types: the plain leaved and the curled-leaf types. These are used for garnish and flavoring. The turnip root or Hamburg type is used mainly in stews and in soups. Parsley is a cool season crop, which grows well in the temperature range of 7°–16°C (45°–60°F). The outer leaves are harvested and the inner leaves continue to grow and mature for other harvests. Turnip rooted parsley is winter-hardy and is ready for harvest 85–90 days from seeding.

JAPANESE HONEWORT, MITSUBA: *Cryptotaenia japonica*

Japanese honewort or mitsuba (three leaflets in Japanese) is a very important minor crop in Japan; it is cultivated in Korea, Taiwan, and on the mainland of China. The origin is probably Japan as it is reported that wild forms can be found there. A cool season crop, the main production is in March through May, but is grown throughout the year in Japan. Mitsuba in Japan is forced in plastic greenhouses and in hydroponic culture throughout the year; the main production is from October through March. The petioles are 10–15 cm (4–6 in.) long; and when blanched the petioles elongate to 30 cm (12 in.) or more in length. Like the celery and parsley, the plant has a distinct aromatic flavor and is used in soups and raw in salads.

WATER DROPWORT, WATER CELERY, SERI: *Oenanthe javanica (Oenanthe stolonifera)*

Water dropwort, a native of Southeast Asia has been used as a potherb for thousands of years. Evidence in China indicates that it was used as food as early as 770–2180 BC; the culture in Japan has been dated prior to 750 AD. A cool season aquatic perennial plant, it produces stolons that may be used for propagation; however, propagation is usually by seeds. In Japan seedlings are planted after the rice crop has been harvested in the early fall. The crop is grown in flooded culture into the winter and harvested in January and February when the plants are about 30 cm (12 in.) in height. Water dropwort resemble celery except the petioles are more numerous, slender, rounded, and hollow.

BIBLIOGRAPHY

General

MACGILLIVRAY, J.H. 1953. Vegetable Production. McGraw-Hill Book Co., New York.
THOMPSON, H.C., and KELLY, W.C. 1957. Vegetable Crops. McGraw-Hill Book Co., New York.

Carrot

BANGA, O. 1957. Origin of the European cultivated carrot. Euphytica *6*, 54–63.
BOSWELL, V.R. 1963. Commercial growing of carrots. USDA Leaflet No. 353. U.S. Dept. Agric. Washington, DC.
BOSWELL, V.R., and BOSTELMANN, E. 1949. Our vegetable travelers. Nat. Geographic Mag. *96*, 145–217.
HOWARD, F.D., MACGILLIVRAY, J.H., and YAMAGUCHI, M. 1962. Nutrient composition of fresh California-grown vegetables. Bull. No. 788, Calif. Agric. Exp. Stn., Univ. of California, Berkeley.
MAGRUDER, R., BOSWELL, V.R., EMSWELLER, S.L., MILLER, J.C., HUTCHINS, A.E., WOOD, J., PARKER, M.M., and ZIMMERLEY, H.H. 1940. Descriptions of types of principal American varieties of orange-fleshed carrots. USDA Misc. Pub. No. 361. U.S. Dept. Agric., Washington, DC.
YAMAGUCHI, M., ROBINSON, B., and MACGILLIVRAY, J.H. 1952. Some horticultural aspects of the food value of carrots. Proc. Am. Soc. Hort. Sci. *60*, 351–358.

Celery

SIMS, W.L., WELCH, J.E., and LITTLE, T.M. 1963. Celery production in California. Circ. No. 522, Div. of Agric. Sci.,Univ. of California, Berkeley.

Miscellaneous Umbellifers

BUTTERFIELD, H.M. 1964. Growing herbs for seasoning food. Pub. AXT-112, Univ. of California Agric. Ext. Serv. Berkeley.

HERKLOTS, G.A.C. 1975. Vegetables in South-East Asia. Hafner Press, New York.

LOWMAN, M.S., and BIRDSEYE, M. 1946. Savory herbs: culture and use. USDA Farmers' Bull. No. 1977. U.S. Dept. Agric. Washington, DC.

WILLIAMS, L.O. 1960. Drug and condiment plants. USDA Agric. Handb. No. 172. U.S. Dept. Agric., Washington, DC.

21

Vegetable Legumes

Family: Leguminosae (pea)
Subfamily: Papilionoidaceae

These are dicotyledonous annuals and perennials. There are some 480 genera and over 12,000 species.

Both *Pisum* and *Phaseolus* genera are very important as vegetables in the immature stage, as well as staple food crops in the mature stage (seeds). Many of the other species, although grown for the dry seeds, are used in the immature stage (pods and/or seeds) as vegetables. In some, the leaves, tender shoots, or the roots are harvested and used as vegetables. A list of legumes used in the world as vegetables is given in Table 21.1. Each is discussed following the general section.

Definitions

Gram: Word used in India for whole seed of legumes. *Pulse*: Word used in India for seeds of legumes without seed coat and cotyledons split into half. *Dal*: The split pigeon pea in India.

In India the following legumes, called pulses, are the important ones: *Cicer arietinum*—garbanzo bean, chick pea; *Canjanus cajan*—pigeon pea, red gram; *Phaseolus aureus*—mung bean; *Phaseolus mungo*—black gram, urd; *Lens esculenta*—lentil; *Lathyrus sativus*—chickling pea, grass pea.

LEGUMES

Origin

Legumes have been cultivated for over 6000 years in various parts of the world. The wild form of many of the present-day legumes cannot be

TABLE 21.1. LEGUMES USED AS VEGETABLES

Crop	Scientific name	Edible part	Toxic substances[a]	Other uses	Regions grown, misc. information
Peas	*Pisum sativum* $n = 7$	Immature seeds; pods also used, (var. *macrocarpon*); leaves used.	–	Forage and green manure	Cool, relatively humid climate (13°–18°C). Grown as winter crop in India. Above 1200 m in tropics; 1800–2600 m in Uganda.
Beans Common, snap, string	*Phaseolus vulgaris* $n = 11$	Immature pods. Leaves as pot herbs in some areas of tropics. Mature seeds.	–	Forage	Warm season crop, can be grown in temperate as well as in tropics, not suited in ever-wet tropics such as Malaya, because of disease and poor fruit set. Short-day or day-neutral varieties.
Mat bean, moth bean	*P. aconitifolis* $n = 11$	Green pods. Mature seeds.	–	Forage and green manure	Native of India, Pakistan, and Burma where it grows wild and is also cultivated. Grown in Ceylon and China. Drought resistant, not suitable for very wet tropics.
Adzuki bean	*P. angularis* $n = 11$	Immature pods. Mature seeds.	–		Warm season, short-day plant. Important dry beans in Japan.
Mung bean, green gram	*P. aureus* $n = 11$	Green pods. Bean sprouts. Mature seeds.	–	Forage and green manure	India probable origin, 0–1800 m on dry land following rice. Short- and long-day varieties grown in India, Burma, China, Indo-China, and Java. Introduced to East and Central Africa, West Indies, and United States.
Rice bean	*P. calcaratus* $n = 11$	Young pods and leaves. Dried beans used in place of rice.	–	Fodder and green manure	India, China, and Southeast Asia. Short-day, tolerates high temperature.
Scarlet runner bean	*P. coccineus* $n = 11$	Immature pods, green and dry seeds, fleshy tubers.	– Tuberous roots may be poisonous.	Sometimes as ornamental	Central America, humid uplands of tropics (1,800 m). Requires coolness, grows well in Britain. Long-day plant. Perennial.
Lima bean	*P. lunatus (limensis)* $n = 11$	Green shelled seeds. Young leaves and pods sometimes used. Mature seeds.	– Wild types contain cyanogenic glucoside.	Fodder, green manure, and cover crops	Guatemala and Central America origin. 0–2500 m elevation. Fruit set reduced above 27°C. Some are short-day, others are day-neutral; warm season crop.
Blackgram, urd	*P. mungo* $n = 11$	Green pods. Mature seeds.	–	Forage and green manure, hulls and straw fodder	India, most highly prized of pulses of India, vegetarian diet of high caste Hindus. 0–1800 m, drought resistant, not suited for wet tropics.
Peanut	*Arachis hypogea* $n = 10$	Mature seed, tender shoots, and leaves.	–	Fodder	Warm regions of temperate zone, tropics, and subtropics.
Jack Bean, horse bean	*Canavalia ensiformis* $n = 11$	Young pods and immature seeds.	±	Green manure, fodder	Native of Central America and West Indies.
Sword bean	*C. gladiata* $n = 11$	Young pods and beans. Ripe seeds may be poisonous.	+	Green manure, forage	Wild in tropical Asia and Africa. Cultivated in India. Perennial.

(continued)

TABLE 21.1. *(continued)*

Crop	Scientific name	Edible part	Toxic substances[a]	Other uses	Regions grown, misc. information
Pigeon pea, red gram	*Cajanus cajan* $n = 11$	Young green seeds. Green pods.	–	Forage and fodder. Green manure firewood, thatching, and baskets in India.	Probably native of Africa. Warm-season crop sensitive to frost; short-day plants widely adapted to climate and soils. Drought resistant, not suitable for very wet tropics. Perennial.
Chick pea, Bengal gram, garbanzo bean	*Cicer arietinum* $n = 8$	Green pods. Tender shoots. Dried seeds.	± May contain cyanogenic glucoside.	Dry stems and leaves; fodder acid liquid from glandular hairs collected and used as medicine or as vinegar (94% malic acid and 6% oxalic acid).	Most important in India. Crop grown in ancient Egypt, Palestine, and Greece. Grows well in tropical Africa, Americas, and Australia. Grows well in cool dry climate; cannot tolerate heavy rains in wet, hot tropics.
Cluster bean, guar	*Cyamopsis tetragonolobus* $n = 7$	Young tender pods.	–	Fodder and green manure	Indigenous to India, grown in dry tropics. Flour from seeds has five to eight times thickening power of ordinary starch.
Soybean	*Glycine max* $n = 20$	Immature seeds used as vegetable. Mature seeds eaten whole or sprouted. Products: Soy milk, soy curd. Fermented beans. Oil + proteins.	+ Contains digestive enzyme inhibitors.	Fodder and green manure overcrop	Indigenous to eastern Asia. Cultivated in China 2800 BC. Grown in China, Japan, Korea, and Indonesia; grows well in central United States. Subtropical plant, tropics to 52°N. Short-day plant (8–10 hr daylight).
Hyacinth, lablab	*Lablab niger* (*Dolichos lablab*) $n = 11$	Young pods and tender beans. Beans sprouted. Dried bean.	+ White seed coat type contains no cyanide.	Forage and green manure	Asian origin, cultivated in India and Africa, grown in dry areas of low rainfall (640–760 mm) at 0–2100 m. Long-day and short-day varieties. Perennial.
Grass pea, chickling pea	*Lathyrus sativus* $n = 7$	Seeds eaten by poor in India. Leaves are eaten as pot-herbs.	+ Seeds eaten over a long period can cause lathryism, a paralysis of lower limbs.	Fodder	Native of southern Europe and western Asia; cool season crop in India and northern Sudan.
Lentil	*Lens esculenta* $n = 7$	Young pods. Dry seeds.	–	Fodder	Cultivated in ancient Egypt, southern Europe, and western Asia. Spread to India, China, and Ethiopia. Grown at 0–3400 m. As winter crops in northern India and Pakistan. Not suited for hot, wet tropics.
Egyptian lupine	*Lupinus termis* $n = ?$	Mature seed.	+ Seed detoxified by soaking in water.	Green manure	Warm season. Crop important in Egypt, used as snack food.

Yam bean	*Pachyrrhizus erosus* $n = 11$	Succulent roots eaten, raw or cooked immature pods eaten. Starch obtained from tuberous roots.	± Roots and mature seeds contain rotenone (insecticide).		Wild in Mexico and northern Central America. Grown successfully in hot, wet tropical regions. Perennial.
Potato bean	*P. tuberous* $n = 11$	Tubers eaten as above. Young pods have irritant hairs so they are not eaten.			Origin in western South America probably near headwaters of Amazon. Perennial.
Goa bean, winged bean	*Psophocarpus tetragonolubus* $n = 9$	Immature pods. Tuberous roots. Young leaves, shoots, mature seeds. Flowers eaten as vegetable, edible oil from seeds.	–	Green manure, cover crop, fodder	Probably originated in tropical Asia. Grown in New Guinea, Philippines, Southeast Asia. Wet, hot climate, short-day. Perennial.
Fenugreek	*Trigonellia foenum-graceum* $n = ?$	Mature seed, leaves of young plant.	–	Fodder and green manure	Annual winter crop. Important in India. Seeds used as spice in curry.
Fava bean, broadbean, Horse bean, Windsor bean	*Vicia faba* $n = 6$	Green-shelled beans. Dry beans.	Favism, consumption affects only certain people.	Cover crop, fodder	Originated in Mediterranean or Southwest Asia. Temperate zone crop, cool season. Winter crop at edge of tropics in northern Sudan and Burma. High altitudes in Latin America. Grows well at 1800–2500 m.
Catjung, cow pea	*Vigna unguiculata* $n = 11$	Immature pods. Young shoots and leaves as pot-herb. dry beans.	–	Fodder and green manure crop	Tropics and subtropics, in ancient Africa and Asia. Also in India.
Common cow pea	*V. sensis* $n = 11$	Immature pods. Young shoots and leaves as pot-herbs. Dry beans.	–	Fodder and green manure	Cultivated in Africa. Short-day plant.
Yard-long bean, asparagus bean	*V. sesquipedalis* $n = 11$	Immature pods.	–	Fodder	Most widely cultivated in the Far East, dry-neutral. All heat-tolerant and relatively dry conditions. Cold-sensitive and killed by frosts.
Bambara groundnut	*Voandzeia subterranea* $n = 11$	Immature seeds. Mature seeds used after soaking or ground into flour.	–		Found wild in West Africa. Pods produced under soil as in peanut. Cultivated throughout tropical Africa. Crop is self-pollinated.

[a] **Toxic substances: –, none present in cultivated varieties; ±, may be present; + present, must be inactivated or removed before consumption. Except for peas, cluster bean, and fava bean, all contain trypsin inhibitors.**

found. All the main centers of origin according to Vavilov have contributed cultivated legumes of today.

Botany

Most of the cultivated crops used as food are annuals. There are a few perennials (see Table 21.1). Leaves are usually alternate and mostly compound, pinnate, trifoliate, or digitate. The flowers are perfect (have both ovaries and stamens in the same flower) and are characteristically "butterfly"-shaped. Self-pollination and cross-pollination occur. Bees are the main agent for cross-pollination.

Legumes may be classified according to the position of the cotyledons on the germinated seedlings:

Hypogeal: Cotyledons remain underneath the surface of the soil due to the limited elongation of the hypocotyl. The epicotyl is fully differentiated prior to germination; it pushes through the soil. The cotyledon remains in the seed coat. Example: pea *(Pisum sativum)*.

Epigeal: Cotyledons are pushed above the surface by the rapid elongation and straightening of the hypocotyl. The seed coat may be left in the soil or may be carried above the soil and shed by the unfolding of the cotyledons. Example: common bean *(Phaseolus vulgaris)*.

There are certain types where the cotyledons move to the surface of the soil and stop; the length of hypocotyl depends on the depth of the planted seed. Example: peanut *(Arachis hypogaea)*.

The orientation of epigeal seeds is important in their germination. Seeds planted with the long axis vertical and the hypocotyl at the top (the hypocotyl higher than the hilum) will emerge through the soil faster than other seed orientations with the poorest being vertical and the hypocotyl at the bottom. When the hypocotyl is at the bottom, the elongating hypocotyl must during germination turn from an upward direction downward (positively geotropic) before the cotyledons are turned and pushed through the soil surface. Such gyrations often cause difficulties in emergence. Formation of soil crusts add to the difficulties of emergence. Delays in emergence is a factor in weak seedlings.

Nitrogen Fixation

The bacteria of the *Rhizobium* genus can fix nitrogen gas of the atmosphere when living symbiotically in the root nodules of legumes.

The enzyme in the bacteria responsible for nitrogen fixation is nitrogenase. Minor elements, Fe and Mo, are required for the enzyme to be active. These organisms not only supply sufficient nitrogen to the plant, but under certain conditions "excrete" soluble nitrogenous compounds to the soil. The fixed nitrogen in the nodules enriches the soil when the plants die.

The *Rhizobium* organisms are free-living and mobile in the soil and in this state are not able to fix nitrogen from the air. The bacteria enter the root hairs of the leguminous plant and then change their form before they can fix nitrogen.

There are certain host–organism relationships which must be met for this symbiotic relationship. Table 21.2 shows the *Rhizobium* species and plant group relationships. One bacterial strain (within a particular *Rhizobium* species) is able to infect the root system and produce nodules on any of a group of related legumes, but not usually on legumes of another group. There is no overlapping from one plant to another. The plant grouping is also called *cross-inoculation group.* With one cross-inoculation group, strains of one bacterial species vary in their power to form *effective*, *ineffective*, or *intermediate* association in their nitrogen-fixing ability. Effective and intermediate only are symbiotic, and the highly ineffective ones are parasitic. An example of this is *Rhizobium trifolii* strains on different kinds of clover (Table 21.3).

Because the proper *Rhizobium* species may not be in the soil at all, or in adequate amounts, it is recommended that the seeds be inoculated with the proper strain for the particular crop. Soil inoculation by

TABLE 21.2. RHIZOBIUM SPECIES AND PLANT GROUP RELATIONSHIPS

Rhizobium spp.	Plant group	Plant genus
a. *R. meliloti*	Alfalfa (lucerne)	*Medicago*
	Sweet clover	*Melilotus*
		Trigonella
b. *R. trifolii*	Clover (red, white, crimson, and related spp.)	*Trifolium*
c. *R. leguminosarum*	Peas	*Pisum*
	Vetches	*Lathyrus*
	Lentil	*Lens*
	Fava beans	*Vicia*
d. *R. phaseoli*	Bean	*Phaseolus*
e. *R. lupini*	Lupines	*Lupinus*
f. *R. japonica*	Soybean	*Glycine*
(i) strain	Cowpea	*Vigna*
(ii) strain	Pigeon pea	*Cajanus*
	Jack and sword beans	*Canavalia*
	Guar beans	*Cyamopsis*
	Hyacinth	*Dolichos*
	Chickpea	*Cicer*

Source: Whyte *et al.* (1953).

TABLE 21.3. EFFECT OF *Rhizobium trifolii* STRAINS ON NITROGEN FIXATION OF CLOVER

	mg of nitrogen-fixed per plant		
	Strain of *R. trifolii*		
Legume	RT-1	RT-8	RT-13
Red clover	6.43	0.35	0.24
White clover	7.45	1.31	0.88
Subterranean clover	0.99	17.31	19.13

Source Whyte *et al.* (1953).

drilling peat granules mixed with rhizobial bacteria was introduced recently in order to increase nodulation with the proper organism.

Culture

Climate Requirement

Each legume species has strains or varieties adapted to a particular range of conditions.

Temperature

Peas and fava beans are cool season crops having mean optimum growing temperature of 13°–18°C (55°–65°F). Blossoms and pods are more susceptible to frost than leaves and stems. High temperature (27°C/80°F) or higher, especially at harvest, can lower the quality of peas.

Growing degree-days (Chapter 6) have been successfully used to predict harvest dates of peas for the canning and freezing industries. Early varieties require 1200–1300 degree-days (about 60 days), while late varieties require 1400–1600 degree-days (about 75 days) to maturity. Snap beans require an average of 27,000 growing degree hours (50°F base) from planting to optimum maturity; soil moisture stress increases time to maturity.

Beans and most legumes used for food are warm season crops intolerant of frost. In the temperature regions legumes can be grown after all danger of frost has passed and soil is warm enough, 18°C (65°F), for seed germination. Seeds germinate slowly at 16°C (60°F), and if the soil is too damp, they are susceptible to soil-borne diseases. For good healthy seedlings, the soil temperature should be at least 18°C (65°F). High temperature and high humidity are conducive to infection with pathogens.

Photoperiod Requirements

Most of the legumes are short-day or day-neutral for flowering. The early flowering cultivars of peas are generally day-neutral and not influenced by temperatures. However, the late flowering varieties are quantitatively long-day plants; they flower quicker in long days than under short days; high temperatures delay flowering.

Many legumes, which require short days for flowering when planted under long-day conditions of the summer in the temperature zone, will not flower until late summer or early fall; winged bean *(Psophocarpus tetragonolobus)* is such a leguminous vegetable. The scarlet runner, *Phaseolus coccineus*, is a long-day plant.

Soils

Most legumes require light, airy soils, with good drainage and a pH of 5.5–6.7. For optimal nitrogen fixation, good soil aeration is required as oxygen is necessary for the reaction. Adequate soil moisture from rains or irrigation is needed for good growth.

Harvest and Storage

Harvest

Peas and lima bean are harvested for the fresh market, freezing, or canning when the seeds are still immature, succulent, and before the seeds are too starchy. The optimum harvest period for peas and lima beans is a few days when the weather is cool and only a day or two in hot weather.

With edible pod peas, string or snap beans, and other legumes, the pods are harvested at the immature stage before the seeds are starchy and the pods are succulent and "strings" are absent.

Growth of an annual legume plant is depicted in Fig. 21.1. Seed/pod harvest is made at 70–75% of the total growth.

Storage

Peas should be stored at 0°C (32°F) as quickly as possible after harvest to prevent loss of sugars. High temperatures result in poor quality. Shelled peas and lima beans suffer loss of quality from bruising and loss of moisture much faster than the immature seeds in the pod.

Beans and other warm season leguminous crops need to be stored at high humidity (not less than 85% humidity). The best storage temperature is about 5°C (41°F). Lower or higher temperatures decrease the storage life. Shelled beans may require storage colder than 5°C.

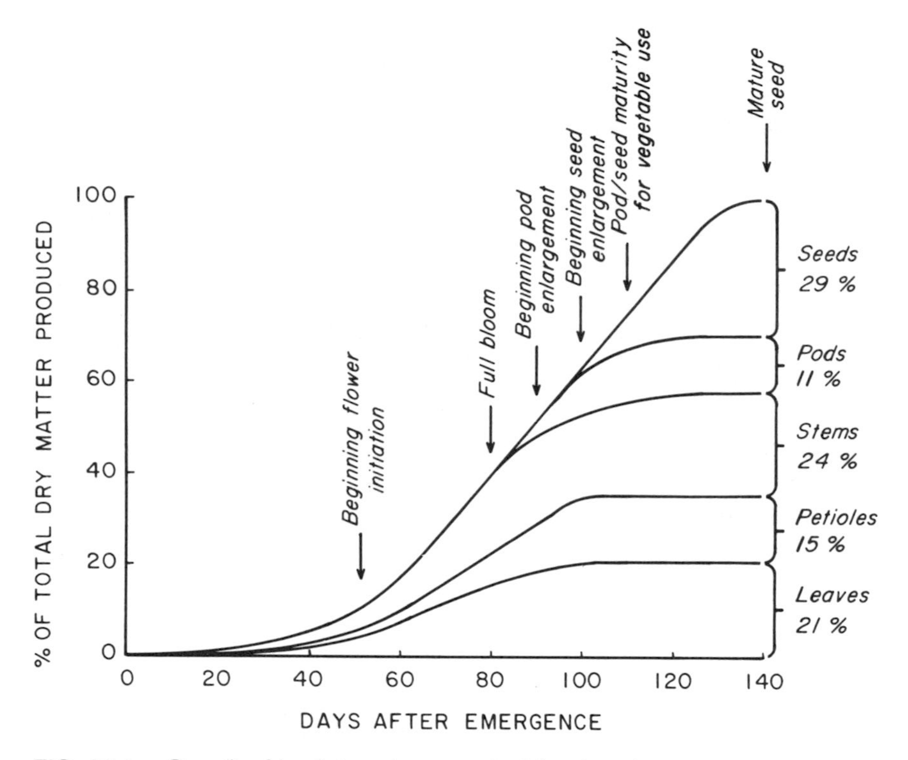

FIG. 21.1. Growth of bush type legume plant (soybean).
Data from Hanway and Thompson (1967).

Nutritive Value

Fresh mature pods and immature seeds are high in vitamins A and C. Although in the dry mature legume seeds the vitamin C content is nil, the sprouted seeds contain large amounts of this vitamin. Soy beans and lima beans are high in phosphorus. The protein content compares favorably with other high-protein foods. Fresh legume vegetables and pulses are the main protein source for the majority in India and Pakistan. The proximate composition of some vegetable legumes are given in Table 21.4 and the essential amino acids in Table A.2. Sprouted legume seeds are rich in vitamins (Table 21.5).

Toxins in Legumes

There are a number of poisonous leguminous plants, over a hundred in the United States alone. All parts of some plants are more or less

TABLE 21.4. PROXIMATE COMPOSITION OF SOME LEGUMES USED AS VEGETABLES IN 100 g EDIBLE PORTION

		Macroconstituents						Vitamins					Minerals (mg)					
			g						mg									
Crop	% refuse	Energy (cal)	Water	Protein	Fat	Total sugar	Other CHO	A(IU)	B_1	B_2	Niacin	C	Ca	Fe	Mg	P	K	Na
Beans																		
Blackeye pea (cowpea), pod	6	40	85	3.3	0.3	3.0	3.3	634	0.37	0.06	1.3	38	32	1.7	54	78	220	5
Fava, immature	70	53	81	5.6	0.6	2.8	3.8	350	0.17	0.11	1.5	33	22	1.9	38	95	250	50
Lima, baby	60	90	69	7.1	1.4	3.1	9.2	390	0.27	0.13	1.6	31	62	3.3	—	175	—	—
Lima, Fordhook	60	80	71	6.3	1.8	2.5	7.5	250	0.29	0.11	1.4	27	28	2.5	30	145	260	5
Lima, Dry			8	18.1	4.3	4.4	22.5	Trace	0.56	0.14	1.5	0	139	7.7	—	454	—	—
Snap, bush, pod	6	34	86	2.7	0.2	2.1	3.6	540	0.09	0.08	0.9	21	35	1.2	51	78	330	9
Snap, pole, pod	5	21	91	1.6	0.1	2.3	1.2	450	0.21	0.07	0.6	16	50	0.8	37	41	200	4
Soybean	40	106	73	9.0	5.0	2.8	4.1	640	0.57	0.14	1.6	33	66	2.5	—	178	—	—
Yard-long, pod	3	30	89	2.8	0.4	3.1	0.7	1,400	0.13	0.11	1.0	32	50	1.0	51	59	210	4
Bean sprouts (Mung bean)	0	25	92	2.7	0.1	2.1	1.4	25	0.11	0.03	0.6	12	20	0.6	16	35	130	2
Peas																		
Edible (podded)	5	35	88	2.8	0.2	4.0	1.8	580	0.15	0.08	0.6	60	43	0.9	22	53	170	6
Garden(green)	70	68	79	5.9	0.3	5.6	5.4	1,000	0.30	0.08	1.5	40	35	1.6	31	110	260	10
Hyacinth bean, pods[a]	20	47	87	3.0	0.5	7.9		460	0.08	0.09	1.1	13	50	1.0	—	47	285	2
Winged bean, pods[a]	4	29	91	2.7	0.3	5.6		545	0.24	0.09	1.2	15	64	0.1	—	36	205	3
Pigeon pea, pods[a]	44	143	62	8.4	0.7	26.5		285	0.44	0.10	1.5	29	78	1.8	—	198	622	5
Bean, leaves[b] (*P. vulgaris*)		36	87	3.6	0.4	6.6		3,200	0.18	0.06	1.3	110	274	9.2	—	75	—	—
Cowpea, leaves[b] (*V. unguiculata*)		44	85	4.7	0.3	8.3		8,000	0.20	0.37	2.1	56	256	5.7	—	63	—	—

Source: Howard *et al.* (1962).
[a] Intengan *et al.* (1964).
[b] FAO-USPHS. 1968.

TABLE 21.5. NUTRIENTS IN SEEDS AND SPROUTS OF SOME LEGUMES

Sample	Water (%)	Food energy (cal)	Protein (g)	Fat (g)	Fiber (g)	Ash (g)	Calcium (mg)	Iron (mg)	Zinc (mg)	Thiamine (mg)	Riboflavin (mg)	Niacin (mg)	Ascorbic acid (mg)
Alfalfa													
Seed	7.4	389	35.1	12.6	7.9	3.1	136	12.9	6.9	1.08	0.58	1.8	26
Sprouts													
Raw	88.3	41	5.1	0.6	1.7	0.4	28.0	1.4	1.0	0.14	0.21	1.6	16
Cooked	87.5		5.1	—	1.7	0.4	28.3	1.4	1.0	0.12	0.20	0.8	11
Lentils													
Seed	9.6	340	26.1	1.6	4.6	2.6	33.0	12.8	4.6	0.72	0.29	3.2	7
Sprouts													
Raw	72.7	104	8.4	0.3	1.1	0.8	12.0	3.0	1.5	0.21	0.09	1.1	24
Cooked	68.7		8.8	—	1.1	0.8	13.7	3.1	1.6	0.22	0.09	1.2	24
Mung beans													
Seed	10.1	334	22.9	1.4	4.9	3.4	83	11.6	3.8	0.70	0.47	1.8	—
Sprouts													
Raw	85.9	53	4.3	0.2	0.6	0.6	13.0	1.9	0.9	0.14	0.18	1.1	20
Cooked	84.3		4.3	—	0.7	0.6	13.1	1.9	0.9	0.14	0.18	1.2	16
Soybeans													
Seed	8.4	428	38.2	20.1	5.1	4.6	220	1.5	6.2	1.19	0.23	3.0	—
Sprouts													
Raw	73.2	105	12.0	2.6	2.3	3.0	75.0	0.4	1.6	0.32	0.16	1.1	12
Cooked	67.2		13.1	—	2.5	3.2	81.7	0.4	2.1	0.42	0.19	1.1	12

Source: Kylen and McReady (1975). Reprinted from J. Food Sci. *40*; 1008–1009. Copyright by Inst. of Food Technologists.

toxic. In others, only the seeds, leaves, roots, or the cortex of the stem may contain a toxic substance.

The roots of plants belonging to the *Derris* genus contain rotenone, a powerful insecticide. The roots are used by natives of some regions to poison fish, which are used for food. Roots of the yam bean *(Pachyrrhizus erosus)* also contain rotenone.

Some lupins *(Lupinus termis)* contain the toxic alkaloid lupinine, while others (*L. albus*, *L. angustifolius*, and *L. luteus*) are practically free of this substance.

Colored, but not white, coated seeds of the hyacinth bean *(Lablab niger)* contain cyanogenic glucoside. Consumption of the large amounts of seeds of the *Lathyrus* genus can cause paralysis in man.

Wild lima beans *(Phaseolus lunatus)* contain much cyanogenic glucoside, whereas domesticated cultivars contain nontoxic amounts.

Certain leguminous plants, when grown in soil high in selenium or molybdenum, can absorb large quantities of these elements. Excessive amounts of these elements are toxic to both man and animal.

Economics

Legumes for human consumption constitute about 5% of the cultivated crops. Table 21.6 gives the world production of vegetable legumes. Most of the vegetable legumes are used fresh. Only peas, limas, and snap beans are canned and frozen to any large extent.

The United States is the world leader in soybean production with 50.1 million MT in 1978, or about 63% of the world production. The other soy bean producing areas are China with 17% and Brazil with 12% of the production.

Use of Legumes

Soybean is perhaps the most important of the legumes used for food. Although this crop is endemic to northern China, soybeans are the most important bean crop in the Middle West of the United States. It is used as a protein supplement for feeding of livestock and used in preparation of many manufactured foods. It is an important source of oils.

In the Orient, soybeans are often eaten as shelled beans, immature and mature. The seeds are sprouted as "bean sprouts," used to make a product of precipitated proteins called tofu and fermented into specialty foods.

TABLE 21.6. WORLD PRODUCTION OF GREEN PEAS AND GREEN BEANS

	Green peas			Green beans	
	Area (10^3 ha)	Production (10^3 MT)		Area (10^3 ha)	Production (10^3 MT)
World	802	4,699		391	2,527
Continent					
Africa	20	102		39	291
North and Central America	174	1,384		51	248
South America	89	144		31	116
Asia	146	604		102	636
Europe	288	2,108		159	1,189
Oceania	24	158		9	47
Eastern Europe and U.S.S.R.	139	596		36	141
Leading countries					
1. United States	145	1,282	1. China	35	345
2. France	56	528	2. Italy	40	322
3. U.K.	53	511	3. Egypt	25	243
4. Italy	41	266	4. Spain	23	203
5. Hungary	38	250	5. United States	32	164
6. India	86	250	6. U.K.	12	113
7. China	41	233	7. Japan	11	95
8. U.S.S.R.	62	200	8. Chile	14	81
9. Belgium–Luxembourg	10	129	9. France	11	80
10. Australia	14	121	10. Greece	10	72

Source: FAO (1979).

PEAS: *Pisum sativum*

Peas have their origin somewhere in the eastern Mediterranean region and Near East: there are wild species of *Pisum* which are closely related to the cultivated species. Carbonized seeds discovered in Switzerland have dated peas back to about 7000 BC. The crop was grown by the Greeks and Romans long before Christ. During the course of cultivation, there came about a gradual separation of types: those grown for vegetable use and those grown for seed and fodder; the edible podded types evolved in more recent times.

Botany

Peas are viny annual herbs, with leaves that are alternate, and the tip of the compound leaf is a tendril (a modified leaflet). Peas are indeterminate (climbing) type or determinate (bush or dwarf) types. The flowers are borne at the axils of the leaves and are usually self-pollinated. Early cultivars bloom after 9 to 10 leaves and the late ones bloom after 15 to 16 leaves have formed. Pods vary in size according to the number of seeds which ranges from 2 to 10.

Seeds are either smooth or wrinkled; the smooth types are more adapted to cool weather conditions than the wrinkled types. Smooth seed coated types are starchy, whereas the wrinkled seed coated types are sweeter. The seed coat may be dark green pigmented or light green to colorless; the former type is used for freezing and the latter is used for canning.

The edible podded cultivars (*Pisum sativum* var. *macrocarpon*, and *P. sativum* var. *saccharatum*), grown for the immature pods, are becoming very popular (Fig. 21.2). Unlike the regular peas, of which the pods swell first, with this group the developing seed distends the pod wall and form a distinct bulge at each seed location. In the Orient the tender shoots of peas are cut and used as greens.

Culture

Peas, a cool season crop, can be grown in a wide range of soil types, from light sandy loams to heavy clays, but in any soil, there must be good drainage as peas do not tolerate soggy or water-soaked conditions. The soil pH optimum is 5.5–6.5.

In the temperate climates the crop is usually planted in the spring where winters are severe and in the late fall and early winter where

FIG. 21.2. Edible podded pea *(Pisum sativum)*. Four pods on the left are frost damaged.

there are little or no frosts. In the tropics and subtropics, peas are planted at high elevations where the temperatures remain cool.

Seeds are planted 2–5 cm (1–2 in.) apart in rows 38 cm (15 in.) in width or in double rows on beds 75 cm (30 in.) in width. They are sown 2–4 cm (1–1½ in.) deep; planting seeds deeper than 5 cm (2 in.) is not recommended. The seeding rate is from 70–110 kg/ha (60–100 lb/acre). Pea seeds germinate over a wide range of temperatures. Table 21.7 shows that germination rate increases with increasing temperature, but at temperatures greater than 18°C (64°F) the percentage of seeds germinating decreases. If the field has not been used for cropping of peas in recent years, the seeds should be inoculated with the proper *Rhizobium* species (Vetch group IV) immediately before planting. The indeterminate or climbing types require staking or trellising; lodging reduces yield.

TABLE 21.7. GERMINATION OF PEA SEEDS AT DIFFERENT SOIL TEMPERATURES

	Soil temperature (°C)						
	5	10	15	20	25	30	35
Percent normal seedings	89	94	93	93	94	89	0
Day to emergence	36	14	9	7	6	6	—

Source: Harrington and Minges (1954).

Although the pea nodules fix N when the plants are large, some N (about 90 kg/ha (80 lb/acre) and a slightly larger amount of P is recommended as preplant application. For soils low in K, a balance fertilizer mix such as 4–12–4 can be used.

Peas grow best at mean temperatures of 13°–18°C (55°–65°F). The crop, depending on the cultivar, matures for vegetable use in 57–75 days after planting. Heat units (see Chapter 6) are extensively used for predicting harvest dates for the freezing and canning industries. Cultivars can vary from 660°C-days (1200°F-days) to 940°C-days (1700°F-days). These values vary slightly from year to year at the same location. For the same cultivar, the values can vary quite a bit between locations, but the relative heat units between cultivars for comparison do not change appreciably. Before bloom, the crop can withstand some frost, but the flowers and pods are susceptible to freezing conditions.

Harvest and Storage

Harvest

For vegetable use the pods should be well filled and contain no hard and starchy seeds. Under cool temperatures, the peas do not mature rapidly and remain at the proper condition for several days. However, under warm temperatures, the seeds accumulate starch rapidly, and the optimum quality lasts for only a day or two. Ordinarily, two harvests are made when picked by hand. With once over machine harvest, there will be very immature as well as starchy seeds harvested; however, these can be removed by brine flotations at two specific gravities.

Storage

Harvested pods or shelled peas should be cooled as quickly as possible to 0°C (32°F) to prevent conversion of sugars into starch and to reduce respiration rate. Shelled peas should be washed with ice water to remove field heat and to prevent decay organisms from growing. Peas in the pod may be stored at 0°C (32°F) for 3 weeks and at 5°C (41°F) for 2 weeks. Storage in controlled atmosphere (CA) of 5–7% CO_2 at 0°C (32°F) give better quality after 3 weeks than storage in air at the same temperature.

Pests and Diseases

Lygus bugs, pea aphid, pea leaf miner, and pea weevil damage the plant and the pod. Wire worms and centipedes attack germinating seeds and nematodes can infest the roots. (Nodules should not be confused with nematode galls on the roots.)

Ascochyta blight is seed-borne and is characterized by purplish to black, streaky, and irregularly shaped lesions on the stem. Septoria blight, a fungus, causes the leaves to appear yellowish and shrunken. Bacterial blight produces water-soaked lesions on all parts of the plant, which may appear creamy and slimy under highly humid conditions. Powdery and downy mildew cause leaves to turn yellow under cool moisture conditions. Warm dry weather checks mildew growth.

Pea mosaic, a viral disease, induces severe stunting and mottling of leaves with streaks of yellowing on the stems. Early infection causes the plant to die. Crop rotation is recommended for control.

SNAP OR STRING BEAN, FRENCH BEAN, FRIJOLES: *Phaseolus vulgaris*

Snap or the common bean is the most widely used of about the 150 species belonging to the genus *Phaseolus*. Its origin is thought to be in Central America (southern Mexico, Guatemala, and Honduras). Evidences of the common bean were found at two widely separated places, at Callejon de Huaylas in Peru for the large seed races, and in the Tehuacan Valley in Mexico for the small seed races, both dated over 7000 years ago. Since post-Columbian times, the bean has spread all over the world. This crop is associated with the corn and squash culture of the tropical Americas.

Botany

Phaseolus vulgaris is an annual twining vine with alternate trifoliate leaves, grown in the tropics, subtropics, and during the warm months in the temperature regions of the world. Besides the climbing forms, the plant may be dwarf or bushy; there are intermediate forms. For vegetable use, the crop is grown for the fleshy pods and immature seeds and the tender shoots in some countries as pot herbs. Large areas are devoted to the production of dry mature seeds.

The indeterminate cultivars grow 2–3 m (6–10 ft) in height, and the determinate cultivars are 20–60 cm (8–24 in.) tall with the stems after four to eight leaves terminating in an inflorescence. The flowers are self-fertilized, fertilization occurring when they open. Pods are long and narrow varying in length from 8 to 20 cm (3 to 8 in.) and from 1 to 1½ cm (⅜–⅝ in.) wide. Depending on the cultivar and conditions at bloom, the number of seeds in a pod varies from 4 to 12. Seed size can vary from 0.7 to 1.5 cm (⅜ to ⅝ in.) in length, weigh from 0.2 to 0.6 g each, and the form may be globular to kidney-shaped. Seed coat colors can be white, yellow, greenish, pink, red, purple, brown, or black and the color may be solid, striped, or mottled. It is interesting that different countries of Latin America prefer a certain seed coat color: black seed coat, Brazil, Mexico, El Salvador, and Venezuela; red, Columbia and Honduras; yellow, Peru; and white, Chile.

The string formation is governed by a recessive gene. Cultivars are often classified into three types according to string formation:

1. No string formation, not influenced by temperature
2. Incomplete string formation, strings stronger when grown at high temperatures
3. Completely developed strings, string formation not influenced by temperature

Culture

A warm season crop, bean seeds should not be planted in cold soils.

The optimum soil temperature for germination is 30°C (86°F). At 10°C (50°F), germination is nil, but at 15°C (59°F) germination is over 90%; however, over 16 days are necessary for the seedlings to emerge. At 20°C (68°F), germination percentage is very high and it takes 11 days for emergence. Only 6 days are necessary for emergence at 30°C (86°F). At temperatures greater than 35°C (95°F), the percentage germination is very much reduced. Seedlings can be injured when the air temperature falls to 10°C (50°F).

Snap bean seeds are planted 5–10 cm (2–4 in.) apart in rows 60–90 cm (24–36 in.) apart and 2–4 cm (¾–1½ in.) in depth. When planted two rows per bed, the bed width should be 100–150 cm (3–5 ft) depending on bush or pole type cultivars. The seeding rate varies from 22 to 55 kg/ha (20 to 50 lb/acre). A fungicide treatment of seeds is recommended before planting. Bean seeds should be handled gently as cracked seed coat and cotyledons decrease germination and seedling vigor.

Beans can be grown on many soil types; a light airy soil with good drainage is best for optimum nitrogen fixation by the nodules of the roots. Inoculation of the seed with the proper nitrogen fixing bacteria is recommended if the field had not been used for the crop in recent years. A preplant fertilization of 65–110 kg N/ha (60–100 lb/acre) may be made. The fertilizer is placed 7.5 cm (3 in.) below and 7.5 cm (3 in.) to the side of the seed row. Also, about 100 kg P/ha (90 lb/acre) and 70 kg K/ha (60 lb/acre) can be applied at the same time. Well-decomposed manure can be used also.

In the temperate climates, plantings of snap beans are made in the spring after all danger of frost has past. The optimum mean temperature for growth is 16–30°C (60°–86°F). In the hot humid tropics, snap bean is difficult to grow because of problems with disease; it grows well in areas of medium rainfall.

Pole beans require support; the vines grow counterclockwise around the support. In the tropics pole beans are grown after corn or okra as the stalk can be used to support the vines.

Snap beans are very sensitive to soil moisture stresses. Yields are decreased substantially when moisture stress occurs during flowering and subsequent pod development period. In loamy sand soils, yields can be significantly reduced when the soil moisture tension goes down to 0.50 bar. High temperatures at flowering can cause flowers to abort; also, hot dry winds are very destructive to the delicate flowers. Cultivars such as 'California Red', which are adapted for high temperatures, have a high percentage of pod set when the maximum day temperatures exceed 38°C (100°F) 1 day before to 1 day after anthesis.

Harvest and Storage

Harvest

Depending on the cultivar and temperature, 45–80 days are required from planting to pod harvest. Generally, the bush (dwarf) types require less time than the climbing (pole) types. Pods can be harvested as they reach optimum stage. For best quality the pods should be about half-grown to about three-fourths of maximum length (before the pods reach full size and while the seeds are still succulent and not starchy).

Storage

Snap beans can be held to a fair condition at 5°C (41°F) for about 25 days; at temperatures below this temperature, 0°–2.5°C (32°–37°F), chilling injury can occur after about 10–12 days. Above 5°C (41°F), the storage life decreases with increase in temperature; at 25°C (77°F), the pods are at a fair condition after only 4–5 days.

Pests and Diseases

Aphid, diabrotica, red spider, thrip, and white fly all attack the bean plant. Appropriate insecticide for each insect should be used. Root knot nematodes attack the roots.

Anthracnose cause circular sunken spots with black edges and pinkish centers on pods. The lesions on the stem and leaves are not noticeable. Disease-free seed is recommended for control.

Bacterial blight spread by water is due to a fungus infection, which causes lesions of dead sunken red spots on the stems and pods. In severe cases the disease may defoliate the plant.

Curley top and mosaic are viral diseases. The former cause plants to be stunted and the leaves deformed and curled; the latter, transmitted by aphids, forms necrotic mosaic patterns on the leaves. Powdery mildew forms white areas on the foliage; it is controlled by dusting with sulfur. Rust is a problem in cool regions. Rust-resistant cultivars should be used.

LIMA BEANS, BUTTER BEANS, SIEVA BEAN: *Phaseolus lunatus (P. limensis)*

There are two types of lima beans, the large seeded type, which can be traced to about 5000–6000 BC in Huaca Prieta in Peru and the small seeded type, or the sieva bean, which dates back to 300–500 BC in Guatemala and Mexico. Wild types can still be found from Central

America to the Andes of Peru and Argentina. Spanish and Portuguese explorers distributed lima beans to other parts of the world.

Botany

The so-called climbing baby or sieva lima is often classified as *P. lunatus* and the bush type as *P. lunatus* var. *lunonnus*. The large seeded type or the "potato lima" is described as *P. limensis* var. *limenanus*.

The lima bean growth habit is similar to the common bean *(Phaseolus vulgaris)*. Large seeded types are considered a perennial but are grown as an annual; small seeded types are annuals. The climbing types range from 2 to 4 m (6 to 12 ft) in height, while the bush types grow 30–90 cm (12–36 in.) tall. Flowers are self-fertile, but cross-pollinations occur frequently in the field. The fruit is a curved oblong pod containing from two to six seeds. Commercial cultivars bear from three to four seeds per pod. Seeds are smooth, flat to round, and 1–2½ cm (⅜–1 in.) long. The seed coat varies from white, cream, red, green, purple, brown, to black. The U.S. cultivars have seed coats that are white, creamy, buff, to light green.

The wild types contain considerable amounts of cyanogenic glucosides (see Chapter 5) and must be cooked and leached before being used as food. The U.S. cultivars contain nontoxic amounts, are safe, and need no treatment for use as food.

Culture

Lima beans require slightly warmer weather for growth than snap beans *(Phaseolus vulgaris)*. Mean monthly temperatures of 15°–24°C (59°–75°F) and a frost-free period is necessary to grow the crops. The large seeded types are more sensitive to the temperature than the small seeded type, which can tolerate hotter and drier conditions. At flowering when the temperatures are too high, above 27°C (80°F), and the humidity too low (less than 60–65% RH), the large seed types fail to set fruit. There are only a few places in the world where the climate is optimal for growing of the large seeded lima beans. The small seeded types do not require such restrictive climates. The growth regulator NAA, at 5 ppm, has been reported to help set lima bean flowers.

Lima bean seeds germinate at 15°–30°C (60°–85°F) with an optimum at 27°C (80°F) soil temperatures. Germination is nil at 10°C (50°F) or lower and above 35°C (95°F). At 15° and 20°C (59° and 68°F) it takes 30 and 18 days, respectively, for emergence; it takes about 1 week at 25° and 30°C (77° and 86°F). Spacing is 10–15 cm (4–6 in.) and in rows 60–75 cm (24–30 in.) apart.

Other than temperature restrictions, the culture of lima beans is similar to snap beans with respect to soil, fertilizer, and general culture.

Harvest and Storage

Harvest

Determination of when to harvest lima beans is difficult. As the bean matures, the seed turns pale green, then white. The beans should be well filled, about 20–30% solids for fresh use. From seeding to harvest, depending on cultivar and temperature, it can vary from 65 to 90 days and sometimes in early planted fields up to 110 days because of cool weather. All pods do not mature to the proper stage at the same time because flowers do not bloom at the same time and conditions for fruit set often times is not right. For vegetable use limas are harvested when the seeds contain about 65–70% moisture. In a once-over harvest for freezing, there should not be more than 10% white beans. As with peas for freezing, the different maturities of lima beans can be separated by floating in different specific gravities of brine.

During seed maturation, the pod expands and bulges as the seed enlarges and the pod color changes from green to yellow to brownish grey (dry); the seed enlarges, changing from dark green, to green, to light green, and when mature is white.

Storage

Lima pods store for 2–3 weeks at 5°C (41°F) and 90% RH. Shelled beans store at or slightly above 0°C (32°F) and high humidity for about 10 days in good condition; at 5°C (41°F), it will hold only about 4 days.

Pests and Diseases

Wire worms, seed corn maggot, nematodes, cut worms, lygus bugs, leaf miner, black bean aphid, spider mite, and lima bean pod borer are the principal pests attacking the plant.

Pythium, *Fusarium*, and *Rhizoctonia* all attack the germinating seed and seedlings. Because these organisms attack seeds, the seeds should be treated with fungicide before planting.

In areas of high rainfall, bacterial spot, pod blight, and *anthracnose* attack the plant. Lima bean scab *(Elsinoe phaseoli)* occurs in Middle America.

Downy mildew and powdery mildew occur under high humidity and cool nights. Lima bean mosaic transmitted by green peach and melon aphids causes a mottling of leaves. Curly top transmitted by the sugar beet leafhopper causes young leaves to dwarf, curl downward, and twist.

SOYBEAN, SOYA: *Glycine max*

Soybean is perhaps the most important leguminous crop in the world as the chief source of plant proteins; it also ranks high as an important oil crop. Although the main production is for dry seeds, soybean is included among the vegetable legumes because of its use as a vegetable at the immature bean stage.

Glycine soja, the wild soybean, can be found in East Asia, but *Glycine max* is a cultigen from the same region. The use of soybean dates back to antiquity in Eastern China; the crop spread to parts of Southeast Asia and Japan. Presently, the United States, China, and Brazil lead in soybean production.

Although soybean is a warm season crop, it can be grown in far northern latitudes during the summer. Soybeans are short-day plants for flowering; low temperatures delay flowering. In the United States, soybeans are classified into nine maturity groups (day lengths or photoperiods), each for a particular range in latitudes. Group 0 for the long days of Maine, North Dakota, northern Minnesota, northern Wisconsin, Oregon, and Washington. Group I for southern Minnesota, South Dakota, Iowa, etc.; Group VIII is adapted for the latitudes of Louisiana and Florida. Figure 21.3 shows the adapted regions in the United States of the different groups. Cultivars of soybeans which flower under long-day conditions will also flower under short-day conditions, but short-day cultivars will not ordinarily flower under long days. There are the so-called "day-neutral" cultivars; the seed maturation of many of these types is delayed under long-day conditions. Depending on cultivars, soybeans can be grown as far north as 58°N latitude, where July mean temperature ranges from 16° to 18°C (61° to 64°F) and require from 115 to 165 days (average of 133 days) from seeding to maturity.

Figure 21.4 shows the temperature requirement of soybean at different stages of growth. Generally, colder temperatures than optimum, especially at floral initiation, anthesis, and seed growth decrease yields.

Soybeans are self-fertile and are normally self-pollinated unless cross-pollinated by insects.

In the tropics soybean is often intercropped with corn. As a vegetable,

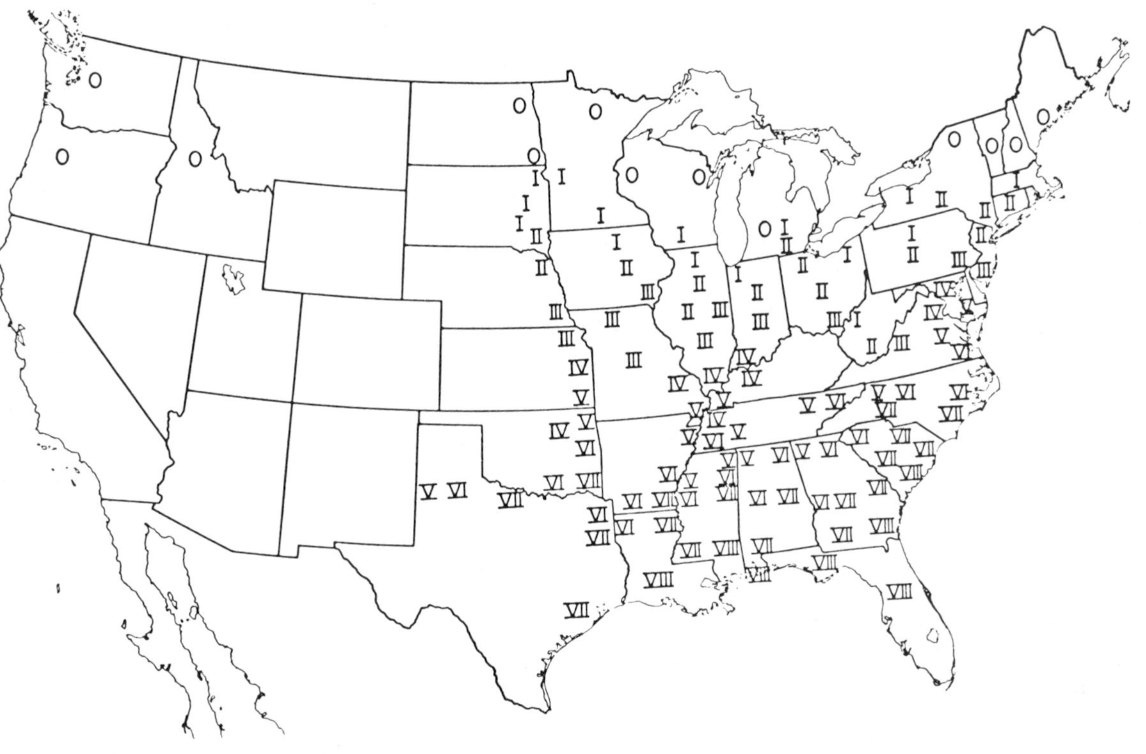

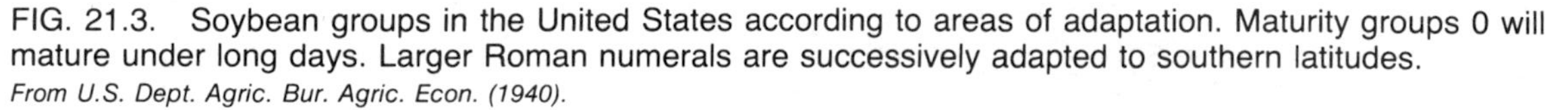

FIG. 21.3. Soybean groups in the United States according to areas of adaptation. Maturity groups 0 will mature under long days. Larger Roman numerals are successively adapted to southern latitudes.
From U.S. Dept. Agric. Bur. Agric. Econ. (1940).

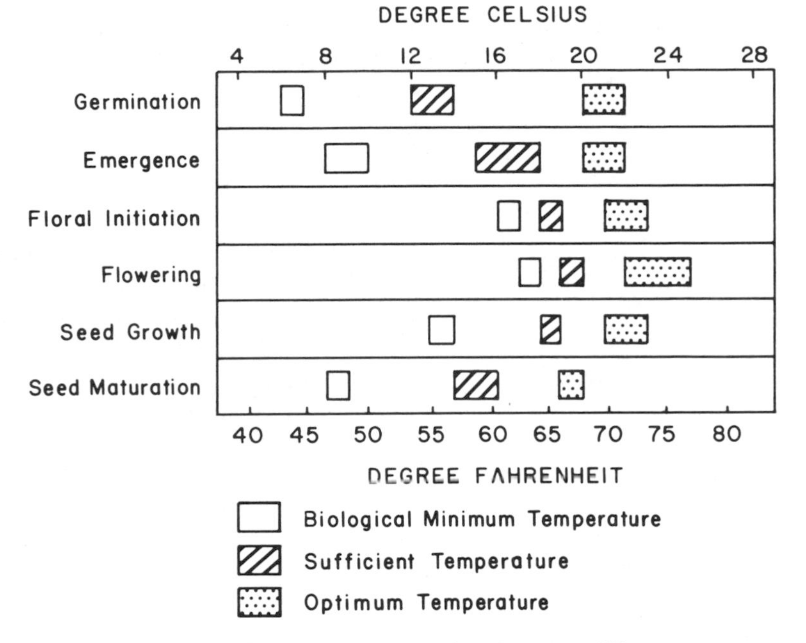

FIG. 21.4. Temperature requirements of soybeans at different stages of growth.
Data from Holmberg (1973).

the crop is harvested while the beans are still succulent and have not begun to accumulate much starch and oil.

PEANUT, GROUND NUT, GOOBER, MONKEY NUT: *Arachis hypogaea*

The peanut is a native of South America, probably Bolivia or southwestern Brazil. Prior to Columbus's discovery of America, it had spread to various parts of the continent and into Central America.

The plant is a spreading annual growing from 25 to 60 cm (10 to 24 in.) high with pinnate leaves, each leaf having two pairs of broad oval leaflets. Germinating seeds are neither hypogeal nor epigeal; the two cotyledons are pushed to the soil surface by the elongating hypocotyl and remain at the surface. Further growth of the seedling above ground is due to the epicotyl elongation.

The flowers are borne in leaf axils, and after self-pollination, the stalk bearing the fertilized ovary called the "peg" becomes geotropic and grows downward until it penetrates the soil about 2–6 cm (1–2½ in.). It is reported that if the lower nodes of the stems are covered with loose soil or mulch, the flowers form at these nodes, self-fertilize, and fruits develop beneath the soil surface.

There are two distinct types: (1) The Virginia types, which have true runners, produce lateral branches readily on which reproductive structures form, have long growing seasons (120–160 days), usually two seeds per pod, and seeds have a long rest period ranging from a month to almost a year. (2) The Spanish and Valencia types, which are erect, have both main and lateral branches bearing reproductive structures, have various growth periods ranging from 90 to 150 days, and the seeds have no rest period. The Spanish types are mainly two seeded and the Valencia types are multiseeded, usually from three to four.

The crop is grown during the warm season in regions with at least 500 mm (20 in.) of rainfall during the growing season. A loose, well-drained soil, sandy loam types are preferred. It can be grown over a wide range of soil pH's from 4.5 to 7.5. Hot dry weather is needed during seed maturation. Depending on the cultivar and growing conditions, yields can vary from 800 to 3000 kg/ha (700 to 2700 lb/acre).

Tender shoots and leaves of the plant are used as a vegetable. The mature seeds contain 21–31% protein, 11–27% carbohydrates, and 41–52% oil. The nuts are consumed raw, boiled, steamed, or roasted; also, oil is extracted from the nuts. The thin seed coat (skin) is reported to be very rich in thiamine (vitamin B_1) and niacin.

PIGEON PEA, RED GRAM, CONGO PEA: *Cajanus cajan (C. Indicus)*

Pigeon peas, probably a native of Africa, were cultivated in Egypt before 2000 BC. There are many cultivars of pigeon peas in India where it still is a very important crop. The immature seeds and sometimes the green pods are used as vegetables. The main use is the mature dry seeds prepared in many ways as food.

Pigeon peas are short-lived woody perennials, but usually grown as an annual in some countries (India). There are botanical varieties: (1) *flavus*, called "tur," which mature early with green three-seeded pods, and (2) *bicolor*, called "arhar," which matures late with red or maroon pods containing four to five seeds. It is a short-day plant.

The crop can be grown under semiarid condition and in various soil types but not under waterlogged conditions.

The crop is usually propagated by seeds but stem cuttings can be made. Pods set in about 3–4 months after sowing, but 5–6 months are required before seeds mature. Late types require almost a year to mature.

SWORD BEAN: *Canavalia gladiata*; AND JACK BEAN: *Canavalia ensiformis*

Sword bean, a viney perennial, is found growing wild in tropical Asia and Africa. Jack bean, an annual bushy plant, is of New World origin, a native of Central America. Immature pods and seeds from both sword beans and jack beans are used as vegetables in the tropics and subtropics. Also, they are important as fodder and green manure crops.

Sword bean has pods in the range 15–40 cm (6–16 in.) long and 3½–5 cm (1¼–2 in.) wide and contain from five to ten dark red seeds. Jack bean pods are in the range of 10–35 cm (4–14 in.) in length and 2–2½ cm (¾–1 in.) in width. Depending on the cultivar, the pods may contain from three to 18 white seeds.

Mature seeds of both species are reported to be poisonous and must be prepared by boiling with several changes of water before consumption. Mature seeds contain 23–26% protein, 49–62% carbohydrates, and 2–3% oil. Commercial preparations of the enzyme, urease, is made from jack bean.

CHICK PEA, GARBANZO BEAN, INDIAN OR BENGAL GRAM: *Cicer arietinum*

Chick pea is reported to have originated in the Middle East and spread east to India and west to southern Europe. Whole dried seeds are used for food; also, green pods and succulent shoots are eaten as a vegetable.

Chick pea is a bushy, upright annual 25–50 cm (10–20 in.) in height and covered with glandular hairs. These hairs exude an acrid liquid composed mainly of malic acid and a small amount of oxalic acid; the exudate is collected in India and used as vinegar or as medicine. The mature pods are short and oblong, 2–3 cm (¾–1⅛ in.) long and 1–2 cm (⅜–¾ in.) in width containing one to two angular seeds.

The crop grows well in cool, semiarid regions and not in the hot, humid tropics with high rainfall. The crop is grown in the dry coastal areas and interior valleys of California. Mature seeds are canned in brine and used in salads. Seeds contain 23–25% protein, 53–62% carbohydrates, and 1–5% oil.

CLUSTER BEAN, GUAR: *Cyamposis tetragonobus (C. psoralioides)*

Cluster bean is indigenous to India. The immature tender pods are eaten in India, Pakistan, and Indonesia; also the plants are used as fodder and green manure crop. The seed is high in mannogalactans, which are very mucilaginous; because of this property, the seeds have industrial uses.

Cluster bean is an annual bushy plant with trifoliate leaves growing 1–3 m (3–10 ft) high. Pods grow 4–10 cm (1½–4 in.) long and contain five to 12 oval or quadrangular seeds. Seed coat color may be whitish, gray, or black.

The seeds are usually broadcasted 11–22 kg/ha (10–20 lb/acre) when the crop is grown for seeds and double these rates when planted for fodder. Cluster bean is suited for dry land culture as the crop is drought resistant; however, yields can be doubled if there is adequate moisture during growth.

Tender pods can be harvested in about 120 days from planting and an additional 30–45 days are necessary for mature dry seeds. In Brazil, 200–220 days are required from planting to harvest of mature seeds. Seeds contain about 30% protein, 46% carbohydrates, and 1½% oil.

HYACINTH BEAN, LABLAB BEAN, INDIAN AND EGYPTIAN BEAN: *Lablab niger (Lablab vulgaris, Dolichos lablab)*

Probably of Indian origin, hyacinth bean has been cultivated there since ancient times. The immature pods and tender seeds are a very popular vegetable in India and other regions in the tropics, particularly South and Central America, China, and Africa. Dry ripe seeds are an important food in India; the sprouted seeds are also used.

Hyacinth bean is a short-lived twining perennial grown particularly in areas of low rainfall and high temperatures. The crop can be grown from sea level to 2200 m (7000 ft) and on relatively poor soils but cannot tolerate excessively high soil moisture. The plant grows 50–100 cm (1½–3 ft) in height; the large starchy root is edible. There are reported to exist long-day and short-day types for flowering. The pods are flat, oblong, often curved, and tuberculate at the margins. Each pod contains four to six rounded seeds colored white, red, brown, black, or speckled, with a prominent long white hilum. Hyacinth bean is sown in the fall usually drilled 2–5 cm (1–2 in.) in depth at 15–50 cm (6–20 in.) spacing at a rate ranging from 20 to 80 kg/ha (18 to 70 lb/acre). When seeds are broadcasted the rate is about 90 kg/ha (80 lb/acre). Depending on the season and growing conditions, from 90 to 150 days are required for seeding to harvest.

Seeds with colored seed coat contain cyanogenic glucosides, while the white seed coat cultivars contain nontoxic amounts of this substance. Excellent production of both maize and hyacinth bean in a multiple cropping system was reported in Brazil. The seeds contain 13–20% protein, 50–60% carbohydrates, and 1–2% oil.

CHICKLING PEA, GRASS PEA: *Lathyrus sativus*

Chickling pea is indigenous to parts of southern Europe and the Middle East. In India the crop is grown for animal feed and the seeds for human food. Leaves of the plant are used as potherb. In times of famine, the poor people consume large quantities of the seeds; prolonged diet of these seeds cause a disease of the limbs called lathyrism (see Chapter 5).

Chickling pea is a hardy cool season crop, which grows on land too low in soil moisture for other crops; also, it will tolerate waterlogging and grows in generally poor soil conditions. It is a much branched, herbaceous annual with alternate pinnate leaves ending in tendrils. Seeds are broadcasted at 40–50 kg/ha (35–45 lb/acre) in the late fall. The pods, containing three to five seeds, are oblong and flattened 2½–4 cm (1–1½ in.) long. Angled seeds are colored white, brown, gray, or mottled. Seeds contain 25–28% protein, 43–58% carbohydrates, and ½–9% oil.

LENTIL: *Lens esculenta (Lens culinarus) (Ervum lens)*

Lentil is considered to be among the first crops domesticated. It was cultivated in the near East, 8000–10,000 years ago. The crop is mainly grown for seeds used for food. In India the immature pods are used as a vegetable. Lentil is divided into two types according to seed size: the large seeded cultivars (subspecies *macrospermae*) and the small seeded cultivars (subspecies *microspermae*).

A cool season crop, it is grown in northern India during the winter. It tolerates a wide range of soil conditions, even moderate alkaline soils in India and Pakistan. It can be grown up to 3400 m (11,000 ft) in elevation. Lentil cannot be grown in the hot humid tropics.

Lentil is a much branched, annual herb with pinnate leaves ending in a tendril growing to heights of 25–40 cm (10–16 in.). The oblong pods contain one to two seeds. In the late fall or early winter seeds are broadcast at the rate of 45–130 kg/ha (50–150 lb/acre) or drilled 2–5 cm (1–2 in.) deep at 20 cm (8 in.) spacing at the rate of 80 kg/ha (70 lb/acre). The crop matures from 150 to 180 days from planting. Mature seeds contain 24–28% protein, 59–67% carbohydrates, and 0.7–1.0% oil.

EGYPTIAN LUPINE: *Lupinus termis*

This crop is cultivated in Egypt and Sudan for the mature seeds. The crop has been used as a green manure crop in the Mediterranean region for centuries. It is an erect cool season annual 1–1½ m (3–4½ ft) in height, grown during the winter season. The crop is moderately frost and drought resistant and fairly resistant to high temperatures. Seeds are usually broadcasted at the rate of about 100 kg/ha (90 lb/acre) in October and November. The crop matures 6–7 months after plantings. Pods are 8–12 cm (3–5 in.) in length containing three to four seeds, which are oval and flat shaped similar to baby lima beans.

The seeds contain a bitter poisonous alkaloid, lupinine, which is destroyed by soaking the seeds in water. Daily changes of water over a period of several days are required to render the seeds nonpoisonous for consumption as snack food. Many cultivated species of the genus *Lupinus* are practically free of this alkaloid. Mature seeds contain about 14% protein, 5½% carbohydrates, and ½% oil.

MAT OR MOTH BEAN: *Phaseolus aconitifolius*

Mat bean is indigenous to India. It is an annual, warm season crop and quite drought resistant. In India the green pods are used as a vegetable and the mature seeds used cooked, whole or split. It is also used for fodder and green manure crop.

A short-day plant, the small yellow flowers are borne on axillary racemes. The small rounded pods are 2½–5 cm (1–2 in.) long and ½ cm (⅜ in.) wide and contain four to eight small seeds.

ADZUKI BEAN, AZUKI: *Phaseolus angularis*

Adzuki bean is probably a native of Japan and used mainly as dried beans cooked with rice or in confections. A short-day plant, it is intolerant of waterlogged soils and is reported to be rather drought resistant. Pods are cylindrical 6–12 cm (2½–5 in.) long and ½ cm (¼ in.) wide containing five to 12 dark red coated seeds. At the immature stage, the pods can be used as a vegetable. Mature seeds contain about 22% protein, 65% carbohydrates, and 0.3% oil.

MUNG BEAN, GREEN OR GOLDEN GRAM: *Phaseolus aureus (Vigna aureus)*

Mung bean is an ancient crop of India used mainly as the dry beans and the green pod as a vegetable. There are two cultivars: the green seed coat (green gram) and the yellow or golden seed coat (golden gram). The green gram is used for the production of mung bean sprouts. One gram of seeds will produce 6–8 g of sprouts.

Mung bean is an annual plant with trifoliate leaves. There are short-day and long-day cultivars. The mature pods are long and slender 5–10 cm (2–4 in.) long and 0.5 cm (¼ in.) wide, containing 10 to 15 small rounded seeds. Seeds contain 22% protein and less than ½% oil.

RICE BEAN, RED BEAN: *Phaseolus calcaratus*

Boiled rice bean is used instead of boiled rice in Southeast Asia. The immature pods and leaves are used as vegetables and the whole plant used as fodder or as green manure crop. Rice bean is reported to grow wild in the Himalayas to central China and into Malaysia.

A twiny annual short-day plant, rice bean grows 1–3 m (3–10 ft) in length. The long slender pods bear six to 12 seeds, which may be yellow, red, brown, black, or speckled.

SCARLET RUNNER BEAN: *Phaseolus coccineus (P. multiflorus)*

Scarlet runner bean is a perennial and a native of the highlands of Central America 1800 m (6000 ft). It is a long-day plant suited for growth during the summers at high elevations of the tropics. However, it is killed by frosts.

The plant is a twiny vine growing to lengths of 4 m (12 ft) or more. The flowers are scarlet, white, or variegated. The immature pods are used like snap beans. In Central America green seeds and mature dry seeds are used for food. The tuberous roots are starchy and are also used; however, they are reported to be poisonous. Mature seeds contain about 17% protein, 65% carbohydrates, and 2% oil.

BLACK GRAM, URID: *Phaseolus mungo (Vigna mungo)*

Black gram is not to be confused with mung bean or green gram *(Phaseolus aureus)*; black gram is named for the black seed coat. It is a very important crop in India. Beside the mature seeds, the green pods are used as vegetables. Black gram can be grown in areas of moderate rainfall, in heavy soils, and from sea level to 1800 m (6000 ft) elevation. The plant has hairy stems with a trailing habit growing 25–80 cm (10–30 in.) high. The mature pods are hairy 5–8 cm (2–3 in.) long and contain six to eight seeds. Seeds contain 19–24% protein, 5–59% carbohydrates, and 1–2% oil.

WINGED BEAN, GOA BEAN, FOUR-ANGLED BEAN: *Psophocarpus tetragonolobus (Tetragonolobus purpureus)*

The winged bean has been cultivated for centuries in Asia, from southern India and Sri Lanka through the Malayan peninsula to New Guinea and the Philippines; its origin is in southeastern Asia, perhaps New Guinea. In recent years, there has been a considerable amount of interest in this crop because it can be grown in the hot humid tropical and subtropical regions not suitable for soybean *(Glycine max)*. The entire plant can be used for food: the immature pods, the mature seeds, the tender shoots, flowers, leaves, and the tuberous roots. It is also used as fodder and green manure crop.

The winged bean is a twining perennial with trifoliate leaves usually grown as an annual (Fig. 21.5). It thrives in hot wet climates but cannot be grown in waterlogged soils. It grows well under irrigation in dry regions. Soils low in nitrogen can be used as the crop can obtain this element from the nitrogen-fixing bacteria in the root nodules. A short-day plant for flowering, winged bean cannot be successfully grown in the temperate climates as the pods cannot mature after flowering (12 hr days) before the onset of cold weather. Scientists are searching for day-neutral types.

Winged bean propagation is usually with seeds, but it can be started with stem cuttings under mist. If treated with IAA or NAA, there is an increase in rooting. Winged bean has a hard seed coat, which delays germination; scarfication of the seed coat increases the percentage germination.

FIG. 21.5. Winged bean *(Psophocarpus tetragonolobus)*.

In the tropics, seeds are sown at the start of the rainy season, but they can be planted in any season where rainfall or water is not a problem. Plants are spaced about 1 m (3 ft) apart and 1.5 m (4½ ft) between rows. Under short-day and warm climatic conditions, flowering can commence in about 50 days following emergence, and as long as 3–4 months at high elevations.

The inflorescence is an axillary raceme bearing two or more white to light blue or purple flowers. Fully mature pods, depending on the cultivar, vary from 5 to 35 cm (2 to 14 in.) in length and 1.2 to 4 cm (½ to 1½ in.) in width. The characteristic four-sided pods have wings extending the full length of the fruit. Pods can be harvested a few days before they reach maximum length. In the Philippines, the winter crop pods are ready for harvest 15–20 days after anthesis. At 25 days, the pods are still green but quite tough and cannot be used as a vegetable. Seeds are mature 60 days after anthesis.

FENUGREEK, FENUGREC, METHA (INDIA): *Trigonella foenum-graceum*

Fenugreek, a native of the Mediterranean region, is an annual herb that has been grown in the Middle East and India as food, fodder, and green manure crop for thousands of years. In India the leaves of young plants are used as greens and sun-dried for use when the crop is not available.

The plant has long petioled trifoliate leaves and grows to heights of 40–90 cm (16–36 in.). Flowers are whitish in color blooming 50–80 days from planting and develop into long slender pods 8–15 cm (3–6 in.) long containing 10 to 20 seeds. In Egypt seeds are broadcasted in October to November at the rate of 100 kg/ha (90 lb/acre); the crop matures in 180–210 days. In Ethiopia seeds are sown in July and harvested in November, and in Russia it is a spring or fall crop. A winter crop in India, fenugreek seeds are sown in October and November at the rate of about 45 kg/ha (40 lb/acre) with two or three irrigations to grow the crop and at the rate of 17 to 22 kg/ha (15 to 20 lb/acre) without irrigation.

The mature seeds are pungent and are used in India in the preparation of curry powder. Seeds contain about 25% protein, 50% carbohydrates, and 6% oil; they are reported to have medicinal qualities.

FAVA BEAN, BROAD BEAN, HORSE BEAN, WINDSOR BEAN: *Vicia faba (Faba vulgaris)*

Possibly of Mediterranean origin, fava bean is one of the oldest legumes under cultivation. Important since the Stone Age of man, it was cultivated in ancient Egypt, Greece, Italy, and the Middle Eastern countries.

A cool season temperate zone crop, fava beans can be grown at high elevations in the tropics up to about 2500 m (8000 ft); it cannot be grown in the low elevations of the tropics.

The immature and mature seeds are eaten; sometimes the immature pods are used as vegetable. In the temperate region, it is grown in the winter as fodder or green manure crop.

Fava bean is an erect annual growing 45–180 cm (1½–6 ft) in height. The leaves are pinnate rather than trifoliate and the inflorescences are on short axillary racemes, and the flowers are self-fertile. The pods are large, 5–15 cm (2–6 in.) in length, and contain one to several seeds.

Some people, particularly of the Mediterranean region, are adversely affected by eating the beans or inhaling the fava bean pollen (see Chapter 5, Favism). Mature seeds contain 17–25% protein, 48–64% carbohydrates, and 1–2% oil.

YARDLONG BEAN, COWPEAS: *Vigna* spp.

There are three distinct types (species) of cowpeas (*Vigna* genus), characterized by growth habit and pod character. The immature pods of all three species are used as vegetables.

Cow pea (common), southern bean, blackeye pea or bean: *Vigna sinensis (Dolichos sinensis)*

Plant is erect or semierect; pods are pendant 10–30 cm (4–12 in.) long; seed has "eye," which may be black, dark purple, brown, or maroon around the hilum.

Catjung, catjang, Bombay cowpea: *Vigna unguiculata (V. cylindrica, V. catjung, Dolichos unguiculata)*

Plant is erect or semierect; pods are erect and short 7–12 cm (3–5 in.) long; seeds are small and cylindrical.

Yard-long bean, asparagus bean, tau kok: *Vigna sesquipedalis* (*Vigna sinensis* subspec. *sesquipedalis, Dolichos sesquipedalis*)

Yard-long bean is grown especially for the long immature pods and for the dry mature seeds. The origin of yard-long bean is rather obscure; it is possibly of tropical African origin, as wild species of *Vigna* can be found there. Vavilov puts the origin as China.

Yard-long bean is an annual trailing vine with trifoliate leaves (Fig. 21.6). The plant is day-neutral for flowering. The pods range from 30 to 75 cm (12 to 30 in.) in length; seeds are kidney-shaped and long, 8–12 mm (5⁄16–½ in.). The crop can be grown in a wide range of climatic conditions but is quite sensitive to cold temperatures. The plant can tolerate heat, low rainfall, and acid soils. However, in poor soils the yield is reduced, and under low rainfall or low soil moisture, the pods are short and fibery.

Vine support increases yield; in the Philippines yard long beans are often planted after okra so that the okra stalks supports the vines. Cow peas contain about 24% protein, 55% carbohydrates, and 2% oils.

FIG. 21.6. Yard-long bean, asparagus bean *(Vigna sesquipedalis)*.

BAMBARA GROUND NUT, MADAGASCAR PEANUT: *Voandzeia subterranea*

Bambara ground nut, of West African origin, has been cultivated in tropical Africa for centuries. The crop is grown for the seeds, which are usually harvested green. The mature seeds are hard and difficult to cook. The mature seeds are very nutritious; they contain about 15% protein, 57% carbohydrates, and 7% oil.

Bambara ground nut is an annual creeping or erect herb with branching stems growing to heights of 20–30 cm (8–12 in.). The leaves are pinnate trifoliate on long erect petioles. Following self-fertilization, the peduncle of the flower grows downward, and the developing pod is buried in the soil as with peanuts *(Arachis hypogaea)*. The mature pods vary from 1 to 3 cm (3/8–1 1/8 in.) in length containing one to two seeds, which may be spherical or elliptical, 7–15 mm (1/4–1/2 in.) in diameter. Seed coat color may be white, yellow, brown, red, black, or mottled. The crop matures 130–150 days from planting.

Bambara ground nut is resistant to high temperatures and can be grown on poor marginal soils not suitable for other leguminous crops.

BIBLIOGRAPHY

Leguminous Crops

ANON, 1959. Tabulated information on tropical and subtropical grain legumes. Plant Production and Protection Division, FAO, Rome, Italy.

ANDERSON, A.J., LONERAGAN, J.F., MEYER, D., and FAUCETT, R.G. 1956. The establishment of legumes on acid soils. Rural Research in CSIRO, Div. Plant Industry, Canberra, Australia.

BRITTINGHAM, W.H. 1946. A key to the horticultural group or varieties of southern pea, *Vigna sinensis*. Proc. Am. Soc. Hort. Sci. *48*, 478–480.

FAO-USPHS. 1968. Food Composition Table for Use in Africa. FAO-US Dept. Health, Education and Welfare.

HERKLOTS, G.A.C. 1972. Vegetables in South-East Asia. Hafner Press, New York.

HOWARD, F.D., MACGillivray, J.H., and YAMAGUCHI, M. 1962. Bull. No. 788, California Agric. Exp. Stn., Univ. of California, Berkeley.

KAISER, W.J. 1981. Diseases of chick pea, lentil, pigeon pea and tepary bean in continental United States and Puerto Rico. Econ. Bot. *35*, 300–320.

KNOTT, J.E., and DEANON, J.R., Jr. 1967. Vegetable Production in Southeast Asia. Univ. of Philippines Press, Manila.

KYLEN, A.M., and McCREADY, R.M. 1975. Nutrients in seeds and sprouts of alfalfa, lentil, mung bean and soy bean. J. Food Sci. *40*, 1008–1009.

LIENER, I.E. 1962. Toxic factors in edible legumes and their elimination. Am. J. Clin. Nutrition *11*, 281–298.

PATWARDHON, V.N. 1962. Pulses and bean in human nutrition. Am. J. Clin. Nutrition *11*, 12–30.
PURSEGLOVE, J.W. 1968. Tropical Crops: Dicotyledons. John Wiley & Sons, Inc., New York.
SMITH, C.R., Jr., SHEKLETON, M.C., WOLFF, I.A., and JONES, Q. 1959. Seed protein sources—Amino acid composition and total protein content of various plant seeds. Econ. Bot. *13*, 132–1150.
WHYTE, R.O., NILSSON-LEISSNER, G., and TRUMBLE, H.C. 1953. Legumes in Agriculture. FAO Agric. Studies No. 21. Rome, Italy.

Peas *(Pisum sativum)*

BERRY, G.J., and AITKEN, Y. 1979. Effect of photoperiod and temperature on flowering in pea (*Pisum sativum* L.) Aust. J. Plant Physiol. *6*, 573–587.
HALL, H., and VOSS, R.E. 1976. Growing green peas. Leaflet 2913, Div. Agric. Sci., Univ. of California, Berkeley.
HARRINGTON, J.F., and MINGES, P.A. 1954. Vegetable seed germination. Leaflet, Div. Agric. Sci., Univ. of California, Berkeley.

Bean *(Phaseolus vulgaris)*

ALLARD, H.A., and ZAUMEYER, W.J. 1944. Responses of beans *(Phaseolus)* and other legumes to length of day. USDA Tech. Bull. No. 867. U.S. Dept. Agric., Washington, DC.
GENTRY, H.S. 1969. Origin of the common bean, *Phaseolus vulgaris*. Econ. Bot. *23*, 55–69.
KAPLAN, L. 1981. What is the origin of the common bean. Econ. Bot. *35*, 240–254.
KISH, A.J., OGLE, W.L., and LOADHOLT, C.B. 1972. A prediction technique for snap bean maturity incorporating soil moisture with heat unit system. Agric. Meteorol. *10*, 203–209.
SIMS, W.L., HARRINGTON, J.F., and TYLER, K.B. 1977. Growing bush snap bean for mechanical harvest. Leaflet 2674, Div. Agric. Sci., Univ. of California, Berkeley.
WATADA, A.E., and MORRIS, L.L. 1966. Post-harvest behavior of snap bean cultivars. Proc. Am. Soc. Hort. Sci. *89*:375–380.

Lima Beans *(Phaseolus lunatas)*

ALLARD, R.W. 1953. Production of dry edible lima beans in California. Circ. 423. California Agric. Exp. Stn., Univ. of California, Berkeley.
YAMAGUCHI, M., MACGILLIVRAY, J.H., HOWARD, F.D., SIMONE, M., and STERLING, C. 1954. Nutrient composition of fresh and frozen lima beans in relation to variety and maturity. Food Res. *19*, 617–626.

Peanut *(Arachis hypogaea)*

LEPPIK, E.E. 1971. Assumed gene centers of peanut and soy beans. Econ. Bot. *25*, 188–194.

OCHSE, J.J., SOULE, M.J., JR., DIJKMAN, M.J., and WEHLBURG, C. 1961. Tropical and Subtropical Agriculture, Vol. 2. Macmillan Co., New York.

Soybeans *(Glycine max)*

DOVRING, F. 1974. Soy bean. Sci. Am. *230*(2), 14–21.

HANWAY, J.J., and THOMPSON, H.E. 1967. How a soy bean plant develops. Spec. Rep. 53, Iowa State Univ., Ames, Iowa.

HOLMBERG, S.A. 1973. Soybean for cool temperate climates. Agric. Hort. Genet. *31*, 1–20.

JOHNSON, H.W., BORTHWICK, H.A., and LEFFEL, R.C. 1960. Effects of photoperiod and time of planting on rates of development of the soy bean in various stages of the life cycle. Bot. Gaz. (Chicago) *122*, 77–95.

LEPPIK, E.E. 1971. Assumed gene centers of peanut and soybeans. Econ. Bot. *25*, 188–194.

POLSON, D.E. 1972. Day-neutrality in soy beans. Crop Sci. *12*, 773–776.

STEPHENSON, R.A., and WILSON, G.L. 1977. Patterns of assimilate distribution in soybean. I. The influence of reproductive developmental stage and leaf position. Aust. J. Agric. Res. *28*, 203–209.

STEPHENSON, R.A., and WILSON, G.L. 1977. Patterns of assimilate distribution in soybean. II. The time course changes in ^{14}C distribution in pods and stem sections. Aust. J. Agric. Res. *28*, 395–400.

WEISS, M.G., WILSIE, C.P., LOWE, B., and NELSON, P.M. 1942. Vegetable soybean. Bull. P32, Iowa Agric. Ext. Serv., Ames, Iowa.

Pigeon Pea *(Cajanus cajan)*

GOODING, H.J. 1962. The agronomic aspects of pigeon peas. Field Crop Abstr. *15*, 1–4.

KRAUSS, F.G. 1932. The pigeon pea *(Cajanus indicus)* its improvement, culture and utilization in Hawaii. USDA Bull. No. 64. Hawaii Agric. Exp. Stn. U.S. Dept. Agric., Honolulu, Hawaii.

MORTON, J.F. 1976. The pigeon pea, a high protein tropical bush legume. HortScience *11*, 11–19.

Hyacinth Bean *(Lablab niger)*

SCHAAFHAUSEN, V., and REIMER, 1963. *Dolichos lablab* or hyacinth bean: its uses as feed, food and soil improvement. Econ. Bot. *17*, 146–153.

Lentil *(Lens esculenta)*

YOUNGMAN, V.E. 1968. Lentils—A pulse of the Palouse. Econ. Bot. *22*, 135–139.

Egyptian Lupine *(Lupinus termis)*

ABOU-CHAAR, I. 1967. The alkaloids of *Lupinus terminus*. I. Isolation of 13-hydroxy lupanine from Lebanese-grown lupine seed. Econ. Bot. *21*, 367–370.

Adzuki Bean *(Phaseolus angularis)*

SACKS, F.M. 1977. A literature review of *Phaseolus angularis*—The adzuki bean. Econ. Bot. *31*, 9–15.

Rice Bean *(Phaseolus calcaratus, Vigna umbellata)*

ARORA, R.K., CHANDEL, K.P.S., JOSHI, B.S., and PANT, K.C. 1980. Rice bean: Tribal pulse of eastern India. Econ. Bot. *34*, 260–263.

Winged Bean *(Psophocarpus tetragonolobus)*

ANON. 1975. The winged bean, a high protein crop for the tropics. Natl. Acad. Sci., Washington, DC.

DATA, E.S., and PRATT, H.K. 1980. Patterns of pod growth, development and respiration in the winged bean *(Psophocarpus tetragonolobus)*. Trop. Agric. (Trinidad) *57*, 309–317.

HYMOWITZ, T., and BOYD, J. 1977. Origin, ethonbotany and agricultural potential of the winged bean *Psophocarpus tetragonolobus*. Econ. Bot. *31*, 180–188.

KHAN, T.N. 1976. Papua New Guinea: a centre of genetic diversity in winged bean *(Psophocarpus tetragonolobus)*. Euphytica *25*, 693–706.

LAWHEAD, C.W., BENNETT, J.P., and YAMAGUCHI, M. 1979. Propagation of winged bean *(Psophocarpus tetragonolobus* (L.) D.C. by stem cuttings. Trop. Agric. (Trinidad) *56*, 271–276.

MARTIN, F.W., and DELPIN, H. 1978. Vegetables for the hot, humid tropics. Part 1. The winged bean, *Psophocarpus tetragonolobus*. Mayaguez Inst. Trop. Agric., U.S. Dept. Agric., Mayaguez, Puerto Rico.

THOMPSON, A.E., and HARYONO, S.K. 1980. Winged bean: Unexploited tropical food crop. HortScience *15*, 233–238.

Fenugreek *(Trigonella foenum-graecum)*

LADIZINSKY, G. 1979. Seed dispersal in relation to domestication of middle-east legumes. Econ. Bot. *33*, 284–289.

MALIK, H.C., and BATRA, P.C. 1958. Studies with fenugreek *(metha)* in Punjab, Indian J. Agric. Sci. *28*, 157–165.

Bambara Ground Nut *(Voandzeia subterranea)*

DOKU, E.V., and KARIKARI, S.K. 1971. Bambara ground nut. Econ. Bot. *25*, 255–262.

22

Solanaceous Fruits: Tomato, Eggplant, Peppers, and Others

Family: Solanaceae (nightshade)

Principal genera/species	Common name
Solanum melongena	Eggplant, brinjal (India), berenjana (Spain)
Solanum macrocarpon	African eggplant
Solanum muricatum	Pepino, melon pear
Solanum quitoense	Naranjillo, lulo
Solanum nigrum	Garden huckleberry, wonderberry (note: unripe fruits and leaves may be poisonous)
Solanum gilo	Jilo
Lycopersicon esculentum	Tomato
Lycopersicon pimpinellifolium	Red currant or grape tomato
Capsicum annuum	Pepper
Physalis ixocarpa	Tomatillo, ground cherry
Physalis pruinosa	Husk tomato
Physalis peruviana	Cape-gooseberry
Cyphomandra betacea	Treetomato, tamarillo

In the Solanaceae are important fruit vegetables: tomatoes, eggplants, and peppers. Potato and tobacco also belong to this family. The family includes about 75 genera and over 2000 species of herbs, shrubs, and small trees distributed in the temperate and tropical regions of the world.

TOMATO, TOMATE: *Lycopersicon esculentum (Lycopersicon lycopersicum)*; AND CURRANT TOMATO: *Lycopersicon pimpinellifolium*

Origin

Tomato's origin has been designated as the Andean region of South America because of the widespread distribution of the genus *Lycopersicon* in this area. Recent archeological evidence indicates the site of domestication to be Mexico and Central America. Since the discovery of America, tomato has become distributed throughout the world. At first it was a curiosity and erroneously reported as having poisonous fruits. It was not until the beginning of the twentieth century that tomato became increasingly popular and in recent years has become one of the most important vegetables produced worldwide.

Botany

Tomato is an annual shrubby plant in the temperate zones and a short-lived perennial in the tropics. Yellow flowers are borne on clusters on trailing stems up to 2 m (6 ft) long. The growth habit in most forms is indeterminated (shoot tips remain vegetative). There are also cultivars with semideterminated and determinated growth habits (shoot tips terminate in a flower cluster) (Fig. 22.1). These latter types have a bushy compact type of growth. Fruits are round, lobed, or pear-shaped, varying in size from 1 to 12 cm (½–5 in.) in diameter.

Fruits of *L. esculentum* and *L. pimpinellifolium* are red when ripe. Fruits of *L. peruvianum*, *L. chilense*, and *L. hirsutum* are green even when ripe. Colors of domesticated tomatoes may be red, orange, or yellow, depending on the genetic makeup of the cultivar. The structure of a tomato fruit is shown in Fig. 22.2.

Climate

The plant is adapted to a wide range of climatic and soil conditions. It can be grown from near the Arctic Circle, under protection, to the equator.

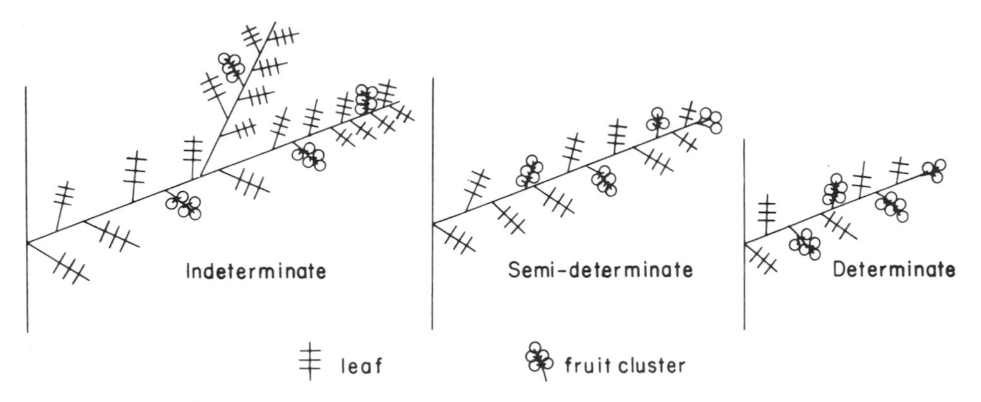

FIG. 22.1. Growth characteristics of tomato types.

The temperature means must be above 16°C (60°F). Any time the temperature drops below 12°C (53°F), the plant gets chilled and, if the temperature remains at or below this temperature for too long a period, chilling injury of the tissues occurs. Short periods of below 12°C (53°F) can be offset by temperatures above this minimum. The optimum range of temperatures is 21°–24°C (70°–75°F). Mean temperatures above 27°C (80°F) are not desirable.

The minimum soil temperature for seed germination is 10°C (50°F); the optimum is 30°C (86°F) and the maximum is 35°C (95°F).

Light intensities of less than 11,000 m candles (1000 fc) can affect plant growth as well as flowering. If the light falls much below this intensity, supplemental lighting is necessary to increase the intensity and to extend the photoperiod. Figure 22.3 shows day-length (energy) effects on the time of planting to flowering.

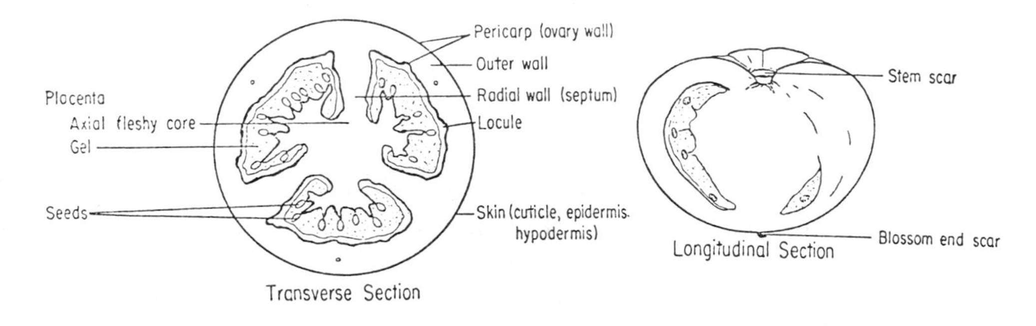

FIG. 22.2. Tomato fruit structure.
(From Spurr 1976.)

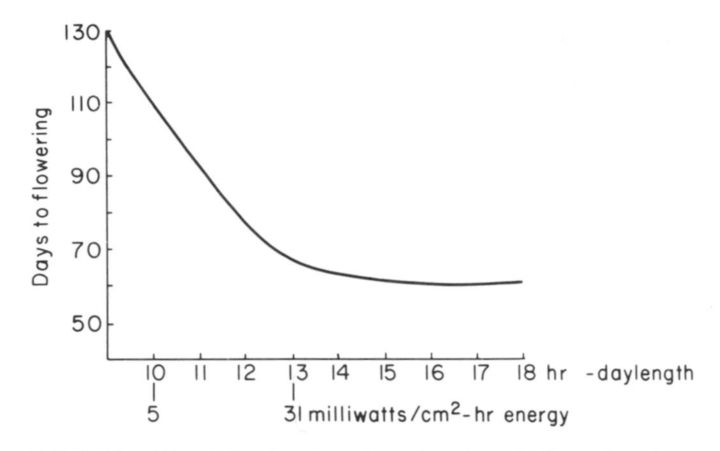

FIG. 22.3. Effect of natural day length on tomato flowering at 51°N lat., greenhouse at 21°C days, 16°C nights.
Data from Cooper (1964).

Flowering and Fruit Set

Poor fruit set occurs if the maximum day temperatures exceed 38°C (100°F) for 5–10 days before anthesis. This is due to destruction of pollen and egg cells. Poor fruit set also occurs if the maximum day temperature exceeds 38°C (100°F) for 1–3 days following anthesis because the embryo is destroyed after pollination. Also, set is poor if the minimum night temperatures are too high, greater than 25°–27°C (77°–80°F), a few days before and a few days after anthesis. The susceptibility to high night temperatures varies with cultivars and there are some tropical cultivars that set fruit at fairly high night temperatures. Hot drying winds can often cause poor fruit set. At 10°C (50°F) or below a large percentage of the flowers aborts.

During times of cool weather, fruit setting hormones such as IAA and parachlorophenoxyacetic acid at 25–50 ppm can be used. If such growth regulators are used, fruits are often parthenocarpic; making fruits puffy. However, addition of gibberellins with IAA reduce the puffiness.

Fruit Ripening

The optimum temperature for fruit ripening is 18°–24°C (65°–75°F). At temperatures below 13°C (55°F), fruits ripen poorly and slowly, and below 10°C (50°F), chilling occurs and the fruits may not ripen at all.

If the temperature exceeds 32°C (90°F) during the storage of harvested mature green fruits, the formation of red color, lycopene, is inhibited and the fruits are yellowish when ripe.

When the mean temperatures are above 40°C (104°F), the fruits remain green because the chlorophyll degrading mechanism is inactivated.

Culture

Soil

Tomatoes can be grown in a wide range of soil types from sandy loams to clay loams and in soils high in organic matter. The pH can range from 5.5–7.0; higher or lower pH's can cause difficulties, especially in mineral deficiencies or toxicities. Tomatoes cannot withstand long periods of flooding so the soil should drain easily or be planted on high beds where drainage is poor. Land with nematodes and bacterial wilt causing organisms should be avoided. Cultivars with fusarium and verticillium resistance should be used when the soil contains these pathogenic fungi.

Most soils for growing of tomato crops require fertilization, about 112 kg N/ha (100 lb N/acre) and 22 kg P/ha) (20 lb P/acre) and some K if the soil is low in this element. It is recommended that a starter fertilizer of 11 kg N/ha (10 lb/acre) and 17 kg P/ha (15 lb/acre) be placed 3–5 cm (1–2 in.) below the seed in direct seed operations; the rest of the fertilizer applied when plants are 15–20 cm (6–8 in.) tall, about the time for thinning. Manures can be used but supplemental, chemical N may be necessary for high yields.

Propagation and Spacing of Plants

Tomatoes can be either transplanted or direct seeded in the field. Cuttings treated with rooting hormone can be used for propagation. The spacing between plants and distance between rows depends on the growth habit of the cultivar (determinate, semideterminate, or indeterminate), whether the plants are staked or left on the ground, and whether the harvest is to be by hand or mechanical harvester. Staked tomatoes are spaced 30–45 cm (12–18 in.) apart in 1–1¼ m (3–4 ft) rows, and without stakes the spacing is 50–60 cm (1½–2 ft) apart and 1¼–1½ m (4–5 ft) rows.

Moisture

Tomato plants require adequate moisture throughout its growth period; overwatering is detrimental. While the plants are small frequent watering is necessary in the root zone and less frequently as the plants grow and extend the soil volume which the roots explore. Tomato roots grow to depths of 4–5 ft or more unless they are restricted by an impervious layer of hard pan, rock layer, or water table.

Water used by tomato plants depends on the temperature. Under cool temperature, 0.3 cm (1/8 in.) water will be used by evapotranspiration per day in a tomato field, whereas under hot dry weather conditions slightly over three times (1 cm) as much water will be used in a day.

The physiological disorder "blossom end rot" appears as a brown to black necrotic sunken area penetrating into the flesh at the blossom end of the fruit. In severe cases the lesion may cover over half the fruit. Incipient blossom end rot may not be noticeable from the outside, but a necrotic zone of brown dead tissue lies beneath the surface at the blossom end. The cause of this physiological disorder has been attributed to water deficit during fruit enlargement and calcium deficiency.

Harvest and Storage

Harvest

From planting of seed to harvest requires 60–90 days depending on cultivars and climate conditions (temperature and day length).

Fresh market tomatoes are often harvested at the mature green stage and ripened either in transit or in storage. Ripening of mature green fruits must take place at temperatures greater than 10°C (50°F); below this temperature chilling injury occurs and ripening does not occur. Above 10°C (50°F) the rate of ripening increases with increasing temperature. Some exposure to light is necessary for the red pigment, lycopene, formation to occur.

Tomato fruits are classified according to maturity stages given in Table 22.1.

Storage

Mature green tomatoes should be stored at 13°–18°C (55°–65°F) and 85–90% humidity. At this temperature range chilling injury does not occur and the fruits will ripen; ripening is more rapid at temperatures >18°C (>65°F) but at constant temperatures of >30°C (>86°F), the red pigment does not readily form and the ripe fruit is orange to yellow in color.

Ripe (red) fruits should be stored at 3°–4°C (38°–40°F); at higher temperatures, the fruit continues to ripen and quality deteriorates rapidly.

Pests and Diseases

Snails, slugs, cut worms, flea beetles, tomato russel mites, stink bugs,

TABLE 22.1. MATURITY STAGES OF TOMATO RIPENING

Stages	Days from mature green at 20°C	Description
Immature green	—	Fruit still enlarging, angular in shape, color dull green, no skin luster. Seed cut with sharp knife when sliced. Gel not well formed. Seed immature and will not germinate. Fruit will not ripen properly.
Mature green	0	Color bright green to whiteish-green and well rounded. Skin has waxy gloss and cannot be easily scraped off. Seeds embedded in gel and are not cut when fruit is diced; seeds mature and will germinate. Fruit will ripen under proper conditions.
Breaker	2	Turning pink at blossom end; internally the placenta is pinkish.
Turning	4	Pink in color from blossom end covering 10–30% of fruit.
Pink	6	Pink to red in color covering 30–60%.
Light red	8	Pink to red in color covering 60–90%.
Red	10	Red color at least 90% of fruit.

and tomato horn worms attack either the foliage and/or fruits. Nematodes attack and cause knot-like swellings of the roots. There are nematode resistant cultivars.

Verticillium wilt, a fungus, causes yellowing of old crown leaves and sometimes causes mottling of the stem. Fusarium wilt, another fungus, causes yellowing of leaves and the vascular system of the stem turns a reddish-brown color.

Phytophthora root rot, caused by a soil fungus, is due to wet soil and is most prevalent in soil high in clay.

Tobacco mosaic virus-infected leaves show blotchy yellow and green spots. The virus is transmitted by seed and handling especially by tobacco smokers.

Spotted wilt is transmitted by thrips; the virus causes dark brown spots and streaks to appear on the stem near the growing point giving the shoots a bronzing effect. Irregular dark brown streaks appear on newly formed leaves.

Curly top, another virus disease, causes leaves to roll with some twisting. With progress of the diseases the veins turn purple and the leaf color turn yellow. Infected plants cease growth and subsequently die. Some cultivars have naturally curly leaves, which is sometimes confused with curly top.

Yellow leaf curl, a viral disease, is transmitted by white flies.

Nutritive Value

Ripe tomatoes are a good source of vitamins A and C; both vitamins increase as the fruit ripens on the vine but decrease when mature green fruits are ripened off the vine. The ascorbic acid content is low in fruits from plants grown under low light intensities. Table 22.2 give the proximate composition of the tomato fruit.

Economics

Since the European explorers found the tomato cultivated in western South America over 400 years ago, it has become by far the most important of the solanaceous fruits the world over. Table 22.3 shows that the highest production is in Europe. In southern Europe, most of the tomatoes are grown out of doors, but in northern Europe (Belgium, Denmark, Finland, Norway, Sweden, and U.K.), tomatoes are grown mainly or totally under glass. Most of the winter production in the United States is under glass also. Yields as high as 135–180 MT/ha (60–80 tons/acre) are obtained from glasshouses in Europe. In the United States, the average yield of all tomatoes is about 38 MT/ha (17 tons/acre), but in California it is slightly under 56 MT/ha (25 tons/acre). In southern European countries and in California, a large percentage of the production is used for processing.

EGGPLANT, AUBERGINE, GUINEA SQUASH, BRINJAL, BERENJENA: *Solanum melongena*

Origin

Wild eggplant is reported to be found in India; the plant is spiny and the fruit bitter. From India the domesticated nonbitter fruited types spread eastward into China by the fifth century BC and later they were taken to Spain and Africa by the traders. The name eggplant was probably derived from the plant type that bears white fruits resembling the chicken egg.

Botany

Eggplant is an annual in temperate zones and perennial in the tropics. Plants grow to a height of 60–120 cm (2–4 ft) and bear a few large

TABLE 22.2. PROXIMATE COMPOSITION OF SOME SOLANACIOUS FRUITS IN 100 g OF EDIBLE PORTION

Crop	Edible part	% refuse	Macroconstituents: Energy (cal)	g: Water	Protein	Fat	CHO	Vitamins: A (IU)	mg: B_1	B_2	Niacin	C	Minerals (mg): Ca	Fe	Mg	P
Eggplant	Fruit	10	20	93	1.1	0.1	4.0	70	0.09	0.02	0.6	7	7	0.4	16	25
Husk tomato	Fruit	2	25	91	1.4	0.5	4.2	380	0.15	0.03	3.5	4	8	0.3	—	34
Peppers	Fruit															
Bell (green)		15	22	93	0.9	0.3	4.4	530	0.06	0.02	0.4	160	7	0.4	13	22
Bell (red)		15	29	91	0.8	0.6	5.3	5700	0.11	0.08	0.7	220	4	0.3	13	28
Chili (green)		5	43	86	2.0	1.5	5.9	10500	0.08	0.08	0.9	245	17	1.4	23	46
Chili (red)		5	46	84	2.0	5.5	5.8	11000	0.10	0.10	1.0	240	18	1.0	27	45
Pimento		30	35	90	1.2	0.9	5.9	2200	0.05	0.46	0.6	165	9	0.5	4	20
Tomato	Fruit															
Pearson		5	19	94	0.9	0.1	3.7	1700	0.10	0.02	0.6	21	6	0.3	10	16
San Marzano (pear tomato)		5	15	94	0.9	0.1	3.5	770	0.10	0.01	0.7	23	9	0.1	11	20
Wonderberry	Fruit	0	24	89	2.0	1.0	1.1	570	0.10	0.01	0.7	12	24	0.6	40	42
African eggplant[a]	Fruit		40	89	1.4	1.0	8.0	—	—	—	—	—	13	—	—	—
	Leaves	42	42	86	4.6	1.0	6.4	—	—	—	—	—	390	—	—	50
Tree tomato[b]	Fruit	—	50	86	2.2	0.9	10.3	760	0.10	0.04	1.2	29	9	0.8	—	48
Naranjillo[b]	Fruit	—	28	92	0.7	0.1	6.8	170	0.06	0.04	1.5	65	8	0.4	—	14
Pepino, melon pear[b]	Fruit	—	32	92	0.4	1.0	6.3	200	0.08	0.04	0.5	32	18	0.8	—	14

Source: Howard *et al.* (1962).
[a] FAO-US PHS. 1968.
[b] NIH. 1961.

TABLE 22.3. WORLD PRODUCTION OF TOMATOES

	Area (10^3 ha)	Production (10^3 MT)	Yield (MT/ha)
World	2,404	49,201	20.5
Continent			
Africa	354	4,769	13.5
North and Central America	314	9,847	31.3
South America	135	2,854	21.2
Asia	706	11,251	15.9
Europe	490	13,871	28.3
Oceania	10	208	21.3
U.S.S.R.	395	6,400	16.2
Leading Countries			
1. United States	182	7,663	42.1
2. U.S.S.R.	395	6,400	16.2
3. Italy	126	4,294	34.1
4. China	284	3,930	13.8
5. Turkey	108	3,136	29.2
6. Egypt	139	2,421	17.5
7. Spain	64	2,050	32.0
8. Greece	40	1,669	41.8
9. Brazil	56	1,500	26.6
10. Romania	72	1,393	19.3

Source: FAO (1979).

fruits which are oval shaped or an elongated oval. Growth is indeterminate. Oriental cultivars bear more fruits that are long and slender, 15–30 cm (6–12 in.) long and 4–5 cm (1½–2 in.) in diameter (Fig. 22.4). Most cultivars are purple to blackish purple in skin color, but some are white, green, or mottled green skinned with white flesh, the skin turning yellow when the fruit matures.

Climate

A warm season crop, eggplant requires continuous long warm weather during growth and fruit maturation. The optimum growing temperature is 22°–30°C (72°–86°F) and growth stops at temperatures below 17°C (63°F). Therefore, the night temperature should be warm. Pollen deformity increases at temperatures of 15°–16°C (59°–61°F). Eggplants can be grown as far north as 50° N latitude if transplanted after the weather warms in the spring. Cultivars with very long-type fruits are more resistant to extremely high temperatures, whereas the small egg-shaped and oval-type fruits cannot stand extremely high summer heat.

Flowering commences after the sixth leaf has opened in the early cultivars whereas with the very late types up to 14 leaves are borne before flowering. Eggplant appears to be insensitive to day length for flowering.

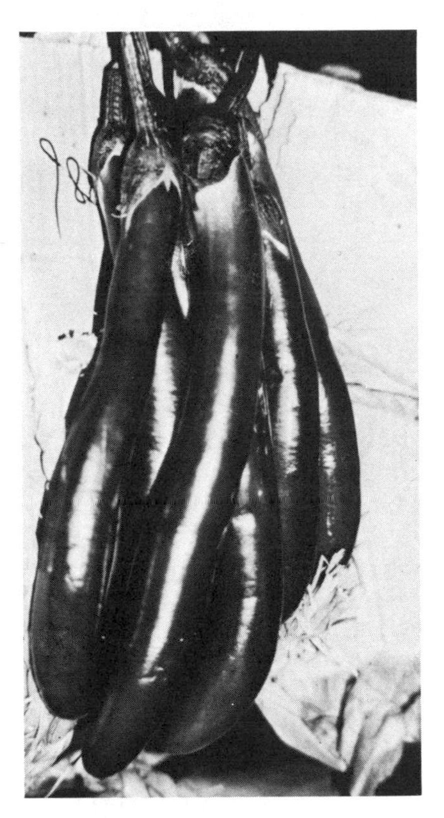

FIG. 22.4. Japanese eggplant *(Solanum melongena)*.

Culture

In the early spring for greenhouse, plastic, or field culture, plants should be placed so that they receive as much light as possible. The crop grows best in light sandy soils in the early spring and in loam for later production. Shallow clay soils are not suitable for eggplant production. The soil should be slightly acid to slightly alkaline (pH 5.5–7.2). Flooding causes root rot of plants, so proper drainage is required. In high rainfall regions, eggplants are grown on raised beds. Poor fruit color can result if the soil moisture is low.

In the tropics (Philippines), eggplants grow satisfactorily up to elevations of 1100 m (3500 ft) and can be grown throughout the year.

The optimum soil temperature for seed germination is 30°C (86°F), and for good germination the range of soil temperatures is 20°–35°C (68°–95°F). Seedlings at the two to three true-leaf stage are usually transplanted from the seedbed to the field. Eggplant stems can be rooted by layering and are sometimes used for propagation. Layering is accomplished by covering the stem with soil while still attached to the plant; adventitious roots form and emerge at the nodes; when well rooted the stem is cut off and transplanted. Cuttings root easily when treated with rooting hormones such as IAA or NAA. High fertility, especially P, is necessary for good growth. About 150 kg/ha (130 lb/acre) each of N, P, and K are recommended.

Harvest and Storage

Harvest

From 75 to 90 days are required from transplanting to harvest of first fruits. Plants continue to produce under favorable conditions for sometime. Plants can be rejuvenated by pruning back unproductive disease-free plants and stimulating new growth; applications of fertilizer may be necessary.

In Japan for early varieties with round or egg-shaped fruits, it takes about 15–20 days from flowering to harvest; for medium, long-fruited varieties, 25–30 days; and for long-fruited varieties, 35–40 days. Fruits are harvested when they are nearly full size and before the seeds mature.

Maturity can be tested by pressing the thumb against the side of the fruit. If the indentation springs back to its original shape, the fruit is too immature. However, with increasing maturity the flesh softens, so at the proper stage the thumb pressure leaves an indentation in the flesh of the fruit. Overmature fruits lose their gloss and become greenish bronze in color.

Figure 22.5 shows the fruit growth of the cultivar 'Black Queen'; harvest is made before the fruit attains full size.

Storage

Fruits can be stored at 10°–15°C (50°–59°F) at 85–90% RH for about 10 days without much loss of quality. Storage below 10°C (50°F) causes chilling injury; and after removal from such storage, pitting and decay occur after several days at room temperature.

Economics

Eggplant is a very important and popular crop in Asia, since it is not difficult to grow in the tropics and it is relatively hardy at high temperatures. Table 22.4 shows that eggplant is most important in Asia and the

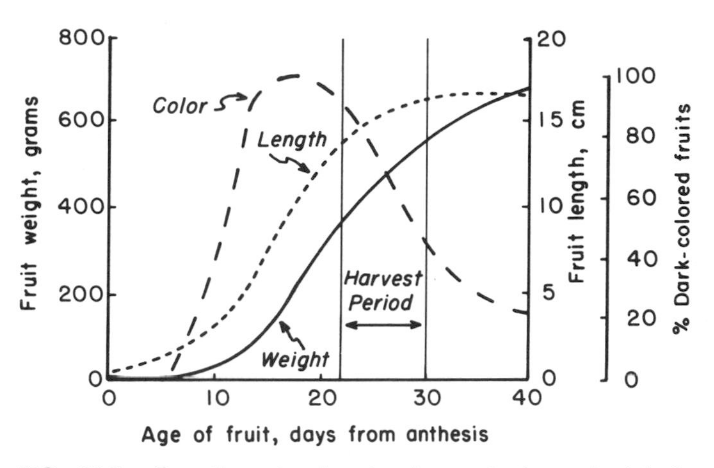

FIG. 22.5. Growth and color development of eggplant fruit 'Black Queen.'
Redrawn from Nothmann et al. (1978).

leading country is China followed by Japan with about half the production.

Pests and Diseases

Eggplant lace bug, aphids, spider mites, leafhoppers, flea beetles, fruit borer, termites, and nematodes attack the plant in the tropics. In the temperature region flea beetles, potato beetles, aphids, spider mites, and nematodes feed on eggplants.

Bacterial wilt, phomopsis rot, anthracnose, verticillium wilt, and mosaic virus infect the plant in the topics. Fruit rot, a seed transmitted and soil borne disease, is the most troublesome in the temperate region.

PEPPER, PIMIENTO, CHILI, AJI: *Capsicum annuum*

Origin

Peppers are a native of the new world (Mexico and Central America, and the Andes of South America). *Capsicum annuum* is the most widely cultivated species; depending on the cultivar it may be pungent or nonpungent. Other species, *C. baccatum, C. chinense, C. pubescens*, are distributed in parts of South America. *Capsicum frutescens*, a

TABLE 22.4. WORLD PRODUCTION OF PEPPER (CHILI AND GREEN) AND EGGPLANT

	Pepper (chili and green)			Eggplant	
	Area (10^3 ha)	Production (10^3 MT)		Area (10^3 ha)	Production (10^3 MT)
World	890	6,671		323	4,273
Continent					
Africa	148	1,032		29	414
North and Central America	80	760		3	72
South America	18	155		1	7
Asia	492	2,566		267	3,215
Europe	152	2,177		24	565
Oceania	—	—		—	—
U.S.S.R.	—	—		—	—
Leading Countries					
1. China	142	1,340	1. China	152	1,350
2. Nigeria	68	620	2. Japan	23	675
3. Spain	28	537	3. Turkey	36	550
4. Italy	20	486	4. Italy	13	335
5. Turkey	47	480	5. Egypt	14	290
6. Mexico	53	474	6. Syria	8	148
7. Yugoslavia	40	340	7. Iraq	8	116
8. Bulgaria	17	260	8. Spain	5	98
9. United States	22	248	9. Philippines	17	95
10. Indonesia	104	200	10. Greece	4	84
11. Hungary	18	200			

Source: FAO (1979).

perennial, has a wide distribution in tropical American and a secondary distribution in southeast Asia; the tobasco pepper is the only one cultivated.

Botany

Like the tomato and eggplant, peppers are perennials in the tropics and annuals in the temperate zones. Pepper plants have been domesticated and used as a condiment in Central and South America many centuries before the arrival of Columbus to the New World. The flowers are self-pollinated and the fruit is a pod-like berry with a cavity between the placenta and fruit wall. Most wild types are pungent. The bell, pimiento, and paprika types are nonpungent and have thick-wall fruits.

Temperature requirements for growing of peppers are slightly higher than for the tomato. Peppers are more sensitive to cool, wet weather. Chili and tobasco hot peppers are more tolerant of high temperatures than the sweet varieties.

Fruit set of peppers does not occur at temperatures below means of 16°C (60°F) or above 32°C (90°F). Maximum set of bell peppers occurs between constant temperatures of 16°–21°C (60°–70°F).

Culture

Peppers grow well in warm climates and a long growing season; they can tolerate extreme hot weather better than tomatoes and eggplants. They grow best in light fertile well-drained soil.

Peppers may be either direct seeded or transplanted in the field after all danger of frost has past. Seeds germinate very slowly in cool soils (15°C/59°F); the optimum is 30°C (86°F) and the maximum is 35°C (95°F). Plants are spaced 45–60 cm (18–24 in.) in rows 75–90 cm (2½–3 ft) apart. About 170–220 kg N/ha (150–200 lb/acre) and 22 kg P/ha (20 lb/acre) is used to grow the crop. Potassium is applied when the availability of the element is low.

Harvest and Storage

Harvest

From 60 to 90 days from transplanting the fruits are ready for harvest. Bell peppers are harvested when they have attained full size and are still green. This stage is comparable to the mature green tomato. The ripening process is similar to the tomato. Fresh chilies are harvested green, but for processing they are harvested ripe (red), as are pimientos and paprikas.

Storage

Peppers of the mature green bell type can be stored for 10–15 days at 7°–10°C (45°–50°F). If stored below these temperatures, chilling injury occurs, causing cells to die and the fruit to decay.

Nutritive Value

Peppers are an excellent source of vitamins A and C. Table 22.2 shows that as bell peppers mature the provitamin A (B-carotene) and ascorbic acid increase.

The pungent principle in peppers is capsaicin, found only in the septa and in the placental tissue of the fruit but not in the fruit wall. The seeds also have small amounts of the pungent compound.

Economics

Peppers are the second most important crop among the solonaceous fruits. From the point of view of nutrients obtained from these crops, peppers are by far superior to both tomato and eggplant in vitamins A and C content. The world production of peppers and chilies is given in Table 22.4.

Pests and Diseases

Flea beetles, cut worms, aphids, vegetable weevils, grasshoppers, wire worms, corn seed maggots, leaf miners, and caterpillars damage both the plant and/or fruit.

Tobacco mosaic virus, tobacco etch virus, potato Y virus, and cucumber mosaic virus infect pepper plants. Western yellow or curly top virus causes old leaves to curl upwards and turn yellow. Spotted wilt is caused by another virus and infection results in the die back of the growing plant. Nematodes attack the roots, causing gall formations.

Phytophthora root rot causes rotting of roots under high temperatures and high soil moisture. It can be controlled by careful regulation of soil moisture.

Seedling damp-off is caused by *Rhizoctonia soloni, Phythum,* and *Phytophthora* attacking the seed and seedling before emergence through the soil.

AFRICAN EGGPLANT: *Solanum macrocarpon*

This perennial crop is important in the Ivory Coast countries of Africa being grown commercially and also in the home gardens. It is also grown in Madagascar. The small fruits are similar to the eggplant (*S. melongena*) and the leaves are edible.

PEPINO, PEPINO DULCE, MELON PEAR: *Solanum muricatum*

Pepino is a native of Peru and not known in the wild. It is an ancient cultivated crop of the Andes, grown principally along the central coast of Peru. It is also cultivated from Colombia to Boliva at elevations of 1000–3000 m (3300–10,000 ft) and on a small scale in northern Argentina and Chile. It was introduced into Central America and Mexico in post-Columbian times. The plant is a bushy shrub bearing fruits 10–15 cm (4–6 in.) long which are ovoid or elliposidal and attached to long stalks. Fruit color varies from light green to pale yellow. Immature fruits are eaten cooked. Mature fruits have a fleshy pulp with an aroma and flavor similar to the cucumber. It also has been described as aromatic, tender and juicy, tasting musky like a melon. Fruits are high in vitamin C (29 mg/100 g).

NARANJILLO, LULO: *Solanum quitoense*

Naranjillo, native to Ecuador, is a small bushy plant attaining 1–2 m (3–6 ft) in height. It is grown at 1100–2400 m (3500–8000 ft) elevations in humid regions and in soils high in organic matter. Propagation is usually by seed as vegetative cuttings are difficult to root because of fungus infections. Naranjillo flowers about 7 months after seeding and fruits ripen 5–6 months after anthesis. The fruits are spherical in shape, 3–5 cm (1 1/6–2 in.) in diameter, yellow when ripe, and rough after the white hairs are shed. The pulp is green, acidic in taste, and contains several white seeds. Grown for the juice, it is cultivated in Ecuador, Peru, Colombia, and Central American countries.

GARDEN HUCKLEBERRY: *Solanum nigrum*

Garden huckleberry is a native to North America with wide distribution in temperate to tropical regions. Plants can withstand considerable frost. The fruit is a berry, about 6 mm (1/4 in.) in diameter, green when unripe and black in color when ripe. Ripe fruits are used for pies and preserves. The immature fruits and leaves are poisonous.

JILO: *Solanum gilo*

Probably a native of central Africa, it is an important crop in Nigeria. Jilo (Fig. 22.6) is a minor crop in central and southern Brazil. The culture of the crop is similar to eggplants. The fruit, spherical to oval in shape about 4 cm (1½ in.) in diameter and 6 cm (2¼ in.) in length, are harvested immature (green in color) about 90–100 days from seeding. On maturation the fruits turn orangish-red. The fruits are quite bitter in taste. In Nigeria, the young shoots, which are very bitter, are finely chopped for use in soups.

TOMATILLO, TOMATE DE CASCARNA: *Physalis ixocarpa*

Tomatillo is of Mexican origin and has been introduced to the United States. The plant is an annual 1–1.3 m (3–4 ft) in height. The fruit is

FIG. 22.6. Jilo (*Solanum gilo*).

enclosed in a husk (enlarged purple veined calyx). The fruit is a large round stickly green or purplish berry, high in ascorbic acid (36 mg/100 g). In Mexico the fruits are used in the making of chili sauce and dressing for meats.

GROUND CHERRY, HUSK TOMATO: *Physalis purinosa*

Ground cherry grows wild from New York to tropical America. The plant is an annual vine trailing close to the ground and sometimes ascending to about 30 cm (1 ft). The prominent five-angled husk, which encloses the fruit, is smooth, thin, and paper-like (Fig.22.7). The fruit is slightly smaller than the husk and is yellowish, nonglutinous, and acidic sweet. It is used for preserves and sometimes for sauces.

CAPE-GOOSEBERRY, UCHUBA: *Physalis peruviana*

Cape-gooseberry is a native of the Andes; it is grown from Venezuela to Chile. The plant is bushy, about a meter (3 ft) in height. The fruit,

FIG. 22.7. Husk tomato (*Physalis pruinosa)*.

about 4 cm (1½ in.) long and 3 cm (1¼ in.) wide, is enclosed in a hairy green calyx with purple spots. The calyx grows at a faster rate than the ovary so that it covers the fruit completely. The fruit is a smooth spherical or ellipsodial berry, greenish yellow in color and 1 cm (3/8 in.) in diameter. The fruits are sometimes eaten raw but generally they are preserved in pickles. A minor crop of the Andes region, fruits are nearly always found in the market places.

TREE TOMATO, TAMARILLO: *Cyphomandra betacea*

A native of subtropical America (Mexico) the tree tomato is a perennial tree-like shrub growing to heights of 2–3 m (6–10 ft). The crop is usually grown in frost-free areas and in light fertile well-drained soils. There are two types of tree tomato grown commercially in the North Island of New Zealand: (1) yellow fruit color with yellowish flesh and light colored seeds, and (2) dark red colored fruit with orange flesh and black seeds.

Tree tomatoes (Fig. 22.8) are propagated in nurseries from seed or they may be propagated from 1- or 2-year-old stem cuttings. Plants from cuttings tend to be bushy while those from seeds are more tree-like.

FIG. 22.8. Tree tomato, tamarillo (*Cyphomandra betacea*).

From the nursery the plants are set out into the field at 3–5 m (10–16 ft) spacing between plants. Plants start to bear fruit within 18 months and are in full bearing in 3–4 years. Fruits are commercially mature 21–24 weeks following anthesis and are fully ripe in 24–25 weeks. Ripe fruits can be kept for a week at room temperature. Fruits, which had been dipped in water at 50°C (122°F) for 10 min., waxed, and then stored at 3°–4°C (38°–40°F) for 12–14 weeks, have a 7 day shelf life at room temperature.

Ripe fruits are slightly acid and tomato-like in flavor. They are high in vitamin C (20–40 mg/100 g) and also a good source of provitamin A (1600–5600 IU/100 g).

BIBLIOGRAPHY

CALVERT, A. 1964. The effects of air temperature on growth of young tomato plants in natural light conditions. J. Hort. Sci. *39*, 194–211.

COOPER, A.J. 1964. The seasonal pattern of flowering of glasshouse tomatoes. J. Hort. Sci. *39*, 111–119.

CURME, J.H. 1962. Effect on low night temperatures on tomato set. Proc. Plant Sci. Symp., Campbell Soup Co., Camden, New Jersey.

FAO-USPHS. 1968. Food Composition Table for Use in Africa. FAO-US Dept. Health, Education and Welfare.

FLETCHER, W.A. 1965. Tree Tomato Growing. Bull. No. 306. N. Z. Dept. Agric. Auckland, New Zealand.

FILGUEIRA, F.A.R. 1972. Manuel de Olericultura. Editora "Ave Maria" Ltd., Sao Paulo, Brasil.

HEDRICK, U.P. (Editor). 1919. Sturtevant's Notes on Edible Plants. 27th Ann. Rep. Vol. 2, Part II, pp. 343–348. New York State Dept. Agric., Albany, New York.

HOWARD, F.D., MACGILLIVRAY, J.H., and YAMAGUCHI, M. 1962. Nutrient composition of fresh California-grown vegetables. Bull. No. 788. Calif. Agric. Exp. Sta., Univ. of California, Berkeley.

KNOTT, J.E., and DEANON, J.R., Jr. 1967. Vegetable Production in Southeast Asia. Univ. of Philippines Press, Manila, Philippines.

LEON, J. 1964. Plantas Alimenticias Andinas, Bol. Tech. No. 6, Junio. Instituto Interamericano de Ciencias Agricolas Zona Andina.

MACGILLIVRAY, J.H. 1953. Vegetable Production. McGraw Hill Book Co., New York.

MARTIN, F.W., and POLLACK, B.L. 1979. Vegetables for the hot humid tropics. Part 5, Eggplant, *Solanum melongena*. Mayaguez Inst. Trop. Aric., Sci. Ed. Admin., US Dept. Agric., Mayaguez, Puerto Rico.

MARTIN, F.W., MARTIN, W., SANTIAGO, J., and COOK, A.A. 1979. Vegetables for the hot humid tropics. Part 7, The peppers, *Capsicum* species. Mayaguez Inst. Trop. Agric., Sci. Ed. Admin., U.S. Dept. Agric., Mayaguez, Puerto Rico.

NIH. 1961. Tabla de Composición de Alimentos para Uso en America Latina. Natl. Inst. Health, Washington, DC.

NOTHMANN, J., RYLSKI, I., and SPIGELMAN, M. 1978. Effects of air and soil temperature on colour of eggplant fruits (*Solanum melongena* L.) Exp. Agric. *14*, 189–195. Cambridge Univ. Press, London.

PRATT, H.K., and REID, M.S. 1976. The tamarillo: fruit growth and maturation ripening, respiration and role of ethylene. J. Sci. Food Agric. *27*, 399–404.

OCHE, J.J., SOULE, M.R., and DIJKMAN, M.J. 1961. Tropical and Subtropical Agriculture. Macmillan Co., New York.

RICK, C.M. 1978. The tomato. Sci. Am. *239* (2), 76–87.

SANTIAGO, A. 1976. Tomato Growing in the Tropics. World Crops, *28*, 89–91.

SCHAIBLE, L.W. 1962. Fruit setting responses of tomatoes to high night temperatures. Proc. Plant Sci. Symp., Campbell Soup Co., Camden, New Jersey.

SPURR, A.R. 1976. Structure and development of the fruit. Proc. 2nd Tomato Quality Workshop. Ser. No. 178, Dept. Veg. Crops, Univ. of California, Davis.

THOMPSON, H.C., and KELLEY, W.C. 1957. Vegetable Crops, 5th Ed., pp. 471–500. McGraw-Hill Book Co., New York.

23

Cucurbits

Family: Cucurbitaceae (gourd)

There are about 96 genera and 750 species in the warmer parts of the world, especially in the tropics.

Principal genera/species	Common name
Cucumis sativus	Cucumber
C. anguria	West Indian gherkin
C. melo	Melon, muskmelon, Persian melon
Citrullus vulgaris (C. lanatus)	Watermelon
C. vulgaris var. *citroide*	Citron, preserving melon
Cucurbita spp.	Squash, pumpkin
Sechium edule	Chayote, vegetable pear, mirliton
Luffa cylindrica	Dishcloth gourd, sponge gourd, loofah, *mo kwa*
L. acutangula	Chinese okra
Benincasa hispida	White gourd, wax gourd, Chinese preserving melon, *don kwa*, Chinese squash, *mo kwa*
Lagenaria siceraria	White flowering gourd, Calabash
Momordica charantia	Balsam pear, bitter melon, bitter cucumber, alligator pear, *fu kwa*
Tricosanthes anguina	Snake gourd
Cyclanthera pedata var. *edulis*	Caihua, wild cucumber

CUCURBITS

Origin

Cucurbits are of widely different places of origin. Most are of the old world, only the *Cucurbita* genus (pumpkins and squashes) and possibly *Cyclantera* (caihua) are of warm temperate, subtropical, and tropical American origin.

Botany

Most of the cucurbits are climbing or prostrate dicotyledonous plants of the tropics, subtropics, and milder regions of the temperate zones of both hemispheres. Most are annuals and some are perennials. All are frost sensitive. Cucurbits are grown mostly for their fruits; however, shoots and flowers of some species are used as food.

In the cucurbits there are several kinds of flower types or sex expressions as follows.

Hermaphrodite—Flower has both functional pistillate and staminate parts, i.e., perfect flower.

Monoecious—Flowers are staminate and pistillate.

Gynomonoecious—Some flowers are hermaphrodite and some pistillate.

Andromonoecious—Some flowers hermaphrodite and some staminate.

Trimonoecious—Three types of flowers: hermaphrodite, pistillate and staminate.

Gynoecious—All flowers pistillate.

Androecious—All flowers staminate.

Dioecious—Pistillate and staminate flowers each type borne on separate plants, i.e., female plant and male plant.

Most cucurbits are monoecious or andromonoecious.

The root system is extensive, and depth of rooting depends on the species. The stems are trailing with laterals branching from the nodes. Some vines of the *Cucurbita* may reach lengths of 12–15 m (40–50 ft). Leaves may be simple but most are three to five lobes with considerable variation in shape, size, and depth of lobes. Tendrils are borne at the axils of leaves; they are absent in dwarf or bush squashes. The flowering habit, depending on the species or cultivar, is monoecious (male and female flowers as separate organs but borne on the same plant) or dioecious (unisexual) (male and female flowers on separate plants). The fruits for which these plants are cultivated vary greatly in size, shape, and color. The fruit is an inferior berry or pepo. The cucurbits produce the largest fruits in the plant kingdom.

Culture

Temperature

Cucurbits are warm season crops with mean optimum temperatures of 18°–30°C (68°–85°F). They cannot tolerate long periods at temperatures below 10°C (50°F). Melons require a long warm season while cucumbers can be grown at the lower range and in shorter time because immature fruits are harvested.

Photoperiod

Most cucurbits are day-neutral. Since cucurbits are tropical or subtropical plants, they seem to grow best at about 12 hr days. Chayote is an exception; this plant is short-day and initiates flowers when day lengths are slightly under 12½ hr. It has been observed that cucumbers flower more profusely in 11 hr photoperiods.

Culture

Soil reaction should be from slightly acid to slightly alkaline; if too acid it should be limed. Soils high in organic matter are not recommended. Because of the extensive root system, the soil should be deep, a minimum of 45 cm (18 in.). It is stated that the extent of lateral roots of cucurbits exceeds the vine growth. Squash and melon roots penetrate the soil 2 m (6 ft) or more. In areas of limited rainfall and where plants cannot be irrigated, a heavy soil with high water holding capacity is preferred. Soil should be well drained; cold and wet soils are not tolerated by cucurbits. They are very sensitive to saline soil conditions.

For a good stand of seedlings, the soil temperature at planting must be above 16°C (60°F); seeds germinate best at 24°–35°C (75°–95°F). The effect of soil temperature on cucurbit seed germination is shown in Fig. 23.1. In cool soils the seedlings are very much susceptible to diseases. In contrast, in warm soils seedlings are vigorous and can overcome infections. In cold climates, cucurbits are started in pots and transplanted with soil surrounding the roots (bare-rooted plants do not transplant well) to the field when the soil has warmed in late spring. Plants are spaced according to vine growth of the particular cultivar. For cucumber spacing between plants is 30–40 cm (12–16 in.) between plants in 90 cm

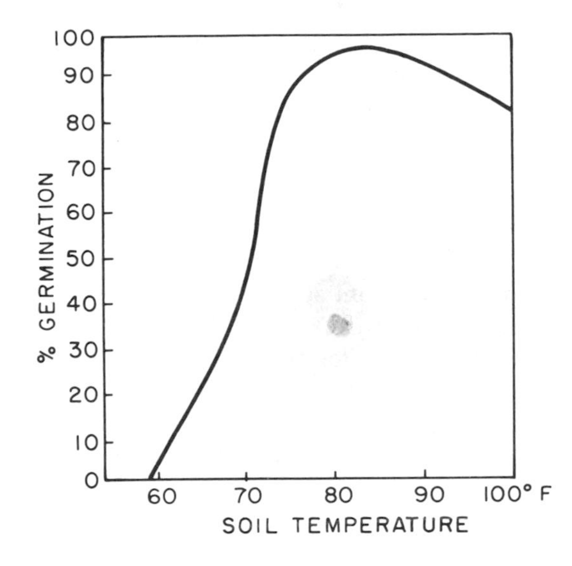

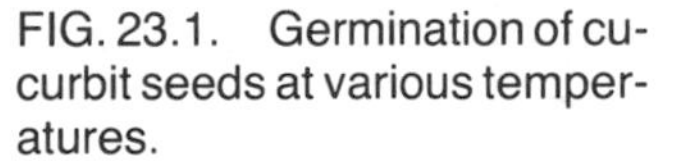
FIG. 23.1. Germination of cucurbit seeds at various temperatures.
Drawn from data of Harrington and Minges, (1954.)

(36 in.) rows while watermelon is spaced 90–130 cm (36–45 in.) between plants in 2.7–3 m (9–10 ft) rows.

Large amounts of water are required to grow cucurbits; 400–600 mm (15–24 in.) of water are necessary in semiarid regions. Lack of water during fruit development can cause misshapen fruits.

Some fertilization is generally necessary to obtain high yields. About 70–110 kg N, 45–90 kg P, and 35–110 kg K/ha (60–100 lb N, 40–80 lb P, and 30–100 lb K/acre) are recommended, depending on the residual fertility of the soil.

Since fruits are the part harvested, increases in perfect or female flowers is desirable. In monoecious species, the following has been observed. Male flowers increase under long days and high temperatures. Male flowers increase with increase in fruit load. Some cucumber breeding lines are gynoecious (all pistillate flowers); in such lines, pollinator plants (those with staminate or perfect flowers) are needed to set fruit. Chemical growth regulators, such as ethephon, have been used to increase female flowers. (Ethephon decomposes releasing the growth regulator, ethylene.)

Seed Storage

For longevity, seeds must be kept at low temperatures and under low humidity. The moisture of the seeds should be 5–8%. In the tropics, if seeds are kept at high temperatures with high humidity, germination of cucurbit seeds can be drastically reduced in 1 month. Under high temperature and high humidity, seeds will not germinate after 6 months. Seeds in airtight containers with silica gel to absorb moisture and at low temperatures of 0°C (32°F) will remain viable for many years.

Time of Harvest (Maturity)

Summer squashes are harvested immature. The fruit grows very rapidly in warm weather, and are ready for harvest 3–7 days following anthesis. Many of the cucurbits, except melons, are harvested immature and used like summer squash in Asia. From seeding, pumpkins and winter squash require 3–4 months to mature.

Melons should be harvested when mature for best quality. For shipment, they are harvested when almost mature and allowed to mature (ripen) before using. Ethylene gas treatment assures uniform ripening in some cultivars (honeydew melon). Immature fruits are undesirable because the sugars, flavor, texture, and aroma have not developed; ethylene treatment will not improve the quality of these fruits. With overripe fruits many biochemical and physical changes have taken place which altered the flavor as well as the texture.

Handling

Harvested fruits should be handled and shipped with care to avoid bruising and exposure to unfavorable conditions. Field heat of the harvested melons should be removed as quickly as possible, to about 10°C (50°F).

Storage

Immature fruits can be stored for short periods of time. Depending on the species, mature fruits can be stored for several weeks to many months provided they are stored under proper conditions (see Table 23.1). Winter squash and pumpkin can be cured prior to storage at

TABLE 23.1. STORAGE OF CUCURBITS: RECOMMENDED CONDITIONS

Crop	Storage (°C)	Temp. (°F)	Relative humidity	Approximate storage life
Cucumber	7–10	45–50	90–95	10–14 days
Summer squash	7–10	45–50	85–95	5–14 days
Melons				
Cantaloupe	2–4	35–40	85–90	5–15 days
Crenshaw and Persian	7–10	45–50	85–90	2 weeks
Honeydew	7–10	45–50	85–90	3–4 weeks
Casaba	7–10	45–50	85–90	4–6 weeks
Watermelon	4–7	40–45	80–85	2–3 weeks
Pumpkin	10–13	50–55	70–75	2–3 months
Winter squash	10–13	50–55	70–75	4–6 months

Source: Ryall and Lipton (1979).

27°–29°C (80°–85°F) at 80–85% RH for about 10 days for healing of wounded tissue.

Cucurbit fruits are subject to "chilling injury" if stored for prolonged periods at temperatures below 10°C (50°F).

Nutritive Value

The nutritive value of some cucurbits is shown in Table 23.2. Of those on the list, the *Cucurbita* genus fruits, both immature and mature, are high in vitamin A value. Leaves, shoots, flowers, seeds, and, in some cases, the roots are also used for food.

Economics

Most cucurbits are used fresh; only the cucumber and gherkin are preserved to any large extent. Cucumbers are made into pickles and some pumpkins and squash are canned or frozen. Citron and lagenaria (calabash) are used to make glace for fruit cakes. Cucumbers, melons, and squash are easily handled and shipped to distant markets. Greenhouse cucumbers are of considerable importance during the winter in Europe and in the eastern parts of the United States. Tables 23.3 and 23.4 shows that, among the cucurbits, watermelon is produced the most and cucumber ranks second.

CUCUMBER: *Cucumis sativus*

Origin

Cucumber has not been found in the wild form but *C. hardwickii*, found wild in the Himalayas, has been suggested as the ancestor of the present-day cucumber. Indigenous in India are many types of cucumber of various fruit sizes, shapes, and color.

Botany

There are about 40 species in the genus *Cucumis*; cucumber *(C. sativus)*, which is cultivated for its immature fruits, is very important. The plant is an annual vine trailing along the ground from 1 to 3 m (3 to 10 ft). Cucumber has rough cordate leaves with three to five lobes and angled stems. It is monoecious. Under long days and high temperatures the plant produces more staminate than pistillate flowers but under

TABLE 23.2. NUTRITIVE VALUE OF CUCURBITS IN 100 g EDIBLE PORTION

			Macroconstituents					Vitamins									
				g					mg				Minerals (mg)				
Crop	Edible part	(%) refuse	Energy (cal)	Water	Protein	Fat	CHO	A (IU)	B1	B2	Niacin	C	Ca	Fe	Mg	P	Ref.[a]
Cucumber	Immature fruit	15	12	96	0.6	0.1	2.2	45	0.03	0.02	0.3	12	12	0.3	15	24	1
Melons	Mature fruit	41–45	26–41	87–92	0.6–1.0	0.1	6.3–10.3	Trace–4200	0.06	0.02	0.4–0.9	19–45	5–10	0.2–0.4	8–17	7–39	1
Watermelon	Mature fruit	50	36	90	0.6	0.1	9.1	300	0.08	0.02	0.2	6	5	0.2	10	14	1
Watermelon	Seeds	65	514	10	40	43	3.1	30	0.02	0.06	1.2	Trace	54	5.6	—	444	2
Pumpkins and winter squash	Mature fruit	15–40	20–40	85–91	0.8–2.0	.1–0.5	3.3–11.0	340–7800	0.07–0.14	0.01–0.04	0.5–1.2	6–21	14–48	0.4–7.0	16–34	21–38	1
Chinese winter melon	Mature fruit	25	15[2]	96	0.2	0.1	3.5[2]	Trace	0.02	0.03	0.5	14	14	0.4	16	7	1
Summer squash	Immature fruit	1–5	13–22	92–95	1.0–1.4	.1–0.2	2.0–4.0	80–340	0.05–0.07	0.03–0.04	0.4–0.6	9–19	15–19	0.4–0.5	20–26	28–38	1
Summer squash	Leaves	51	33	90	3.8	0.7	4.9	2400	0.12	0.18	1.1	19	159	1.6	—	99	2
Summer squash	Flowers	41	29	90	2.0	0.5	5.6	910	0.05	0.11	0.9	24	74	3.1	—	38	2
Chinese okra	Immature fruit	10	20	93	1.2	0.2	4.0	410	0.05	0.06	0.4	12	20	0.4	—	32	1
Chinese squash	Immature fruit	10	18	94	1.0	0.2	3.3	450	0.02	0.04	0.5	57	13	0.4	—	23	1
Sponge gourd	Immature fruit	29	21	94	0.6	0.2	4.9	45	0.04	0.02	0.3	7	16	0.6	—	24	2
Snake gourd	Immature fruit	14	16	95	0.6	Trace	4.0	235	0.02	0.03	0.3	12	22	0.3	—	15	2
Bottle gourd	Immature fruit	16	15	95	0.5	0.1	3.5	10	0.04	0.02	0.4	11	16	0.4	—	14	2
Chayote	Immature fruit	14	26	93	0.9	0.3	5.3	50	0.03	0.04	0.5	11	19	0.4	14	20	1
Chayote	Root	—	—	72	0.1	0.1	—	0	0.07	0.07	1.1	21	4	1.0	—	84	3
Chayote	Leaves	70	34	93	—	0.2	6.4	3000	0.07	0.07	0.8	33	66	0.3	—	47	2

[a] **Key to references:** (1) Howard *et al.* 1962; (2) Intengan *et al.* (1964); (3) Munsell *et al.* (1950); (4) Watt and Merrill (1950).

TABLE 23.3. WORLD PRODUCTION OF CUCUMBERS AND GHERKINS AND PUMPKINS, SQUASH, AND GOURDS

	Cucumbers and gherkins			Pumpkins, squash and gourds	
	Area (10^3 ha)	Production (10^3 MT)		Area (10^3 ha)	Production (10^3 MT)
World	788	10,147		529	5,176
Continent					
Africa	18	263		64	967
North and Central America	98	1,104		38	237
South America	3	45		66	668
Asia	361	5,101		189	2,259
Europe	122	2,218		162	960
Oceania	1	16		10	85
U.S.S.R.	185	1,400			
Leading countries					
1. China	210	2,570	1. China	79	900
2. U.S.S.R.	185	1,400	2. Egypt	24	446
3. Japan	26	1,150	3. Turkey	20	350
4. United States	79	874	4. Romania	132	332
5. Turkey	28	450	5. Argentina	31	315
6. Netherlands	2	385	6. Italy	13	299
7. Poland	32	232	7. Japan	15	265
8. Egypt	15	231	8. South Africa	12	210
9. Spain	7	210	9. Korea, REP	11	175
10. Romania	18	202	10. Syria	13	148

Source: FAO (1979).

short-day conditions the ratio of pistillate to staminate flowers increases. Pollination is usually by bees.

Some European cultivars bear parthenocarpic fruits and require no pollination for fruit set. Gynoecious lines are used in production of hybrid seed. When the chemical ethephon is sprayed on cucumber plants, only pistillate flowers are produced.

Cucumber cultivars are classified according to the color of spines (spicules), either white or black on the fruit. White spines are usually market types as they retain the green color longer than the black spine types. Black spine types are almost exclusively used for pickling because of the better color retention when kept in brine.

Although cucumbers are day-neutral to flowering, under long days and high temperature conditions, the plant produces more staminate flowers than pistillate flowers.

In Japan cucumbers are classified into two groups: (1) North China group: summer cultivars which are day-neutral (2) South China group: during short days (10 hr days) the plant produces mainly pistillate flowers and under long days the plant produces few pistillate and many staminate flowers.

TABLE 23.4. WORLD PRODUCTION OF WATERMELONS AND CANTALOUPES AND OTHER MELONS

	Watermelons			Cantaloupes and other melons	
	Area (10^3 ha)	Production (10^3 MT)		Area (10^3 ha)	Production (10^3 MT)
World	1,965	24,096		481	6,372
Continent					
Africa	121	2,083		34	544
North and Central America	122	1,413		77	1,018
South America	131	1,034		23	280
Asia	819	13,169		228	2,977
Europe	158	2,949		119	1,551
Oceania	5	47			
U.S.S.R.	610	3,400			
Leading countries					
1. China	222	4,070	1. China	83	1,420
2. Turkey	210	4,000	2. United States	46	719
3. U.S.S.R.	610	3,400	3. Spain	67	705
4. Egypt	53	1,344	4. Iran	52	480
5. Japan	35	1,263	5. Italy	14	351
6. United States	68	1,094	6. Japan	14	290
7. Iran	100	930	7. Egypt	13	266
8. Italy	25	800	8. Mexico	25	250
9. Greece	30	707	9. Syria	29	212
10. Iraq	51	648	10. France	15	211

Source: FAO (1979).

Culture

Cucumbers, a warm season crop, grow best in the temperature range of 18°–30°C (65°–86°F); the plants suffer chilling injury at temperatures below 10°C (50°F).

A well-drained fertile loam type soil in a pH range of 6.5–7.5 is recommended for growing the crop. Usually the seeds are planted directly in the field as transplanting of cucumber seedlings is difficult. Soil temperature should be 20°C (68°F) or higher; the optimum for seed germination is 25°–35°C (77°–95°F). At 20°C, it takes 6–7 days for seedlings to emerge, while at 25°C the time is cut in half. The spacing between plants is 30–45 cm (1–1½ ft) on rows 1.2 m (4 ft) apart. Cucumbers are often planted on hills 90–120 cm (3–4 ft) apart, with two seedlings per hill.

About 70 kg N/ha (60 lb/acre), 110 kg P/ha (100 lb/acre), and 70 kg K/ha (60 lb/acre) are recommended for cucumbers. About half of the nitrogen fertilizer is used as preplant application and the remainder is applied as a side-dressing after the plants are thinned at the three to four true-leaf stage.

Although cucumbers are a deep root crop, adequate soil moisture is needed for good yields. In cool regions when the crop is planted in heavy soil near field capacity, there is sufficient moisture to grow a crop. However, yields are generally increased if additional water is applied during the growing season. A minimum of 400 mm (16 in.) of water is necessary to grow cucumbers in a dry climate.

Harvest and Storage

For the fresh market, fruits are harvested before they have fully elongated and the seeds are still succulent. For pickling several stages of immature fruits are harvested. Depending on the temperature and cultivar, 55–70 days are required from planting to first harvest.

Storage

Cucumber fruits held lower than 10°C (50°F) suffer chilling injury, the optimum condition are 12°–13°C (54°–55°F) and 95% RH. Yellowing occurs at 15°C (59°F). Waxing or packaging in plastic film retard moisture loss.

Nutritive Value

The bitterness in cucumbers is caused by terpenes called cucurbitacins. The amount in the fruit is influenced by growing conditions. It has been reported that a mutant lacking the ability to form these bitter compounds has been discovered. The nutritive value of cucumbers is given in Table 23.2.

Pests and Diseases

Cucumber beetle, squash bug, and melon aphids are insects that feed on cucumbers. Nematodes attack the roots which inhibit growth and reduce yields.

Fusarium wilt, anthracnose, downy and powdery mildews, curly top, cucumber mosaic, and squash mosaic virus are the principal diseases of cucumber.

GHERKINS (WEST INDIAN), BUR GHERKIN: *Cucumis anguria*

The gherkin, sometimes called the West Indian gherkin, is not indigenous to the Americans, as was first thought, but is of African origin. It was probably brought to the New World by slaves from West Africa. Gherkin is a nonbitter variant of the normally bitter fruited wild African species *C. longipes*.

The plant is a monoecious trailing vine with tendrils and stiff hairs (Fig. 23.2); the leaves are deeply three to five lobed with rounded stems. At first glance the plant appears somewhat like the watermelon *(Citrullus vulgaris)*. Fruits are oval to oblong, about 3–5 cm (1¼–2 in.) long and borne on slender peduncles 10–20 cm (4–8 in.) in length. The rind is covered with long flexible prickly spines about 5 mm (¼ in.) or more in length, pale green when immature, and turn yellowish when mature. The flesh is pale green in color with white seeds.

The culture of gherkins is similar to that of cucumbers.

MUSKMELONS, CANTALOUPE, HONEYDEW: *Cucumis melo*

Origin

The primary center of origin of *Cucumis melo* is deduced to be in tropical and subtropical West Africa as there are forty or more spe-

FIG. 23.2. Gherkin *(Cucumis anguria)*.

cies of this genus endemic to this region. A secondary center occurs in the region covering Iran, southern Russia, India, and east to China.

Botany

Muskmelons are either monoecious or andromoecious annuals with long trailing vines similar to the cucumber *(Cucumis sativus)* except the leaves are shallow lobed and the more or less rounded.

There are many different forms of *Cucumis melo*. These are classified into several botanical varieties as follows:

Cucumis melo var. *cantaloupesis*—cantaloupe (Europe)
var. *inodorus*—winter or casaba melon
var. *flexucus*—snake melon, serpent melon
var. *reticulatus*—netted or nutmeg muskmelon, cantaloupe (United States), Persian melon
var. *conomon*—Oriental pickling melon
var. *chito*—mango melon, garden melon
var. *dudaim*—pomegranate melon, Queen Anne's pocket melon

There is considerable variation in fruit size and shape; external appearance may be smooth, sutured, or netted; the skin color may be white, green, yellowish green, yellow, yellowish brown, or speckled yellow or orange with green or yellow background. Fruits of some cultivars abscise when ripe. With maturation the fruit accumulates sugars, which are predominately fructose glucose and sucrose. As with many cucurbits the carbohydrate translocated from the leaves to the fruit is stachyose, a tetrasaccharide. There is no stachyose in the developing or ripe fruit; also there is no starch accumulation. Upon ripening the fruit softens and fruity aromatic essences are formed in the fruit.

Culture

Muskmelons are warm season crops requiring 85–120 days from planting to harvest depending on cultivar and growing conditions. The mean optimum temperature for growth is 18°–24°C (65°–75°F). It does best in somewhat arid conditions as it is less susceptible to leaf fungal disease in low humidity.

Muskmelons require deep well-drained fertile soil with neutral to slightly alkaline conditions. The plant is sensitive to acid soil. For early production light sandy loam types are recommended but good yields are obtained from heavier soils but not from heavy clay or peat soils.

About 65–135 kg N/ha (60 to 120 lb/acre) and 28–66 kg P/ha (25–50 lb/acre) are recommended. About half of the fertilizer is placed about 6 cm (2½ in.) to the side and 6 cm (2½ in.) below the seed row; the rest of the fertilizer is side-dressed shortly after thinning when the seedlings are at the four to six true-leaf stage. K may be added to the fertilizer mix if the availability of this nutrient is low in the soil.

Seeds are planted 1½–4 cm (½–1½ in.) deep in rows or in hills 180–210 cm (6–7 ft) apart. The depth of planting depends on soil type and moisture. The spacing between plants in the row should be 30–60 cm (12–24 in.) apart; a single plant for 30 cm spacing and two plants for 60 cm spacing.

Cultural practices for muskmelons is similar to that of cucumbers. Adequate moisture is required until about a week before the fruits begin to ripen.

At spacing of 30 cm, a plant can set one to two fruits; additional fruits will not set but will abort until the seeds in the fruits are mature. At this stage, the perfect flowers will set farther up the vine and will not interfere with the continued development of the fruit(s) set earlier. If more than two fruits are set and develop to maturity on a single plant, these fruits are more likely to be small and the soluble solids lower than those fruits from one or two fruits per plant.

Harvest and Storage

Harvest

Cantaloupe (*C. melo* var. *reticulatus*) maturity can be judged by the abscission zone (slip) which develops between the fruit and the peduncle; also there is a change in ground color from green to yellow and a slight development of aroma from the blossom end. Persian melons do not develop the abscission zone.

Honeydew (*C. melo* var. *inodorus*) also develops a change in color to yellowish white from greenish white and a waxy feel to the touch develops on the rind; also a slight aroma emanates from the blossom end when ripe. Honeydews do not form the abscission zone.

For shipment both cantaloupe and honeydew are harvested before they are eating ripe. If stored a few days at room temperature, the cantaloupe fruit ripens naturally. The honeydew generally requires gassing with ethylene to ripen. After ethylene treatment the melon ripens, softening from the blossom end and forming aromatic compounds. The sugar content of cantaloupes ranges from 8 to 14% whereas in honeydews it ranges from 10 to 16%.

Storage

Following harvest, cantaloupes should be cooled as rapidly as possible to about 10°C (50°F); this is best accomplished by "hydrocooling" with ice water. Cooling rapidly decreases the respiration rate and loss of sugars in the process. Hard ripe cantaloupes should be stored at 4°–5°C (39°–41°F) from 4 to 10 days followed by 1–2 days at 15°–16°C (59°–61°F) to ripen further; the relative humidity should be about 95%. Cantaloupe does not require ethylene treatment to ripen properly.

Honeydews of different ripeness require different treatment (H.K. Pratt, personal communication) as follows:

1. *Unripe but mature melons* (fruit with white background color which may have some light green, surface may be covered with fine fuzzy hairs; blossom end hard, no aroma, not springy). Gassing with ethylene (200 ppm) is essential at temperatures above 20°C (68°F) for fruits to ripen.
2. *Fruits that have initiated ripening* (fruit with white background color and slight waxy surface; blossom end slightly springy; slight development of aroma). Gassing with ethylene (200 ppm) is not essential but is considered beneficial.
3. *Ripe fruit* (fruit creamy white in color and has waxy surface; entire blossom end springy; at room temperature good aroma from blossom end, eating ripe). Needs no ethylene treatment.
4. *Slightly overripe* (early senescence). The surface color creamy white to pale yellow and quite waxy to touch. Flesh is soft and has strong aroma. Quality not good as (3) but still edible.
5. *Overripe* (senescent). Entire surface yellow and soft. Strong somewhat fermented odor. Flesh soft, stringy, and mealy. Usually not edible.

Honeydew melons need no precooling as with cantaloupes. After treatment of maturity (1) with ethylene the fruit should be cooled slowly in 2–2½ days to 16°C (60°F) and further cooled to 7°–10°C (45°–50°F) in the following 3–4 days. Fruits of maturity (2) and (3) can be placed directly into 7°–10°C (45°–50°F) and held for 2–3 weeks at 85–95% RH. Chilling injury occurs when the fruits are held below 5°C (41°F). If melons are not eating ripe from storage, they can be further ripened at 20°–27°C (68°–80°F) to the proper ripeness.

Nutritive Value

The yellow and orange fleshed melons contain B-carotene, provitamin

A; cantaloupe is particularly high in this pigment. Melons are quite high in vitamin C; cantaloupe contains 45 mg and honeydew contain 32 mg ascorbic acid per 100 g edible portion (Table 23.2). In the melon the aromatic flavor compounds are predominantly esters of acetic acid.

Pests and Diseases

Melon aphid, green peach aphid cucumber beetle, leafhopper, leaf miner, and red spider mites are pests of the melon plant. Soil pests include nematodes, wire worms, and corn seed maggot.

Melon seedlings are attacked by pythium and rhizoctonia in cold, wet soil while older plants are additionally attacked by fusarium, phytophthora, alternaria, stem blight *(Mycosphaerella)*, and powdery and downy mildews.

Bacterial wilt is transmitted by the cucumber beetle, and fusarium and verticillium wilts both soil borne can reduce crop yields.

Viral diseases include watermelon mosaic, cucumber mosaic, cantaloupe latent virus, which are aphid transmitted; squash mosaic is seed borne and beetle transmitted, and curly top is transmitted by beet leaf hoppers.

ORIENTAL PICKLING MELON, URI (JAPAN), YUEH KUA (CHINA): *Cucumis melo* var. *conomon*

The origin of the oriental pickling melon is obscure; reference to the plant is found in Chinese literature from 560 AD. Oriental pickling melon in Japan is grown mainly for making of pickles called "tsukemono" or "koko" perhaps according to one authority in Japan the variety name *conomon* was a corruption of the word *koko no mono* meaning material to make *koko*. There are several cultivars of oriental pickling melon, characterized by the color: white, 'shiro uri'; green, 'ao uri'; and striped, 'shima uri.'

The optimum temperature for growth is 25°–30°C (77°–85°F); but at 13°C (55°F) growth is very much retarded.

Culture of Oriental pickling melons is similar to cucumber and melons. As with most cucurbits extra care must be taken not to disturb roots when transplanting. Plants begin to flower about 40 days after planting in the spring. The fruits are usually cylindrical, 20–30 cm (8–12 in.) in length and 6–9 cm (2½–3½ in.) in diameter. For pickling fruits are harvested when full size but still immature. The seed cavity is removed and only the rind used in making of pickles. In Southeast Asia, besides pickling, the fruits are used like summer squash in various ethnic dishes.

WATERMELON, SANDIA (SPANISH): *Citrullus vulgaris (Citrullus lanatus)*

Watermelon is indigenous to south-central Africa, where reports by the early explorers describe the entire countryside covered with watermelon vines. The crop has been cultivated in the Mediterranean region for thousands of years. Watermelon was taken to China from India and it was introduced to the Americas in post-Columbian times.

The citron or preserving melon (*C. vulgaris* var. *citroide*) is white fleshed and not edible raw unlike watermelons. The flesh is used in preparation of glace and in sweet preserves.

Botany

A monoecious spreading annual vine with large pinnately lobed leaves, watermelon is grown for the large juicy sweet fruits. Fruits may be oblong, ellipsoidal, or spherical in shape with a thick but fragile rind and can exceed 10–12 kg (22–26 lb) in weight. The flesh may be white, greenish white, yellow, or pink to red at maturity. Seeds may be white, greenish, yellow, brown, red, or black in color and contain large amounts of carbohydrates, fats, and proteins. Some cultivars are grown especially for the large seeds, which are roasted and eaten.

There are two types: bitter and nonbitter. The bitterness is genetically controlled by a single dominant gene. The bitter substances are called cucurbitacins, tetracyclic triterpenes, found in many cucurbits and reported to be both insect attractants and repellents.

The triploid hybrid watermelon is seedless; it has small empty ovules, remnants of which can be found in the flesh. The triploid (3X) is produced by crossing a tetraploid (4X) plant, which was obtained by colchicine treatment, with a diploid (2X). At pollination, a regular diploid plant, is required to "set" a tetraploid pistillate flower.

Culture

Watermelons are a warm season crop requiring a relatively long growing season, about 4 months of frost-free weather. It grows well at high temperatures, at means of greater than 21°C (70°F).

Soils with good drainage are preferred; sandy loams to loam types are usually selected though heavier soils can be used. Unless the cultivar has resistance to fusarium wilt, continuous cropping of the same land should be avoided. Rotation once every 4–6 years is desirable and once every 10 or more years if fusarium and nematodes are a problem.

Recommended fertilization is 70–110 kg N/ha (60–100 lb/acre) and 30–55 kg P/ha (25–50 lb/acre) and K if the soil is low or deficient in this element. About half of the fertilizers should be applied 6 cm (2½ in.) to the side and 6 cm (2½ in.) below the seeds; the remainder should be applied as a side-dressing shortly after thinning at about the four to five true leaf stage.

Watermelons are planted in hills about 1–2 m (3–6 ft) apart in rows 2½–3 m (8–10 ft) apart; the seedlings are thinned to one or two plants per hill, depending on the spacing. When thinned, the plants should be cut so that the roots of the remaining plant(s) are not disturbed. If seedlings are transplanted, the soil surrounding the roots should not be removed and the root disturbance be kept at a minimum in handling.

The optimum soil temperature range for seed germination is 25°–35°C (77°–95°F). At 20°C (68°F) germination is satisfactory but it takes about 12 days for emergence; at 15°C (59°F) germination is poor.

A minimum of 380 mm (15 in.) of water is required to grow watermelons in light soils. Since watermelon is deep rooted, in heavy soils there may be adequate available water to mature a crop, provided the soil is near field capacity at planting.

Staminate flowers are formed first followed by pistillate flowers; insects, usually bees, are the pollinating agents. Misshapen fruits are caused by poor or inadequate pollination; therefore, to ensure good fruit set, hives are placed in fields. Pruning to two to three fruits per plant is sometimes practiced to gain larger size and to increase the sugar content. The vines should not be pruned. From planting to harvest requires 75 to 130 days, depending on cultivar and growing conditions, especially temperature.

Harvest and Storage

Harvest

The fruit is harvested when the flesh is sweet but not overripe. When overripe the flesh becomes mealy and stringy. The following criteria are often used for ripeness: (1) the ground spot, light colored or white spot where melon rests on ground, changes from white to a light yellow color; (2) by thumping: immature fruits give a high pitched sound; ripe fruits give a low pitched sound; (3) the tendril directly opposite the fruit peduncle changes from green to brown and may appear dry; however, ripe melons with criteria (1) and (2) sometimes have green tendrils. The best way is to plug or cut a few melons in several parts of the field and taste and examine the flesh for ripeness, it can be assumed that most fruits of comparable size in the field are in the same stage of ripeness. The use of refractive index of the juice is a measure of percentage soluble solids. Good quality melons should have a soluble solids reading of at least 10.5% at the center or core of the fruit. This location is important as the soluble solids content varies with the location within the fruit; the highest is at the center, blossom end the next highest, upper side intermediate, and the bottom rind (ground spot) and the stem end the lowest.

Ripe fruits harvested during the early morning hours are turgid and may easily crack while those harvested later in the day are not quite as turgid and less apt to crack when handled.

Storage

Watermelon fruits being a warm season crop, can be subject to chilling and injury. The recommended temperature is 13°–16°C (55°–60°F) for storage up to 2 weeks. For storage of longer than 2 weeks 7°–10°C (45°–50°F) is recommended. The flesh color fades at temperatures below 10°C (50°F). Relative humidity of 80–85% is recommended since water loss from the waxy rind is very low at the recommended temperatures.

Nutritive Value

Watermelon often serves as a source of water in areas of drought and when water is contaminated. The immature fruit can be harvested and used like summer squash. The juice can be made into syrup and also fermented into a wine-like alcoholic beverage.

The toxic bitter compounds, cucurbitancins, in watermelon are normally hydrolyzed by specific enzymes to nontoxic substance in most cultivars. There is a small percentage of the human population which cannot detect bitter flavors. When the combination of bitter watermelon and inability to detect bitterness occurs, poisoning can take place.

Pests and Diseases

Aphids and 12 spotted cucumber beetle *(Diabrotica)* attack leaves, stems, and very young fruits. Spider mites and flea beetle chew on the leaves. Nematodes attack the roots and the fusarium wilt fungus attack the vascular tissue causing the plant to wilt.

Curly-top, a viral disease transmitted by leafhoppers causes old leaves to yellow and new leaves to appear stunted and dark green. Watermelon mosaic virus, carried by aphids, causes the mottling of leaves, stunting of vines, and misshapen fruits.

SQUASHES AND PUMPKINS: *Cucurbita* spp.

Principal genera and species	Common name
Cucurbita pepo	Common field pumpkin, winter and ornamental gourds
C. pepo var. *melopepo*	Summer squash
C. moschata	Winter squash, pumpkin
C. mixta	Pumpkin, winter squash (Mexico and Central America)
C. maxima	'Hubbard,' 'banana,' and 'Boston Marrow'
C. ficifolia	Buffalo gourd

The cultivation of *C. pepo* prior to Columbus ranged from present-day Mexico City north into the Southwest of the United States, northeast into the Middle West and East Coast of the United States to the Canadian border. This species is more tolerant of cold climates than others in the genus. *Cucurbita mixta*'s distribution has been confined from Guatemala north into the Southeast of the United States, as it is less tolerant to cold than *C. pepo*.

Cucurbita moschata is more widespread; it spread from its center of origin north to the Southwest of the United States and south through the Andes to Peru.

The progenitor of *C. maxima* moved into South America where the species evolved in Brazil, Bolivia, Chile, and Argentina.

Cucurbita ficifolia, the only perennial species, spread along the high altitudes of the Andes as far south as Chile. It is tolerant to cool weather and is short day for flowering.

Botany

Except for *C. ficifolia*, all the cultivated species are annual plants which are usually climbing or trailing vines with tendrils. Bushy plants such as summer squash *(C. pepo)* have short internodes and no tendrils.

Cucurbita: leaves are large simple, alternate, and shallow to deeply lobed. Plants are monoecious with large brilliant yellow petaled flowers borne singly in the leaf axils. The filaments of the staminate flowers are free but the anthers are more or less united; the entire structure of stamens is called androecium. The peduncle of the pistillate flower is characteristically different in the five cultivated species. Fruits are inferior berries, varying greatly in size, shape and color. *Cucurbita* fruits are among the largest found in the plant kingdom, some exceed 50 kg (110 lb) in weight.

The characteristics of the five cultivated species are given in Table 23.5 and the peduncles are shown in Fig. 23.3.

The *Cucurbita* genus is important in that the mature fruits can be stored up to 6 months under proper conditions. There is some confusion as to the names of pumpkins and squashes. The following explanation may be helpful.

Pumpkin. Edible fruit of *Cucurbita* genus, utilized when ripe as fodder or for human consumption as in pies. The flesh is somewhat coarse and strong flavored and is usually *not* served as a vegetable.

Squash: Immature fruit of *C. pepo*, baking cultivars of *C. maxima*, and those mature fruit of *C. pepo*, *C. moschata*, and *C. mixta* used as table vegetable or pies and as livestock food. The flesh is usually fine grained and mild in flavor, and suitable for baking.

Summer squash in Great Britain is called "vegetable marrow" and the word "marrow" is used for either mature fruits of *C. pepo* or *C. maxima* served boiled or stewed.

Table 23.6 shows the common names of various *Cucurbita* species.

Culture

Pumpkins and winter squashes are warm season crops requiring 85–120 days from planting to maturity. Summer squashes require 40–50 days as fruits are harvested only a few days after bloom. These crops are adapted to monthly mean temperatures of 18°–27°C (60°–80°F) and are not tolerant to near freezing temperatures. For seed germination soil temperatures should be above 15°C (59°F), the warmer the soil temperature the more rapid the germination rate; germination percentage is highest at 35°C (95°F). Seeds are planted about 2–3 cm (1–1¼ in.) deep in moist soils; in sandy and very light soils, seeds are planted 4–5 cm (1½–2 in.) deep. *Cucurbita* should be grown in fertile well-drained soil in the pH range 6.5–7.5. A balanced fertilizer of about 110 kg N/ha (100 lb/acre), 40 kg P/ha (36 lb/acre) and 90 kg K/ha (80

TABLE 23.5. CHARACTERS DIFFERENTIATING THE CULTIVATED SPECIES OF *Cucurbita*

Species		Foliage	Stem	Androecium	Peduncle	Fruit flesh	Funicular attachment of seed	Seed margin; color
C. pepo	Annual	Spiculate	Hard angular	Short, thick conical	Hard angular, ridged	Coarse grained	Obtuse, symmetrical	Smooth, obtuse; white, buff, or brown
C. moschata	Annual	Non-spiculate	Moderately hard, smoothly angled	Long, slender, columnar	Hard, smoothly angular, flared	Fine grained or coarse with gelatinous fibers	Obtuse, slightly asymmetrical	Scalloped, obtuse; white, buff, brown
C. mixta	Annual	Non-spiculate	Hard angular	Long, slender, columnar	Hard, basically angular, but enlarged by hard cork	Coarse grained	Obtuse, slightly asymmetrical	Barely scalloped, acute; white, buff, or brown
C. maxima	Annual	Moderately spiculate	Soft, round	Short, thick, columnar	Soft, basically round, but enlarged by soft cork	Fine grained	Acute, asymmetrical	Smooth, obtuse; white, buff, or brownish
C. ficifolia	Perennial	Moderately spiculate	Hard, smoothly angled	Short, thick, columnar	Hard, smoothly angled, slightly flaring	Coarse, tough, fibrous	Obtuse, slightly asymmetrical	Smooth, obtuse: black or tan

Source: Modified from Whitaker and Bohn (1950).

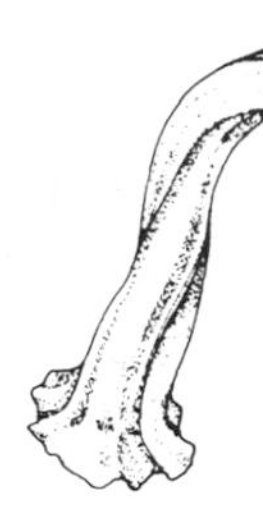

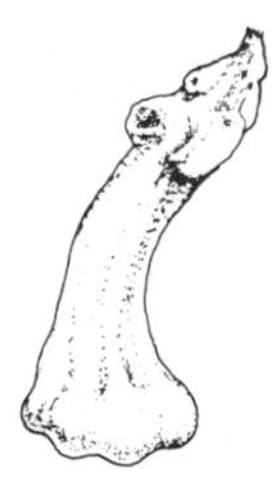

C. maxima C. pepo

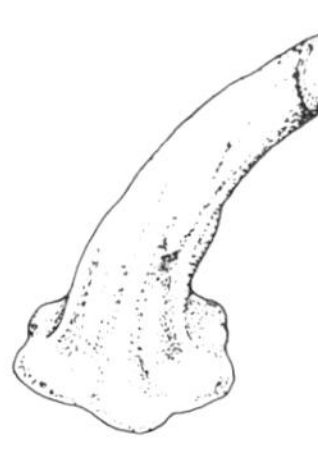

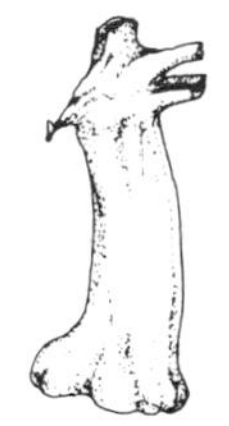

C. mixta C. ficifolia

FIG.23.3. Peduncle or fruit stalk characteristics of *Cucurbita* genus. *C. maxima*: Large, soft, spongy not noticeably furrowed (ridged). *C. pepo*: Hard and woody, distinctly furrowed (five to eight sharp angled ridges) some may have spicules. *C. moschata*: Hard and woody distinctly furrowed (five rounded ridges) some have flared base at attachment to fruit. *C. mixta*: Hard five-edged rounded ridges with little or no enlargement at base. *C. ficifolia*: Small hard, smooth angled and slightly flared at the base.

TABLE 23.6. COMMON NAMES OF SPECIES OF *Cucurbita*

Species	Common name					
	Summer squash	Winter squash	Pumpkin	Marrow	Cushaw	Gourd (yellow flowers)
C. pepo	+	+	+	+		+
C. moschata		+	+			
C. maxima		+	+	+		
C. mixta		+	+		+	

Source: Whitaker and Davis (1962).

lb/acre) is required. About half of the fertilizer is applied at seeding and the remainder at about the four to six true leaf stage as side-dressing.

Pumpkins and winter squashes are spaced about 2–3 m (6–10 ft) apart in hills, two seedlings per hill. Summer squash, being a bush, does not require such wide spacing; it is planted about 1–1½ m (3–5 ft) apart in rows 2 m (6½ ft) apart. Summer squash requires more frequent watering than the vining types as the roots of summer squash penetrate the soil to a depth of 1½ m (4 ft) compared to the latter's depth of 2 m (6½ ft) or more.

Pumpkins and squashes being monoecious (separate male and female flowers) require pollinators for fruit set. Bees are the main pollinating agent. Poor pollination could result in yield reduction and misshapen fruits.

Harvest and Storage

Harvest

Summer squashes are harvested while the fruits are very immature, usually 2–7 days after anthesis, the size depends on the market preference. Growth of summer squash is shown in Figure 23.4. Fruits should be continually picked off since fruits left to mature on the vine suppress pistillate flower formation. Care should be taken to avoid bruising or cutting harvested fruits.

Fruits of pumpkins and winter squash are allowed to mature on the vine. In temperate climates if the fruits are mature, the rind is hard and the vines begin to senesce; they should be harvested before frost. A sharp knife should be used to cut the peduncle leaving about 2–5 cm (1–2 in.) on the fruit.

Curing

Cuts and bruises of mature fruits can be healed by suberization. This is accomplished at 27°–30°C (80°–86°F) and 80% RH for a period of 10 days. The healing of cuts and bruises prevents decay organism from entering the wounds. Following this period the temperatures should be lowered to 10°–15°C (50°–59°F) and lower humidity.

Storage

Summer squash can be held for 1 week at 10°C (50°F) and for only 2–4 days at 4°C (40°F); chilling injury occurs if held for several days at temperatures below 10°C (50°F).

Winter squash and pumpkins can be held 30–180 days, depending on the cultivar, at 10°–15°C (50°–59°F) and 60% RH. If stored at tempera-

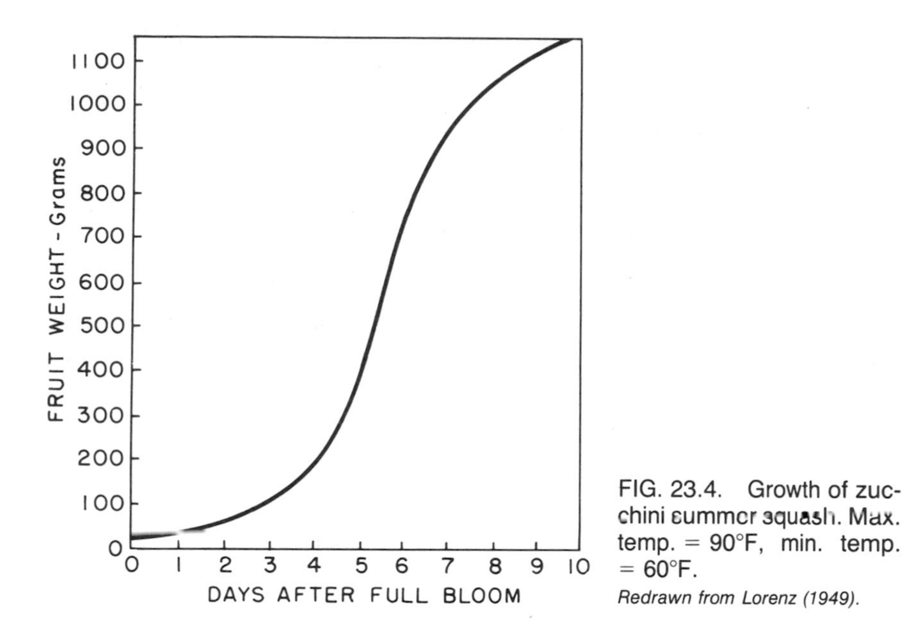

FIG. 23.4. Growth of zucchini summer squash. Max. temp. = 90°F, min. temp. = 60°F.
Redrawn from Lorenz (1949).

tures greater than 15°C (59°F), respiration rate is high, resulting in loss due to shrinkage. The fruits should not be piled on top of each other as humidity increases, which results in decay.

Nutritive Value

Winter squashes and pumpkins are higher in stored carbohydrates (energy) than summer squashes. Cucurbits are good source of vitamin C and the yellow and orange fleshed pumpkins and winter squashes are very high in provitamin A. In the Orient and in Mexico the male squash blossoms are used for food.

The bitterness reported present in certain cultivars of summer squash is undoubtedly due to cucurbitacins also present in some cucumbers and in bittermelons.

Pests and Diseases

Melon aphid, cucumber beetles, squash bug, squash vine borer, pickle and melon worms, leaf miner, and cut worms are pests that attack pumpkins and squashes.

Many diseases are prevalent in the crops; some of these include downy and powdery mildews, and scab, a fungus that attacks leaves, stems, and immature fruits of summer squashes. Bacterial wilt infects leaves and is spread by cucumber beetles. Fusarium root rot starts as a soft rot of the stem at the soil surface and as the disease worsens the plant wilts. *Choanephora*, a soil-borne fungus, attacks flowers and immature fruits. It is common in moist soils under rainy conditions; in well-drained soils there are fewer problems with this disease. Gummy stem blight and black rot of fruits are caused by the same fungal organism; the disease may be seed borne and rotation and fungicides are used to combat it. Cucumber mosaic and watermelon mosaic are spread by aphids, and squash mosaic is transmitted by cucumber beetles; use of virus-free seed and control of vectors can suppress the diseases. Curly top, spread by the beet leafhopper, kills seedlings but infected older plants are dwarfed with leaves turning yellow and curled at the edges.

MATURE FRUITS: WAX, WHITE, OR ASH GOURD, CHINESE WINTER MELON, TON KWA; AND IMMATURE FRUITS: CHINESE SQUASH, MO KWA: *Benincasa hispida*

Wax gourd is reported to grow wild in Java. This crop is grown throughout Southeast Asia, China, and India. The name wax, white, or ash gourd is derived from the mature fruit, which produces a thick layer of white wax on the epidermis.

Wax gourd is an annual hispid (rough with bristle-like hairs) climbing herbaceous plant several meters in length. The leaves are broad with five to 11 angular lobes. Whereas the staminate flowers have long peduncles, the pistillate flowers have densely haired ovary on short peduncles. The corolla has five large yellow petals.

The mature fruit is somewhat spherical to oblong in shape measuring 30–40 cm (12–16 in.) in diameter or length and 15–23 cm (6–8 in.) in width and weighs as much as 45 kg (100 lb) (Fig. 23.5). As the fruit matures, a white waxy bloom forms which thickens with time. The wax continues to form even in storage after harvest.

Chinese squash, called *mo kwa* is a cultivar grown specifically for immature fruits (Fig. 23.6). The immature cylindrical fruit has many bristle-like trichomes on the epidermis. It is harvested when the fruit is about 12–15 cm (4½–6 in.) in length and used like summer squash.

Wax gourd grows best at 24°–27°C (75°–80°F). It is susceptible to cold temperatures but can tolerate drought. As with most cucurbits, it grows best in well-drained soils high in organic matter. The optimum soil pH is

FIG. 23.5. Wax gourd, Chinese winter melon, *don kwa (Benincasa hispida)*.

5.5–6.4. Cultural practices similar to winter squashes and pumpkins are used for growing the crop.

The plant starts to flower about 60–80 days from planting and can be harvested in about 1 week for the immature fruits and after 30–40 days for the mature fruits.

Wax gourds can be stored the longest among the cucurbits. Bruises and cuts on mature fruits are capable of healing by suberization. When stored at 13°–15°C (55°–59°F) and 70–75% RH, fruits remain in good quality for over 6 months. In storage it is best to put each fruit on a shelf and not stacked.

For use the skin is scraped or peeled off and the seed cavity scooped out and discarded. The flesh is cut up into pieces or the whole fruit with various cooked ingredients poured into the center cavity is steamed until cooked. Wax gourd is a delicacy in Chinese soups especially in the winter. The flesh is sometimes candied.

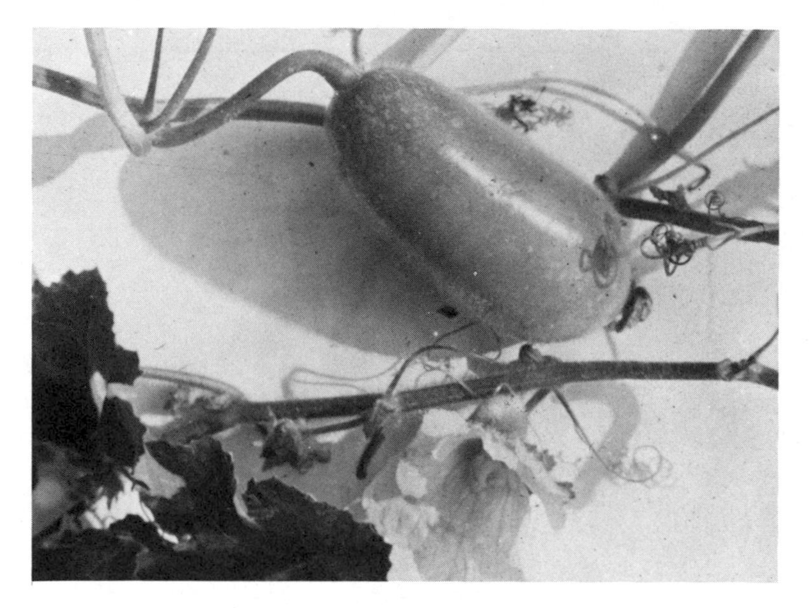

FIG. 23.6. Chinese squash, *mo kwa (Benincasa hispida).*

BOTTLE GOURD, WHITE FLOWERED GOURD, CALABASH GOURD, MO KWA: *Lagenaria siceraria*

The bottle gourd has been utilized as food and utensil for thousands of years by man in the tropics and subtropics of the world. Its origin is probably in the tropical lowlands of south–central Africa. It is postulated that mature fruits with intact seeds drifted on ocean currents from Africa to South America. Archaeological evidence indicates the presence of *Lagenaria* in Peru 12,000 years ago; remains of gourds have been found in Egyptian tombs over 5000 years old.

Bottle gourd is a monoecious annual vine with very large cordate–oval leaves measuring 15–30 cm (6–12 in.) in diameter. The branched vines with branched tendrils spread or climb 3–15 m (10–50 ft). The foliage is pubescent and emits a characteristic somewhat musky and unpleasant odor when bruised. The characteristic large white flowers, borne on slender peduncles 15–20 cm (6–8 in.), open in the evening and may remain open until the following midday. Cultivars are characterized by the shape and size of fruits varying from 10 to 90 cm (4 to 36 in.); the fruits' shapes may be globular, flat, bottle-shaped, club-like, coiled, etc., or variations of these shapes.

The immature fruits (Fig. 23.7) are used for food in various ways. In

FIG. 23.7. Immature bottle gourd, *mo kwa (Lagenaria siceraria)*.

California the flesh of immature fruits is used in making glace for cakes, etc.; in the Orient, the flesh after removal of the hard skin is sliced into very thin strips and air dried into a product called *kampyo* which is used in Oriental (Japanese) cooking. The Chinese use the immature fruits like summer squash; they are called *mo kwa* as are immature *Benincasa hispida*. Immature starchy fruits of *Lagenaria* spp. called *kashi* in Brazil are used as a vegetable. With maturation, the rind becomes very hard and durable; the fruit then is used to make utensils or fabricated into works of art.

The culture of bottle gourd is similar to winter squash and pumpkins.

CHAYOTE, VEGETABLE PEAR, MIRLITON, CHOCHO, CHAYOTL: *Sechium edule*

Chayote is an herbaceous perennial climbing vine, indigenous to southern Mexico and Central America. The plant grows to 15 m (50 ft) or more in length and has large shallow lobed leaves producing both male and female flowers (monoecious) in the same leaf axil (Fig. 23.8). It is grown principally for the pear-shaped fruits, which are cooked in many ways. Unlike the other many-seeded cucurbit fruit, chayote has a single

FIG. 23.8. Chayote *(Sechium edule)*.

large seed. Parthenocarpic fruits can be produced by two applications of 10^{-3} M gibberellic acid (GA_3 and $GA_{4/7}$) in a lanolin paste applied to the stigma of pistillate flowers. The plant produces enlarged starchy roots, which are harvested and used; also the tender shoots and young leaves are used as greens.

Chayote cultivars are classified by the characteristics of the fruit shape and color. Some cultivars have prickly spines or thorns on the fruit. The plant is adapted to the warm tropical and subtropical climates; flowering occurs under slightly over 12 hr or shorter day lengths.

Propagation is either by vegetative cuttings or planting of mature seeds. Young shoots are removed from the crown with a sharp knife, placed in moist sand until well rooted, and then transplanted into pots. When the plants are established, they are planted in the field with as little as possible disturbance of roots. For seed propagation the entire fruit is left on the vine to mature fully; it is harvested before the seed starts to sprout. The entire fruit is placed on its side in a shallow hole and covered with 5–8 cm (2–3 in.) of soil. Fruit for seed should be carefully handled and not be stored below 10°C (50°F) and planted without much delay.

Chayote grows well in loose well-drained soil high in organic matter. Plants or seeds are placed in hills spaced approximately 2 m (6 ft) and in rows 3½ m (12 ft) apart. The vines need support; usually they are allowed to climb trellises. Large "T" supports with a 10 cm (4 in.) heavy wire course meshed fencing over the supports are ideal as the fruits can hang through the fencing for easy harvest.

Besides the tropics and subtropics, this crop can be grown in warm temperate regions where the temperatures are warm into the late fall for fruit maturation after the plant flowers under short days. About a month is required from anthesis to harvest. In temperate regions the roots should be protected against cold weather by placing a layer of mulch on the soil surrounding the plant. Crowns thus protected will send up shoots when the soil warms in the spring.

Harvest fruit should be handled carefully to avoid cuts and bruises. They should be kept at 10°–15°C (50°–59°F); at this temperature fruits will be in good quality for a few weeks. If stored at lower temperatures, chilling injury will take place.

Chayote is attacked by striped and spotted cucumber beetles and in the soil by nematodes. In areas of high rainfall, the plant can become infested with fungus diseases.

BALSAM PEAR, BITTERMELON OR GOURD, ALLIGATOR PEAR, BITTER CUCUMBER, FU KWA: *Momordica charantia*

Balsam pear is indigenous to the Old World tropics but now found in all parts of the tropics and subtropics. It is an important cultivated food crop in India and Southeast Asia. In many regions of the tropics, the vine as well as the fruits are used in folk medicine.

Balsam pear is a rapid growing herbaceous annual vine growing 7–10 m (20–30 ft) in length. The leaves, borne on long petioles, are palmate with five to nine deep lobes. The plant is monoecious; the five yellow petaled flowers are borne solitarily on slender peduncles in the leaf axils. The fruit vary in size from 4 to 30 cm (1½ to 12 in.) in length and from 1½ to 6 cm (½ to 2½ in.) in width. The fruit has about 10 irregular longitudinally rounded ridges and between them smooth pebbled protrusions on the surface give a warty appearance (Fig. 23.9). In addition, some types have pointed projections on the surface.

The green immature fruits are used for food, for at this stage they are reported to be the least bitter. To reduce bitterness fruits before use are often parboiled or steeped in salt water after peeling. With ripening, the fruit turns yellow and orange and when ripe splits open exposing the

FIG. 23.9. Balsam pear, bitter melon, *fu kwa (Momordica charantia).*

scarlet arils covering the seeds. The arils are sweet and eaten by people and birds. Tender shoots and leaves are parboiled to leach the bitterness and used as greens. The bitter substance is probably the poisonous cucurbitacins.

Balsam pear is grown on trellis and cultured much like other climbing cucurbits (loofah). It flowers about a month after planting; the fruits are harvested in about 2–3 weeks after anthesis. Fruits are often protected from fruit flies with paper or plastic bags open at the bottom; evidently the flies do not fly up into the bag covering the fruit.

SMOOTH LOOFAH, DISHCLOTH GOURD, SPONGE GOURD, HECHIMA: *Luffa cylindrica* AND CHINESE OKRA, ANGLED LOOFAH: *Luffa acutangula*

All but one species of the genus *Luffa* is indigenous to tropical Asia; two, *L. cylindrica* (Fig. 23.10) and *L. acutangula* (Fig. 23.11), are cultivated for food and used as a commercial fiber. Wild forms of the angled loofah, which has extremely bitter fruits, grow in India, but the domesticated types are less bitter.

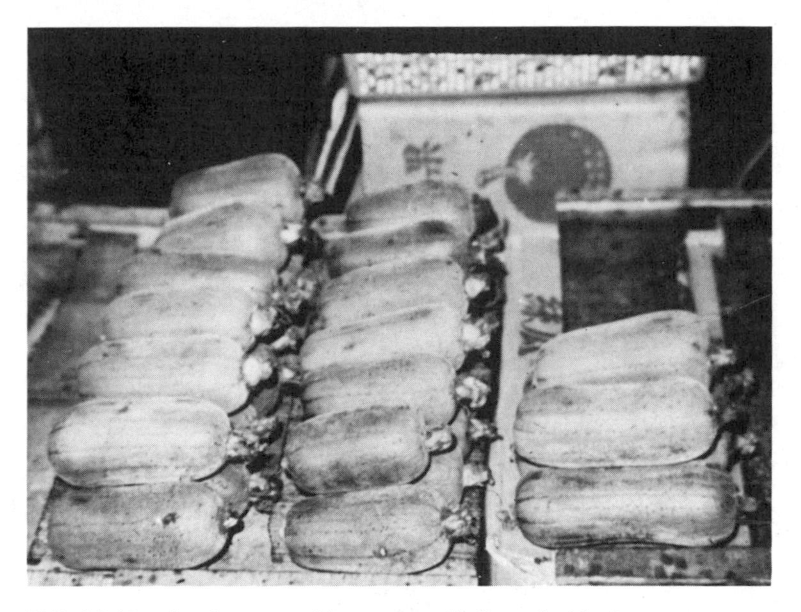

FIG. 23.10. Loofa, vegetable sponge *(Luffa cylindrica)*.

Both species are annual climbing vines with tendrils, and monoecious in flowering habit. However, perfect flowers of the angled loofah have been reported. The two species are easily distinguished by the shape of the fruits. Smooth loofah have long smooth cylindrical shaped fruits with longitudinal lines 30–50 cm (1–1½ ft) in length and 8–12 cm (3–4½ in.) in diameter. The angled loofah fruits have 10 distinct longitudinal acute ribs from the stem to the tip and are about the same size and shape as the smooth type.

In the temperate region loofah are propagated by planting seeds in hills, spaced about 60–200 cm (2–6½ ft) and in rows 3 m (10 ft) apart. Plants should be allowed to climb trellis to support the vines and fruits. For good shaped fruits, the pistillate flowers should be hand pollinated; incomplete pollination results in misshapen fruits.

Staminate flowers are produced in a ratio ranging from 25–40 to 1 pistillate flower. This rate can be reduced by a factor of two by treatment of the grow regulator IAA. The stigma of smooth loofah are reported to be receptive to pollination for 60 hr after anthesis and angled loofah for 36 hr.

Immature fruits can be harvested for food after about 2 months from seeding. Mature fruits required about 4–5 months. For fibers, mature fruits of *L. cylindrica* are processed by submersion in water for 7–10

FIG. 23.11. Chinese okra *(Luffa acutangula)*.

days which results in the disintegration of the outerwalls and pulp. Washing removes seeds, skin, and pulp from the fibers. If fibers are not white, they may be bleached with hydrogen peroxide, then dried.

SNAKE GOURD: *Trichosanthes anguina (T. cucumerina)*

Snake gourd occurs in the wilds from India, Southeast Asia, to tropical Australia. It is a monoecious annual vine with large angular five to seven lobed leaves and climbing by means of branched tendrils. The five or more staminate flowers, forming a pubescent inflorescence, and a single sessiled pistillate flower are borne in the same leaf node. Both flower types have white delicately fringed petals, which open late in the afternoons and are quite fragrant.

When mature the greenish white and sometimes striped fruits are very long and slender, often twisted, 30–180 cm (1–6 ft) in length and 4–10 cm (1½–4 in.) at the largest diameter.

Culture of snake gourd is similar to other climbing cucurbits, such as balsam pear. Plants are grown on trellises about 2 m (7 ft) high and horizontally supported so that the developing fruits are allowed to hang. To induce straight growth, a small stone is tied to the tip of the elongating fruit. After about 3 months from planting immature fruits 30–75 cm (1–2½ ft) can be harvested for use.

CAIHUA, ACHOCCHA, KORILA, WILD CUCUMBER: *Cyclanthera pedata*

The origin of caihua is thought to be in the Caribbean region; it is grown presently from Mexico to Bolivia.

It is an annual crawling vine branched at the lower nodes and spreading as much as 5 m (16 ft) in length. Leaves are 8–18 cm (3–5 in.) in width, normally five lobed; the two lower lobes may be further divided into three or four minor lobes (the three center lobes are undivided) which gives the leaf an appearance of having nine to 11 lobes. The vine climbs with forked tendrils.

FIG. 23.12. Caihua, wild cucumber *(Cyclanthera pedata)*.

Fruits are soft, flattened, oblong with longitudinal lines, and sparsely spined, 10–15 cm (4–6 in.) in length, 5–10 cm (2–2¾ in.) in width and 2–3 cm (¾–1¼ in.) in thickness (Fig. 23.12). The interior of the fruit is hollow; fleshy rind is 3–4 mm (⅛–3⁄16 in.) thick; the inner tissue is a white spongy pulp containing black seeds attached to the placenta.

The crop is cultivated in the mountainous valleys up to 2000 m (6500 ft) in elevation. Seeds are planted in hills spaced 1.5–2.0 m (5–6 ft); the seedling are thinned to one to two plants per hill. Plants are either supported on poles or grown prostrate on the ground and with adequate moisture during the growing season. Fruits are ready for harvest about 3 months after planting and can be continually harvested for an additional 3 months with proper care and growing conditions.

Fully grown fruits are harvested for use either raw or cooked. Fruits are often used for stuffing by removing seeds and pulp and stuffed with ingredients similar to stuffed peppers.

BIBLIOGRAPHY

Cucurbits (Many or Several Crops)

HARRINGTON, J.F., and MINGES, P.A. 1954. Vegetable seed germination. Leaflet, Div. Agric. Sci., Univ. of California, Berkeley.

HERKLOTS, G.A.C., 1972. Vegetables in South-East Asia. Hafner Press, New York.

HOWARD, F.D., MACGILLIVRAY, J.H., and YAMAGUCHI, M. 1962. Bull. No. 788. California Agric. Exp. Stn., Univ. of California, Berkeley.

INTENGAN *et al.* 1964. Food Composition Tables. Handb. No. 1, Food Res. Center, Manila, Philippines.

MACGILLIVRAY, J.H. 1953. Vegetable Production. Blakiston Co., New York.

MANSELL *et al.* 1950. Composition of food plants of Central America, II. Guatemala. Food Res. *15*, 6.

PURSEGLOVE, J.W. 1968. Tropical Crops: Dicotyledons. John Wiley & Sons, New York.

RYALL, A.L., and LIPTON, W.J. 1979. Handling, Transportation and Storage of Fruits and Vegetables, Vol. 1. AVI Publishing Co., Westport, Connecticut.

SHIMIZU, S. (Editor). 1977. Encyclopedia of Vegetable Crops Yokendo, Tokyo, Japan. (*In* Japanese)

TISBE, V.O., DEANON, J.R., JR., and BANTOC, G.B., JR. 1967. The cucurbits. *In* Vegetable Production in Southeast Asia. J. E. Knott and J.R. Deanon, Jr. (Editors). Univ. of Philippines Press, Manila.

WATT, B.K., and MERRILL, A.L. 1950. Composition of Food. USDA Handb. No. 8, US Dept. Agric., Washington, DC.

WHITAKER, T.W., and CUTTER, H.C. 1971. Prehistoric cucurbits from the Valley of Oaxaca. Econ. Bot. *25*, 123–127.

WHITAKER, T.W., and DAVIS, G.N. 1962. Cucurbits. Interscience Publishers, New York.

Cucumber (*Cucumis sativus*)

EAKS, I.E., and MORRIS, L.L. 1957. Deterioration of cucumbers at chilling and non-chilling temperatures. Proc. Am. Soc. Hort. Sci. *69*, 388–399.

JOHNSON, H., Jr. 1972. Greenhouse cucumber production. Agric. Ext. Service, Univ. of California, Berkeley.

LOOMIS, E.L., and CRANDALL, P.C. 1977. Water consumption of cucumbers during vegetative and reproductive stages of growth. J. Am. Soc. Hort. Sci. *102*, 124–127.

MILLER, C.H., and RIES, S.K. 1958. The effect of Environment on fruit development of pickling cucumbers. Proc. Am. Soc. Hort. Sci. *71*, 475–479.

SMITTLE, D.A., and WILLIAMSON, R.E. 1977. Effect of soil compaction of nitrogen and water use efficiency, root growth, yield and fruit shape of pickling cucumbers. J. Am. Soc. Hort. Sci. *102*, 822–825.

WARD, G.M. 1967. Growth and nutrient absorption in greenhouse tomato and cucumber. Proc. Am. Soc. Hort. Sci. *90*, 335–341.

Watermelon *(Citrullus vulgaris)*

CHAMBLISS, O.L., ERICKSON, H.T., and JONES, C.M. 1968. Genetic control of bitterness in watermelon fruit. Proc. Am. Soc. Hort. Sci. *93*, 539–546.

MIZUNO, S., and PRATT, H.K. 1973. Relations of respiration and ethylene production to maturity in the watermelon. J. Am. Soc. Hort. Sci. *98*, 614–617.

SACHS, M. 1977. Priming of watermelon seeds for low temperature germination. J. Am. Soc. Hort. Sci. *102*, 175–178.

SCHWEERS, V.H., and SIMS, W.L. 1976. Watermelon production. Leaflet 2672, Div. Agric. Sci., Univ. of California Berkeley.

Melons *(Cucumis melo)*

FLOCKER, W.J., LINGLE, J.C., DAVIS, R.M., and MILLER, R.J. 1965. Influence or irrigation and nitrogen fertilization on yield, quality and size of cantaloupes. Proc. Am. Soc. Hort. Sci. *86*, 424–432.

McGLASSON, W.G., and PRATT, H.K. 1963. Fruit set patterns and fruit growth in cantaloupe (*Cucumis melo* L. var *reticulatus* Naud.). Proc. Am. Soc. Hort. Sci. *83*, 495–505.

PRATT, H.K., GOESCHL, J.D., and MARTIN, F.W. 1977. Fruit growth and development and role of ethylene in 'Honeydew' muskmelon. J. Am. Soc. Hort. Sci. *102*, 203–210.

YABUMOTO, K., YAMAGUCHI, M., and JENNINGS, W.G. 1978. Production of volatile compounds by muskmelon, *Cucumis melo*. Food Chem. *3*, 7–16.

Squashes and Pumpkins *(Cucurbita)*

ANON, 1968. Growing pumpkins and squashes. USDA Farmers' Bull. No. 2086, US Dept. Agric., Washington, DC.

CASTETTER, E.F., and ERWIN, A.T. 1927. A systematic study of the squash and pumpkin. Bull. No. 244 Iowa Agric. Ext. Sta., Univ. of Iowa, Ames.

LORENZ, O.A. 1949. Growth rates and chemical composition of fruits of four varieties of summer squash. Proc. Am. Soc. Hort. Sci. *54*, 385–390.
WHITAKER, T.W., and BEMIS, W.P. 1975. Origin and evolution of the cultivated *Cucurbita*. Bull. Torrey Bot. Club *102*, 362–368.
WHITAKER, T.W., and BOHN, G.W. 1950. The taxonomy, genetics, production and uses of the cultivated species of cucurbita. Econ. Bot. *50*, 52–81.

Wax Gourd *(Benicasa hispida)*

HERKLOTS, G.A.C. 1972. Vegetables in South-East Asia. Hafner Press, New York.
SRIVASTAVA, V.K., and SACHAN, S.C.P. 1969. Grow ashgourd the efficient way. Indian Hort. *14*, 13, 14, 31 (Oct–Dec).

Bottle Gourd (*Lagenaria* spp.)

HERKLOTS, G.A.C. 1972. Vegetables in South-East Asia, Hafner Press, New York.
MARTIN, F.W. 1979. Vegetables for the hot humid tropics, Part 4, Sponge and bottle gourds *Luffa* and *Lagenaria*. Mayaguez Inst. Trop. Agric. Sci. Ed. Admin. U.S. Dept. Agric., Mayaguez, Puerto Rico.

Chayote *(Sechium edule)*

AUNG, L.H., and FLICK, G.J. 1976. Gibberellin-induced seedless fruit of chayote, *Sechium edule* Swarta. Hortic. Sci. *11*, 460–462.
FILGUEIRA, F.A.R. 1972. Manual de Oleri Cultura Chapter 12. Cucurbitaceas. Editora Agronomica "Ceres" Ltd. Sao Paulo, Brazil. (*In* Portuguese).

Balsam Pear *(Momordica charantia)*

MORTON, J.F. 1967. The balsam pear—an edible, medicinal and toxic plant. Econ. Bot. *21*, 57–68.

Loofah or Sponge Gourd (*Luffa cylindrica* and *L. acutangula*)

MARTIN, F.W. 1979. Vegetables for the hot humid tropics. Part 4. Sponges and bottle gourds *Luffa* and *Lagenaria*. Mayaguez Inst. Trop. Agric., Sci. Ed. Admin., US Dept. Agric., Mayaguez, Puerto Rico.
PORTERFIELD, W.M., JR., 1955. Loofah–the sponge gourd. Econ. Bot. *9*, 211–223.

Snake Gourd *(Trichosanthes anguina)*

HERKLOTS, G.A.C. 1972. Vegetables in South-East Asia, Hafner Press, New York.
KATYAL, S.L. 1977. Vegetable Growing in India. Oxford and IBH Publishing Co., New Delhi, India.

Caihua, Wild Cucumber *(Cyclanthera pedata)*

HERKLOTS, G.A.C. 1972. Vegetables in South-East Asia, Hafner Press, New York.
LEON, J. 1964. Plantas Alimenticias Ardinas. Bol. Tech. No. 6. Instituto Interamericano de Cincias. Agricolas Andina, Lima, Peru. (*In* Spanish).

24

Chenopods

Family: Chenopodiaceae (goosefoot)

Principal genera/species	Common name
Spinacia oleracea	Spinach
Beta vulgaris	Table beet, beet root (red flesh); mangel-urzel, mangold (white flesh)
B. vulgaris (Cicla group)	Swiss chard
Atriplex hortensis	Orach, mountain spinach, French spinach, sea purslane

SPINACH: *Spinacia oleracea*

Origin

Records of the usage of spinach in Europe go back as far as 1351 and in China as early as the seventh century. Vavilov places the region of domestication of spinach in the central Asiatic center.

Botany

Spinach is an annual crop grown for its leaves, which may be either smooth or wrinkled depending on the cultivar. Usually dioecious, seeds are produced on the female plant, and the male plant dies after blooming. The pollen is transferred by wind. The fertilized ovule develops into a one-seeded fruit. Plants are classified according to the flowering characteristic: extreme male, vegetative male, female, and rarely monoecious (both male and female flowers on the same plant). Vegetative male and female types are preferred as the plants are larger, whereas the extreme males are smaller and flower early. There are two types of seeds: prickly and smooth.

Spinach is a cool season crop cultivated where the average temperature range is 16°–18°C (60°–65°F), but it can be grown in regions where the average temperature is about 10°C (50°F). Young plants can stand freezing temperatures of as low as −9°C (15°F) without sustaining much injury.

Spinach is a long-day plant for flowering. At day lengths greater than the minimum, higher temperatures cause earlier bolting. Bolting is more rapid with increase in photoperiods, older plants being more sensitive than young plants. Also, a cold exposure followed by high temperatures and long days induces the most rapid bolting of spinach. Crowding of plants produces bolters sooner than those given ample space. Depending on cultivars, the critical photoperiod ranges from 12½ to 15 hr. Vegetative growth is more rapid when grown just short of the critical photoperiod. Cultivars should be selected for their critical day lengths in relation to the particular latitude and season to be grown.

Culture

Sandy loams to clay loams and muck soils in the pH range 6.0–7.0 are used to grow spinach. The crop can tolerate pH's down to 5.5. Light, sandy soils with good drainage are used in regions of high rainfall.

Seeds are either broadcasted or planted in rows 40–60 cm (15–24 in.) apart, and covered with 1–2 cm (½–¾ in.) of soil. Table 24.1 shows that 10°–15°C (50°–59°F) soil temperature is best for germination. Spinach seeds go dormant at temperatures above 30°C (86°F).

Nitrogen fertilization generally increases yields of winter-grown spinach as little nitrification occurs at low soil temperatures. Rates of

TABLE 24.1 EFFECT OF SOIL TEMPERATURE ON SPINACH SEED GERMINATION

Temperature (°C)	% germination	Days to emergence
0	83	63
5	96	23
10	91	12
15	82	7
20	52	6
25	30	5
30	30	6
35	0	—

Source: Harrington and Minges (1954).

70–135 kg N, 40 kg P and 56 kg K/ha (60 to 120 lb N, 35 lb P, and 50 lb K/acre) are used for winter and early spring spinach.

Harvest and Storage

Harvest occurs when plants have reached good size, usually 35–70 days after planting depending on the season and climate. A plant ready for harvest has five to six fully developed large leaves. From 22–26 leaves develop from seedling to mature plant ready for harvest; the older leaves senesce as the plant grows. For quality spinach there should be no seed stalks or yellow leaves. Whole plants are cut at crown level for the fresh market; for canning, the leaves are cut (mechanically mowed) above the crown.

Harvested spinach stores best near 0°C (32°F) and under very high humidity; under these conditions they remain in excellent condition for 10–14 days.

The concentration of NO_3^- in the petioles of spinach can reach toxic levels for humans. This is especially true when high amounts of nitrates and ammoniacal nitrogen is used to fertilize the crop (see Chapter 5). The nutritive value of spinach is given in Table 25.1.

Pests and Diseases

Aphids, beetles, mites, and leaf miners are some of the insects which attack spinach. Downy mildew, mosaic virus, curly top virus, leaf spot, and pythium are the major diseases of spinach.

TABLE BEETS: *Beta vulgaris*

Origin

The ancient Greeks used both the leaves and the "roots." The name "beta" came from the Romans according to de Candolle and cultivation dates back to about the fourth century BC. The species is thought to be derived from *Beta maritima*, known as sea beet, indigenous to southern Europe. It was taken to northern Europe by invading armies and has been widely used throughout Europe, the Middle East, and India for food and fodder, particularly during the winter.

Botany

Beets are biennial requiring a period of cold temperature of 4°–10°C (40°–50°F) for 2 weeks or longer for flower initiation. Table beets are grown particularly for the enlarged hypocotyl, the so-called "root." The enlarged issue is mainly hypocotyl with a small portion of the tap root incorporated into the swelling. Enlargement of the hypocotyl is due to growth of several concentric vascular cambia which comprises the "rings" of the beet, when viewed in a transverse section. Also, the young tops are used as greens.

The red color of table beets is betacyanin, a nitrogen-containing compound with chemical properties similar to the anthocyanins. Beets also contain a yellow pigment, betaxanthin. The ratio of these two pigments varies with cultivars and changes during growth and with environmental conditions.

The white fleshed mangel-wurzel or mangold is used for animal feed. Sugar beets are a very recent crop being bred for high sugar (sucrose) content. The breeding for high sucrose started in the later 1700s by Achard in Prussia. Beets with a sugar content of 6% were reported in 1775; now sugar beets are grown in California with three times as much sugar, 15–20% depending on growing conditions.

Culture

Cultivation of table beets is very much like spinach. Unlike spinach, which bolts under high temperature and long days, table beets are vegetative under these conditions. It grows best in cool climates, temperature means of 16°–18°C (60°–65°F). High nitrogen fertilization tend to produce "roots" of poor color.

SWISS CHARD, CHARD, SPINACH BEETS, OR SEA KALE BEETS: *(Beta vulgaris)* (Cicla Group)

All the above names have been used for the crop grown for enlarged leaves; the hypocotyl does not enlarge as in the table beet. Swiss chard has a white midrib of the leaf, whereas in the spinach beet the midrib is green. There are cultivars of chard in which the midrib is red, as was noted by Aristotle in the fourth century BC.

As with table beets, seeds are sown in the early spring in temperate regions. When plants are large enough, the outer leaves are pulled away from the plant or they are cut off near the base and the plant allowed to grow for later harvests. Although a cool season crop, Swiss chard can stand some hot weather and can be grown at high elevations in the tropics and semitropics.

Beta vulgaris var. *orientalis* called *Palang sag* or *Palanki* in India, is an annual herb with prominent tap root and erect stem with trowel-shaped long petiolate leaves at the base. The leaves are used as salad and for preparing stews. The plant is used in India as medicine for disease of the liver and spleen and also as a tonic. The nutritive values of table beets and Swiss chard are in Table 25.1.

ORACH(E), MOUNTAIN SPINACH, FRENCH SPINACH, SEA PURSLANE: *Atriplex hortensis*

Orach, its origin reported to be northern India, has an ancient history. It was first used medicinally by the early Greeks and later as a food plant. The crop was introduced from the Mediterranean region to France, then to England and eventually to the United States where it is now popular in the Middle West and mountain regions because it can be grown during times when spinach (*Spinacia oleracea*) readily bolts (goes to seed).

Culture

A cool season crop, the mature plant attains a height of 2–3 m (6–10 feet). It has large cordate to long ovate leaves, 10–20 cm (4–8 in.) long. The cultivars vary in color from pale yellowish to dark green and to deep blood red, some having a grayish bloom. Orach is a hardy plant; it is resistant to drought, to saline and alkaline soils, and in the early spring the emerging seedlings can tolerate a temperature minimum of −3°C (26°F) without apparent injury.

Seeds are planted in 50–60 cm (20–24 in.) rows, about 25 mm (1 in.) apart and from 5–12 mm (¼–½ in.) deep. Germination is very much reduced when planted 25 mm (1 in.) in depth.

Harvest

Commercial harvest is usually when the plants are 10–15 cm (4–6 in.) in height. For home gardening purposes, plants are thinned at this stage with spacings of 25–30 cm (10–12 in.); the thinned plants are used like spinach as a vegetable. When plants are older, the tender leaves from 6 to 8 cm (2½–3 in.) broad are harvested for use; the coarse old leaves at the base are left to sustain the plant for continued growth of new leaves for future harvests.

For seed production the rows are 100 cm (40 in.) apart and plants spaced at 45–60 cm (18–24 in.). A growing season of over 4½ months is necessary for good seed yields in cool climates. The flowers are wind pollinated.

BIBLIOGRAPHY

Spinach

HARRINGTON, J.F., and MINGES, P.A. 1954. Vegetable seed germination. Leaflet No. **XX**, Div. of Agric. Sci., Univ. of California, Berkeley.

HUYSKES, J.A. 1971. The importance of photoperiodic response for the breeding of glasshouse spinach. Euphytica *20*, 371–379.

MacGILLIVRAY, J.H. 1953. Vegetable Production. McGraw-Hill Book Co., New York.

MAGRUDER, R., BOSWELL, V.R., SCOTT, G.W., WORK, P., and HAWTHORNE, L.R. 1938. Description of types of principal American varieties of spinach. USDA Misc. Pub. No. 316, U.S. Dept. Agric., Washington, DC.

ZINK, F.W. 1965. Growth and nutrient absorption in spring spinach. Proc. Am. Soc. Hort. Sci. *87*, 380–386.

Beets and Chard

CAMPBELL, G.K.G. 1976 Sugar beet, *In* Evolution of Crop Plants, N.W. Simmonds (Editor). Longmans-Green, London.

MacGILLIVRAY. J.H. 1953. Vegetable Production. McGraw-Hill Book Co., New York.

NAYER, M.P., and RAMAMURTHY, K. 1977. *Beta vulgaris* var. *orientalis*, a useful green vegetable of northern India. Econ. Bot. *31*, 372–373.

NISSON, T. 1973. The pigment content in beet root with regard to cultivar, growth, development and growing conditions. Swedish J. Agric. Res. *3*, 187–200.

Orach

BABB, M.F., and KRAUS, J.E. 1939. Orach, its culture and use as a greens crop in the Great Plains region. USDA Circ. No. 526, U.S. Dept. Agric., Washington, DC.

25

Other Succulent Vegetables

Miscellaneous Monocotyledons

ASPARAGUS: *Asparagus officinalis*

Origin

Asparagus is believed to be indigenous to the eastern Mediterranean region. It has been in use for over 2000 years, first for medicinal purposes and later as food.

Botany

Asparagus, belonging to the Liliaceae or lily family, is a perennial monocotyledonous plant grown for the tender green shoots (spears), usually harvested in the spring in temperate regions of the world. In these regions the tops die down with the first frost in the fall, and the crown, composed of fleshy roots and rhizomes, overwinters in the soil. When the soil warms in the spring, the shoots emerge from the crowns. (Fig. 25.1).

Asparagus is dioecious (male and female flowers borne on separate plants), and is pollinated by insects. By tissue culture techniques, pure lines of male and female plants can be produced; and using these parents in making crosses, the resulting F_1 hybrid plant can be vigorous, early, uniform, and high yielding.

Culture

A cool season crop, asparagus grows best at mean temperatures of 16°–24°C (60°–75°F). In the cool temperate regions the dormant crowns should be protected from freezing with straw and/or mounded soil. It does better with a rest season, but can be successfully grown in regions where there is no frost or freezing weather in the winter. The crop is planted in well-drained light soils or in muck (high organic) soils, since in these types of soils the spears can emerge without damage. The crop is either direct seeded or 1-year-old crowns obtained from seedbeds are transplanted to the field. Seeds are drilled 2.5–5 cm (1–2 in.) deep in a trench 15–20 cm (6–8 in.) below ground level for direct seeding. Crowns are transplanted 20–25 cm (8–10 in.) below ground level and 30 cm (12 in.) apart. Soil covering the crowns should be 5–7.5 cm (2–3 in.) in depth. Distance between rows is 1.8 m (6 ft) for green asparagus and 2–2½ m (7–8 ft) for white asparagus. It takes 2–3 years from seeding before harvest is made. The first cutting (2-year-old) is very short, from 3 to 4 weeks; the 3-year-old field is cut longer, for 8–10 weeks, after which the ferns (tops) are allowed to grow for the remainder of the season. The soil is mounded for white asparagus.

In the warm frost-free regions of the world (subtropics), asparagus plants have very short dormant periods, or sometimes none at all. Two harvest periods are possible, one in the spring and another in the fall. The ferns (tops) are allowed to grow between harvest periods to supply carbohydrates to underground storage roots.

A new technique has been devised for the growing of asparagus in the tropics. The so-called "mother stalk" method, wherein four or five stalks

FIG. 25.1. Asparagus *(Asparagus officinalis)* crowns *in situ* in soil.

of ferns are allowed to grow in order to provide the food needs of the plant, is being practiced in Taiwan. Every 3 or 4 months, new stalks are allowed to grow out into ferns as the older ones get diseased or start to become senescent. A year-round production of spears can be attained by this method with yields twice that by conventional methods practiced in the temperate region in a 4–5 month harvest period.

Depending on the soil fertility, the amount of N, P, and K application can vary. If the fertility is low, 65–90 kg N/ha (60–80 lb/acre), 35 kg P/ha (30 lb/acre), and 35 kg K/ha (30 lb/acre) can be applied.

Storage

The recommended temperature for the harvested spears is 2°C (36°F) at 90% RH. Under these conditions asparagus will keep for about 3 weeks.

The nutritive values of asparagus and other succulent vegetables are in Table 25.1.

Pests and Diseases

Asparagus rust caused by *Puccinia asparagi* and *Rhizocotonia* are the main diseases. Asparagus beetle and garden centipede are the most troublesome pests. *Fusarium oxysporum* infection of the roots causes the decline of asparagus plants.

BAMBOO SHOOT: *Phyllostachys*, *Denrocalamus*, *Bambusa* and *Sasa* Genera

The emerging shoots of the bamboo, of the *Phyllostachys, Denrocalamus, Bambusa*, and *Sasa* genera, are harvested and used as a vegetable. The bamboo is a perennial belonging to the Gramineae or grass family. The reproductive organs resemble the grasses, particularly the oat *(Avena sativa)*; the anthers are borne on long filaments resembling those of corn *(Zea mays)*. The important edible bamboos are *P. edulus, P. bambusoides*, and *P. dulcis*.

The cultivation of bamboo has been practiced in the Orient for thousands of years. Bamboo is propagated by transfer of clumps or rhizomes. In the life cycle of a clone, the bamboo flowers once, the top dies. Parts of the same clone, even when transferred to different areas and conditions, will flower at the same time. For a period of 2–3 years, the rhizomes send out shoots which flower. It is reported that viable seeds are produced.

TABLE 25.1. PROXIMATE COMPOSITION OF SOME MISCELLANEOUS SUCCULENT VEGETABLES IN 100 g EDIBLE PORTION

Crop	Edible part	(%) refuse	Macroconstituents					Vitamins					Minerals (mg)				Ref[a]
			Energy (cal)	g				A (IU)	mg								
				Water	Protein	Fat	CHO		B_1	B_2	Niacin	C	Ca	Fe	Mg	P	
Asparagus																	
Green	Shoots	40	27	92	2.8	0.2	2.2	980	0.23	0.15	2.2	48	24	1.5	—	52	1
White	Shoots	30	25	93	1.9	0.2	2.5	50	0.11	0.08	1.1	28	16	1.1	—	52	1
Beets, table	Root	65	34	89	1.9	0.1	6.3	Trace	0.05	0.02	0.4	11	13	0.5	19	55	1
Chard, Swiss	Leaves	5	16	92	1.8	0.2	1.5	3300	0.04	0.09	0.4	30	51	1.8	75	46	1
Mushroom		0	13	92	2.9	0.1	0.4	0	0.08	0.30	4.6	8	5	0.5	12	90	1
Okra (gumbo)	Fruit	10	25	88	2.0	0.1	2.7	660	0.20	0.06	1.0	44	81	0.8	59	63	1
Spinach	Leaves	5	20	90	3.6	0.4	0.8	5800	0.12	0.16	0.8	52	107	2.1	103	66	1
Spinach, New Zealand	Leaves	0	10	94	1.5	0.2	0.6	4400	0.04	0.13	0.5	30	58	0.8	39	28	1
Water convolvulus	Leaves	0	25	92	2.6	0.2	3.4	3500	0.03	0.10	0.9	55	95	2.2	49	40	1
Amaranth, edible	Leaves	29	36	87	3.5	0.5	6.5	6090	0.08	0.16	1.4	80	267	3.9	—	67	2
Bamboo	Shoot	71	27	91	2.6	0.3	5.2	20	0.15	0.07	0.6	4	13	0.5	—	59	2
Malabar spinach	Leaves	0	19	93	1.8	0.3	3.4	8000	0.05	—	0.5	102	109	1.2	—	52	2
Purslane	Leaves	0	21	92	1.7	0.4	3.8	2550	0.03	0.10	0.5	25	103	1.4	—	31	2

[a] Key to references: (1) Howard *et al.* (1962): (2) Leung, W.W., Pecot, R.K., and Watt, B.K. 1952. Composition of foods used in Far Eastern Countries, USDA Handb. 34, US. Dept. Agric., Washington DC.

In the spring, the clumps are covered with soil or compost so that the emerging shoots are in the dark. Exposure to light causes bitterness as cyanogenic glucosides are formed by the shoot. Shoots are harvested as they emerge from the mound.

In Japan, bamboo shoots are forced by growing under mulch heated with electric cables placed 6–8 cm (2½–3 in.) below the soil. About 2–3 cm (1 in.) of rice straw is placed on the soil surface and an additional layer of soil 4–5 cm (1¾–2 in.) on top and plastic sheeting on top of all. The soil temperature is kept between 13° and 15°C (55° and 59°F). By this mulching procedure, bamboo shoots can be harvested almost a month earlier than those not mulched.

In the preparation of shoots for use, the sheaths are removed, the shoots cut lengthwise if the diameter is large, and boiled in water for about 1/2 hr; salt is added a few minutes before the end of boiling. Upon tasting, if the shoot is bitter, it is boiled again to remove the bitterness. Raw bamboo shoot is acrid, and boiling removes the acridity. The bitterness is from cyanogenic glucosides.

GINGER: *Zingiber officinale*

Ginger, often called "ginger root," is not a root but a rhizome. Probably a native of India, it has been used for medicinal purposes as well as food since ancient times in Asia and taken to Europe about the time of Christ. The crop is now extensively cultivated throughout the tropics and in the warmer regions of the temperate zone.

Ginger is a monocotyledonous perennial herb growing to heights of 30–100 cm (12–36 in.) depending on the variety. The rhizomes, 2–3 cm (¾–1¼ in.) in diameter, are thick with nodes compressed. The many branched rhizomes grow near the soil surface. The erect shoots, growing in clumps, are formed from long leaf sheaths; the leaf blades are dark green, smooth, slender, broader at the base than the tip, about 1–2½ cm (⅜–1 in.) wide, and 10–25 cm (4–10 in.) long.

There are several cultivars grown; preferences vary according to countries and regions. Cultivars vary in rhizome color: white, pale yellow, yellow, and red; some are more pungent than others. In Japan, the cultivars are grouped according to size: small, intermediate, and large. The crop is grown in rich, well-drained soil high in organic matter with a pH range 5.5–6.5 and free of nematodes and *Fusarium* fungus. Propagation is by use of clean rhizomes about 3–5 cm (1–2 in.) long having three to four "eyes" (buds). The seed pieces are planted to 10 cm (3–4 in.) deep and spaced 35–45 cm (14–18 in.) between plants and in rows 60–90 cm (2–3 ft) apart. Planting is in the early spring and adequate moisture is maintained throughout the growing period of 8–10 months.

Some N and P about 50 kg/ha each (45 lb/acre) are applied about 5–8 cm (2–3 in.) below the seed piece; if well-decomposed manure is used, the amount of fertilizer is reduced. Three additional fertilizer side-dress applications of 25 kg/ha (28 lb/acre) each of N, P, and K are made during the growing season in Hawaii. The first is made a month after planting about 8 cm (3 in.) deep and 15–20 cm (6–8 in.) to the side; the second is made 3 months after planting, 8 cm deep and 25–30 cm (10–12 in.) to the side; and the third is made 6 months after planting, 8 cm deep and 38 cm (15 in.) to the side of the row. Shoot growth is rapid at 30°C (86°F) but almost ceases at 15°C (59°F).

Harvesting of rhizomes is made after the leaves turn yellow and dry down completely; at this stage the skin of the rhizome has set and will not bruise during harvest and washing. Depending on the region grown, the plant may or may not bloom. It sometimes blooms in September in Japan and rarely in Malaysia.

For pickling or manufacturing of preserves, the rhizomes are harvested before the leaves yellow but are still green; at this stage, the skin is not tough, the flesh not fibery and not as pungent.

Pungent principle of ginger is zingerone, $C_{11}H_{14}O_3$, and the odor is from zingiberene, $C_{15}H_{24}$, a sesquiterpene.

MATAI, WATER CHESTNUT: *Eleocharis dulcis (E. tuberosa)*

Matai or water chestnut is a monocotyledonous plant native to the swampy regions of the Old World tropics and belonging to the sedge family (Cyperacea). The plant is grown for the sweetish corms about 2.5–4 cm (1–1½ in.) in diameter. The corms, which can be eaten raw, are considered a delicacy in Chinese dishes because they remain crisp after being cooked. There are two types: the sweet called *hon matai* and the starchy type called *sui matai.*

Eleocharis dulcis is characterized by numerous upright tubular septate stems, which function as photosynthetic organs in place of leaves. Although the plant flowers and produces seeds, it asexually reproduces by means of rhizomes and corms.

Matai is usually grown in rotation with paddy rice in the Orient. Corms are planted in a nursery in the early spring after danger of frost is past. When plants are about 20 cm (8 in.) high, they are transplanted into paddies at 45–70 cm (1½–2½ ft) spacing. The field is kept flooded during growth. The plant produces rhizomes, which form a series of new plants and soon cover the open spaces. Heavy applications of well-decomposed manure and lime are made before and during growth. Corm formation takes place after the daughter plants from the rhizomes have grown to maturity. The corms mature after the tops have died down or are killed by frost in the fall. Harvest is made after the tops are down, and the water has been drained off. Corms are usually picked by hand after the soil has been turned over by spade or plowed about 15 cm (6 in.) in depth and clods broken by rake or other implements. The corms can be left in the ground over winter and harvested as needed.

Matai is different from dicotyledonous *Trapa bicornis*, another aquatic plant, also called "water chestnut." *Trapa bicornis* has a fruit called a nut with two curved horns (Fig. 25.2). *Trapa natans*, called "Jesuit's nut" and "water caltrop," also has edible "nuts" with four horns. The fruits of both *Trapa* species cannot be eaten raw as they contain toxic substances, which can be destroyed by boiling for an hour. Another aquatic crop which has corms similar to *E. dulcis* is *Sagitteria sagittifolia (S. sinensis)* called arrowhead (Fig. 25.3). The leaves are

FIG. 25.2. Water chestnut *(Trapa bicornis)*.

FIG. 25.3. Arrowhead *(Sagittaria sagittifolia)* corms.

arrowhead-shaped and the corms are ovoid and light colored compared to the flat oval and dark colored ones of matai. Arrowhead corms are cooked and not eaten raw.

Miscellaneous Dicotyledons

LOTUS, EAST INDIAN LOTUS, SACRED LOTUS: *Nelumbo nucifera (Nelumbrium nelumbo)*

Lotus rhizomes, often called lotus roots, are a delicacy in Oriental cooking. The rhizome has a crisp texture after cooking. Lotus is indigenous to Asia, from Iran east to Japan and into Southeast Asia and far south to northeast Australia.

An aquatic monocotyledonous plant, lotus is grown in ponds or in paddy culture. Propagation is usually by rhizomes, with three or four nodes, planted in water-saturated, clay soil with the distal end in the mud and the proximal (basal) end above water level after flooding.

Planting is from mid-April to early May in Japan. The rhizome is buried about 18–20 cm (7–8 in.) deep at about a 30° angle into the muddy soil and enough of the basal end out so that it is still in air for aeration after the field has been flooded with 7–10 cm (3–4 in.) of water.

From the nodes near the apex of the planted rhizome, new leaves emerge, and the petiole elongates to such a length so that the first two to three leaf blades float on water. The subsequent leaves emerging from the new nodes of the growing stem have petioles that stand above water; each succeeding leaf is higher and larger than the previous one. In late July, the terminal leaf emerges; it is short and has a small leaf blade. Roots and secondary stems arise from the same node as the leaf.

The character of the stem growth changes to enlarged internodes during growth. These terminate after production of three to four swollen internodes; each internode becomes successively shorter than the previous one (Fig. 25.4). Stems and rhizomes have several tubular longitudinal air passages of two sizes; large and small. Flowers are showy and

FIG. 25.4. Lotus *(Nelumbo nucifera)* rhizomes. Cross section at left shows ovoid air passages in rhizome section.

fragrant, colored pink to rose, and sometimes white which arise from nodes late in the growing season, from late July to September depending on the cultivar and region grown.

The terminal rhizomes with three to four swollen internodes are harvested in late September through early spring after the tops have died down and the water drained off.

Besides the rhizomes used as vegetables, the seeds are washed and eaten. In China, the seed pods are brewed into tea for medicinal purposes.

EDIBLE AMARANTH, AMARANTH, CHINESE SPINACH, TAMPALA: *Amaranthus tricolor* (*A. gangeticus* and *A. oleraceus*)

Amaranth, an important pot-herb of the tropics, is a short-lived annual belonging to the same genus as the important grain crop of the Aztecs of Mexico in pre-Columbian times. It is one of the few crops that has been domesticated in the Old and New Worlds. *Amaranthus tricolor* is an ancient crop used for greens in Southeast Asia and *A. candatus* is an important grain crop in Ethiopia.

Amaranth is an erect plant growing 30–90 cm (1–3 ft) in height producing small flowers on terminal and axillary spikes (Fig. 25.5). Though the small seeds produced in large number are edible, the flowers are not edible. There are many cultivars in Southeast Asia classified according to leaf color and shape. *Amaranthus tricolor* flowers readily under short-day conditions.

The crop is propagated by seeds planted in light, well-drained soils. Seedlings can be transplanted in 2–3 weeks when they are at the two to four leaf stage or thinned to a stand of 8 × 8 cm (3 × 3 in.). The larger of the thinned plants can be used for greens. In 3–4 weeks after transplanting, when plants are 15–20 cm (6–8 in.) in height, the entire plant may be harvested. Plant spacing is increased to 30–40 cm (12–14 in.) when tender shoots are harvested; several cuttings can be made and yields are greatly increased. Frequent cuttings prolong the productive period and also delay flowering. Periodic (once a week) application of fertilizer, 18 kg N, 18 kg P and 36 kg K/ha (16 lb N, 16 lb P, and 32 lb K/acre), is recommended for high yields.

Other than the fact that amaranths contain high amounts of oxalic acid, which may decrease the nutritionally available calcium, they supply large amounts of protein, vitamins A and C, and fiber.

The fungus, *Choanephora*, causes wet rot of leaves and young stems,

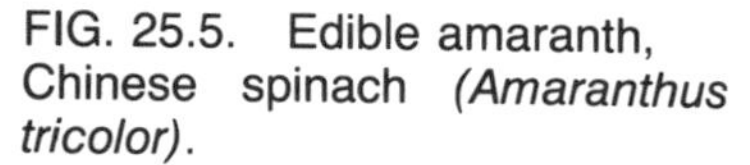

FIG. 25.5. Edible amaranth, Chinese spinach *(Amaranthus tricolor)*.

and *Pythium* causes damp-off of seedlings. Many chewing insects and nematodes attack the plant.

JEW'S MALLOW, JUTE MALLOW, TOSSA JUTE, MOLOKHIA: *Corchorus olitorius*

Probably of south China origin, Jew's mallow, a member of the Tiliaceae (Jute) family, is an important green vegetable of the Middle East, Egypt, and Sudan as well as parts of tropical Africa. Jute is usually grown for the fibers, but the cultivar grown for vegetable use is short and branched. A short-day plant for flowering, it thrives at mean temperatures of 24°–35°C (75°–95°F) and high rainfall conditions of the lowland tropics. The tender mucilaginous leaves are harvested and used as spinach.

MALABAR SPINACH, MALABAR NIGHTSHADE, INDIAN SPINACH, CEYLON SPINACH: *Basella alba* (*B. rubra* and *B. cordifolia*)

Malabar spinach is probably a native to tropical Asia (India and Indonesia) and is now grown in all parts of the tropics and in warm regions of the temperate zone. There is some confusion in the literature between *B. alba* and *B. rubra.* Hortus Third lists *B. rubra* as a cultivar of *B. alba*, and having reddish stems, petioles leaves, and flowers; this cultivar is called Ceylon spinach. The heart-shaped leafed cultivar is often called *B. cordifolia.*

The plant is a rampant vine often extending 3–5 m (10–15 ft) or more. It has long to broad ovate rugose, thick and succulent leaves often 10 cm (4 in.) long and about 8 cm (3 in.) wide, which are green to purplish in color (Fig. 25.6); in some cultivars, the leaves may be acute to slightly obtuse. The flowers are perfect with white to reddish petals and fruits are glossy black and flattened, about 7–8 mm (5⁄16 in.) in diameter. The dark red juice from the fruit was used as a dye and ink in ancient China. Flowering occurs under short days (about 12 hr) and under water stress. Day lengths greater than 13 hr inhibit flowering.

Malabar spinach is adapted to many climates and soil types. It is

FIG. 25.6. Malabar spinach, Ceylon spinach *(Basella alba)*.

propagated by seeds, stem cuttings, or crowns placed in hills about 1 m (3 ft) apart and often grown on stakes or trellises. About 4 weeks after planting if the plant has attained good growth, the succulent young leaves and young shoots 15–30 cm (6–12 in.) are harvested and cooked as greens. The leaves contain a mucilaginous substance. Malabar spinach is a perennial in the tropics, but dies down in the dry regions of the tropics and in the temperate regions.

NEW ZEALAND SPINACH: *Tetragonia expansa (T. tetragonioides)*

New Zealand spinach, which belongs to the carpet weed family (Aizoaceae), was found growing wild in New Zealand by Captain Cook in 1770. Subsequently, it was found along the seacoast of southern and western Australia and in Tasmania. It is now cultivated in most parts of the tropics and temperate regions but is still a minor crop.

New Zealand spinach (Fig. 25.7) is a drought-resistant annual trailing vine with triangular shaped, thick, green and succulent leaves, spreading to a distance of 1–1½ m (3–4 ft). It has yellowish green inconspicuous flowers, which are borne sessile in the axils of leaves.

The crop is propagated by "seeds," which germinate over a long period of time ranging from 2 weeks to 3 months; soaking the "seeds" in water for 24 hr prior to sowing reduces the germination time. The "seed" is actually a hard dry, angular fruit, about 8–10 mm (5⁄16–3⁄8 in.) long, containing several true seeds.

New Zealand spinach is usually grown on beds 1 m (3 ft) wide and spaced 50–100 cm (2–3 ft) between plants. A balanced fertilization program is recommended with adequate soil moisture for rapid growth required for good quality and high yields. From seedling emergence to first harvest requires about 40–50 days. Tender shoots 15–20 cm (6–8 in.) long are cut and used like common spinach and are often put raw in salads. After the first cutting, the plants produce many upright branches which are cut in subsequent harvests.

OKRA, GUMBO, LADY'S FINGER, QUIABO, BHINDI: *Hibiscus esculentus (Abelmoschus esculentus)*

The origin of okra is rather obscure; wild varieties exist in Ethiopia and in the Upper Nile in Sudan and perennial varieties in West Africa.

FIG. 25.7. New Zealand spinach *(Tetragonia expansa)*.

Its presence in the New World has been attributed to the slaves from Africa.

A member of the mallow family, Malvaceae, okra is now grown in all parts of the tropics and during the summer in the warmer parts of the temperate region. It is an herbaceous, shrub-like dicotyledonus annual plant with woody stems growing to heights of 1–2 m (3–6 ft). It has alternate palmate broad leaves and the flowers have five large yellow petals with a large purple area covering the base. Fruits, which are harvested immature, are pale green, green, or purplish pods and in many cultivars are ridged. When mature they are dark brown dehiscent or indehiscent capsules.

Okra is a warm season crop, growing best where the minimum and maximum mean temperatures are 18°C (65°F) and 35°C (95°F), respectively. It can be grown in a wide range of soil types, but is intolerant of wet and poorly drained and acidic soils.

A hard seed coat prevents good germination of okra seed; treatment of seeds with concentrated sulfuric acid for 2–3 hrs improves germination. Germination is stimulated by soaking seeds in a water bath at 45°C (113°F) for 1½ hr. The optimum soil temperature for seed germination is 24°–32°C (75°–90°F). Germination is poor at 20°C (68°F). Okra seeds are planted 2–3 cm deep in rows 1 m (3 ft) apart. Seedlings are spaced at about 30 cm (12 in.) after thinning. About 30 kg/ha (27 lb/acre) each of N, P, and K are recommended at time of seeding and an additional 35–50 kg N/ha (45 lb/acre) is applied after the plant first blooms. Adequate moisture is required for growth and yield, but excessive moisture should be avoided.

Short-day lengths stimulate flowering of most cultivars. Flowering begins at a very early stage of growth at day lengths of less than 11 hr; under long days, the flower buds tend to abort. The cultivar 'Clemson Spineless' is not photoperiod responsive and has been widely accepted for growing in temperate regions where day lengths are long during the summer.

About 35–60 days after emergence, the plant begins to flower; the flower remains open for a day and is either self- or cross-pollinated. Immature pods (Fig. 25.8), 8–9 cm (3–3½ in.) long, are ready for harvest about 4–6 days after anthesis. Pods are either snapped or cut from the plant. Protective coverage for the hands and arms are worn by workers to prevent irritation from the plants. Harvest is recommended at least every other day or every day during hot weather for size and quality. About 35–40 days are required from anthesis to mature seeds. If pods are allowed to mature, plant growth declines and few flowers develop, but with continued harvesting, the plant continues to set fruit.

Immature okra fruits remain in usable quality for 8–10 days if held at 2°–13°C (36°–55°F) and 90% RH. However, due to chilling injury, those held for 4–6 days at 0°–10°C (32°–50°F) deteriorate very rapidly when transferred to 20°C (68°F).

Phythium, Rhizoctonia, Fusarium and *Phytophtora* are diseases that attack okra especially under wet conditions. The cotton aphid is a serious pest of the crop.

PURSLANE, GARDEN PURSLANE: *Portulaca oleracea*

Purslane, belonging to the Portulacaceae or Purslane family, is a common weed in many parts of the world; since ancient times it has been

FIG. 25.8. Okra *(Hibiscus esculentus).*

used as a vegetable, eaten raw in salads or cooked as greens. Purslane is indigenous in the region west of the Himalayas to southern Russia and Greece. It was transported to America from Europe. It is now believed that purslane is also indigenous to America and that it did not spread from material brought from Europe. The tender shoots are collected from the fields (weeds) in China, Philippines, Java, and Malaysia for consumption either raw or cooked. It contains oxalic acid and accumulates nitrates. The crop is cultivated commercially in Egypt and Sudan and listed in seed catalogues in England.

WATER SPINACH, WATER CONVULVULUS, SWAMP CABBAGE, KANG KONG (SOUTHEAST ASIA), ONG TSOI (CHINESE): *Ipomoea aquatica (I. reptans)*

Possibly of East Indian origin, water spinach, a member of the morning glory family, Convolvulaceae, has a long history in China. Reference was made of its uses in Chinese literature of about 300 AD and it was listed in the *Materia Medica* of the eleventh century AD. An herbaceous perennial aquatic and semiaquatic plant of the tropics and subtropics, it is grown for the long tender shoots, which are used as green vegetable, and the entire plant, which is harvested for fodder.

Water spinach has hollow stems, which allow the vines to float on water; adventitious roots are formed at nodes, which are in contact with water or moist soil. The plant is adapted for climates with mean temperatures above 25°C (77°F), growth being retarded at temperatures below 10°C (50°F). The crop is grown year-round in the tropics. Flowering occurs under short-day conditions. Flowers are large and white with a purplish center similar to the sweet potato *(Ipomoea batatas)*.

There are two principal cultivars of water spinach: (1) a narrow pointed leaf cultivar called *Ching Quat* (Fig. 25.9), which is adapted for moist soil culture, and (2) a green, broadleafed cultivars called *Pak Quat*, which is adapted for aquatic culture. Other cultivars are variations of these two types.

FIG. 25.9. Water spinach, *kang kong*, *ong tsoi (Ipomoea aquatica)*.

In the moist soil culture, the crop is grown on raised beds 60–100 cm (2–3 ft) wide. Seeds are sown directly on the bed or nursery grown seedlings about 15 cm (6 in.) tall are transplanted on to the beds at spacing of about 12 cm (4½ in.) between plants. Also, stem cuttings 30–40 cm (12–15 in.) long are planted about 15 cm (6 in.) in depth for propagation. The furrows between the beds are flooded very soon after seeding or transplanting. When rainfall is low, frequent heavy irrigations are necessary for high quality shoots.

For aquatic culture, the broadleaf cultivar cuttings are transplanted into puddled soil, similar to planting of rice in paddies. The cuttings are about 30 cm (12 in.) long with seven to eight nodes and are planted 15–20 cm (6–8 in.) deep and spaced 30–40 cm (12–16 in.) apart in blocks. After planting the land is flooded to 3–5 cm (1½–2 in.) in depth; the water is kept flowing continuously. After the plants are established 45–55 kg of ammoniacal nitrogen/ha (40 to 50 lb/acre) application is recommended; then the water level is raised to 15–20 cm (6–8 in.) depth.

In semiaquatic culture, the crop is ready for harvest 50–60 days after sowing; entire plants are pulled, washed, and bundled. More than one harvest can be made if shoots are cut above ground level, allowing secondary shoots to grow from nodes below the cut.

In aquatic culture, the first harvest can be made after about a month of good growth. The upper part of the main shoot, about 30 cm (12 in.) long, is cut about 5 cm (2 in.) above water level; from eight to 10 shoots are bundled and marketed. Removal of the main shoot stimulates horizontal shoot growth. These new shoots can be harvested in 4–6 weeks, depending on plant vigor and temperature. About 40–90 MT/ha can be harvested from three or more cuttings in a year.

For seed production in the fall, harvesting is ceased after flowering commences and the field is drained of water. The mature pods are harvested, dried in the sun and gently trodded to free the seeds.

Fungi

MUSHROOMS: *Agaricus bisporus (Psalliota campestris, Agaricus campestris)*

Mushrooms are usually cultivated using horse manure or other artificial medium such as mixtures of alfalfa hay, corn cobs, wheat or rice straws, sources of nitrogen (urea, cotton, or soybean meal), and inorganic materials (gypsum and phosphates) composted together.

The compost, after proper preparation, is brought into mushroom houses or in caves and allowed to further decompose. Temperatures rise naturally or by addition of steam to pasteurize at 60°C (140°F) for 3–4 hr and then is lowered to 55°–58°C (131°–136°F) for about 36 hr and then cooled to 48°–50°C (118°–122°F) for 4–7 days. There should be no ammonia odor in the compost and it should have a sweetish odor.

The temperature of the compost is lowered to 25°–30°C (77°–86°F) as quickly as possible and keeping the air spore free. The compost is then inoculated with the properly prepared spawn, the surface made smooth and tamped, followed by light watering. The optimum growing temperature is 25°C (77°F) but satisfactory growth takes place at 20°C (68°F); growth is slow below 15°C (59°F) and the maximum temperature tolerated is between 33° and 35°C (92° to 95°F). Moisture is maintained during mycelial growth.

After the mycelium has spread throughout the compost in about 10–14 days, the top of the bed is covered with casing material, which increases CO_2 concentration in the substrate, causing fructification (formation of reproductive fruiting bodies). Casing material may be sand, clay, peat, lime, or mixtures or combination of these materials at pH of 7.0–7.5 and free of pests and diseases. Casing can be treated with live steam or with formaldehyde to achieve sterility.

The casing layer, from 3 to 4 cm (1¼ to 1½ in.) deep, should be moistened to about field capacity and not over watered. In 6–8 days the mycelium grows into the casing and "pinheads" form at the surface which develops further into mushrooms. Too high a CO_2 concentration (0.6–0.4%) may cause "pinheads" to die, while concentrations of 0.4–0.2% give mushrooms that are tall and caps that open early; the ideal concentration is between 0.2 and 0.06% CO_2 for best quality mushrooms.

The optimum temperature for mushroom storage is 0°–3°C (32°–37°F) and 95% RH.

The odor of mushrooms has been identified as 1-oceten-3-ol and 1-octen-3-one.

STRAW MUSHROOM, PADI STRAW MUSHROOM, CHINESE MUSHROOM: *Volvariella volvacea (Volvariella esculenta)*

The culture of straw mushroom (Fig. 25.10), no doubt, started in China centuries ago, but the first mention of this crop in the Chinese

FIG. 25.10. Straw or Chinese mushroom *(Volvariella volvacea)*, Taipei, Taiwan.

literature dates only to 1822. The substrate of this mushroom is ordinary rice straw, which is composted, but it can be cultivated in the same artificial compost of the common mushroom, *Agarius bisporus*, prepared from alfalfa hay, corn cobs, and nitrogen supplement. Light is required for at least equivalent to 15 min of full sunlight per day for fruiting bodies to appear; the inoculated compost kept totally in the dark does not produce fruiting bodies. The optimum temperature is 30°C (86°F). The beds are kept damp throughout the incubation period in which the mycelium grows, and after 2–3 weeks, the fruiting bodies appear. Harvests can be made for about 2 months.

Special precautions should be taken in the preparation of straw mushrooms for eating. They should be thoroughly cooked with complete softening of all tissues as raw or undercooked mushrooms can cause an unpleasant peppery or numb feeling of the mouth and throat.

The world production in 1975 was estimated at 38,000 MT (42,000 tons), China producing almost 75% of the total and Taiwan and Thailand each producing about 12%.

SHIITAKE, BLACK MUSHROOM, FOREST MUSHROOM: *Lentinus edodes (Cortinellus Berkeleyanus)*

Shiitake is a mushroom grown commercially principally in Japan and smaller productions in China, Taiwan, and South Korea. It is produced on wooden logs of hardwood trees such as *shi, kumegi, kashi*, and chestnut in Japan. Holes are drilled into moist logs (logs are to be soaked in water if dry) and shiitake spawn, which are grown on short wooden pegs, are gently tapped into predrilled holes (Fig. 25–11). The inoculated logs are stacked vertically, preferably not on soil as contamination by undesirable organisms can enter the wood. There should be adequate ventilation between logs and the logs kept under fairly humid conditions. Logs should be moistened if they begin to dry out.

The optimum temperature for mycelial growth is 24°C (75°F) and fairly humid conditions. At temperatures of 40°C (104°F) or higher, the organism is killed.

Shiitake has been successfully grown on artificial substances of organic matter such as combinations of oak chips, rice hulls, and other supplements.

FIG. 25.11. Shiitake, black mushroom *(Lentinus edodes)* culture on wood logs, Japan.

The fruiting bodies are used fresh like common mushrooms but have, in the opinion of many, a much superior flavor. Shiitake can be dried and stored.

BIBLIOGRAPHY

Asparagus *(Asparagus officinalis)*

BENSON, B.L., and TAKATORI, F.H. 1980. Partitioning of dry matter in open-pollinated and F_1 hybrid cultivars of asparagus. J. Am. Soc. Hort. Sci. *105*, 567–570.

HASEGAWA, P.M., MURASHIGE, T., and TAKATORI, F.H. 1973. Propagation of asparagus through shoot apex culture. II. Light and temperature requirements transplant ability of plants and cyo-histological characteristics. J. Am. Soc. Hort. Sci. *98*, 143–148.

HOWARD, F.D., MACGILLIVRAY, J.H., and YAMAGUCHI, M. 1962. Bull. No. 788. California Agric. Exp. Sta., Univ. of California, Berkeley.

HUNG, L. 1975. Annotated bibliography on asparagus. Dept. of Horticulture, National Taiwan Univ., Taipei, Taiwan, Republic of China.

MACGILLIVRAY, J.H. 1953. Vegetable Production. McGraw-Hill Book Co., New York.

TAKATORI, F.H., SOUTHER, F.D., STILLMAN, J.I., and BENSON, B. 1977. Asparagus culture in California, Bull. No. 1882, Div. of Agric. Sci. Univ. of California, Berkeley.

Bamboo (*Phyllostachys, Dendrocalamus, Bambusa,* and *Sasa* genera)

HERKLOTS, G.A.C. 1972. Vegetables in South-East Asia. Hafner Press, New York.

KENNARD, W.C., and FREYRE, R.H. 1957. The edibility of shoots of some bamboo. Econ. Bot. *11*, 235–243.

McCLURE, F.A. 1956. Bamboo in the economy of oriental peoples. Econ. Bot. *10*, 335–361.

YOUNG, R.A. 1954. Flavor qualities of some edible oriental bamboos. Econ. Bot. *8*, 377–386.

Ginger *(Zingiber officinale)*

NAKAGAWA, Y. 1963. Edible ginger culture. Ext. Mimeo. Univ. of Hawaii, Honolulu, Hawaii.

PURSEGLOVE, J.W. 1972. Tropical Crops: Monocotyledons. John Wiley & Sons, New York.

TEWSON, L. 1966. Australian ginger. World Crops *18*(3), 62–65.

Lotus *(Nelumbo nucifera)*

HERKLOTS, G.A.C. 1972. Vegetables of South-East Asia. Hafner Press, New York.

KUMAZAWA, S. 1965. Vegetable Crops. Yokendo Publishers, Tokyo, Japan. (*In* Japanese.)

Waterchestnut *(Eleocharis dulcis)*

HODGE, W.H. 1956. Chinese waterchestnut or Matai—a paddy crop of China. Econ. Bot. *10*, 49–65.
HODGE, W.H., and BISSET, D.A. 1955. The Chinese waterchestnut. USDA Circ. No. 956, US Dept. Agric., Washington, DC.

Amaranth *(Amaranthus tricolor)*

GRUBBEN, G. (Editor). 1980. Proc. 2nd Amaranth Conf., Sept. 1979. Rodale Press, Inc., Emmaus, Pennsylvania.
GRUBBEN, G.J.H. 1976. The cultivation of amaranth as a tropical leaf vegetable. Comm. to 67. Dept. of Agric. Res., Royal Tropical Institute, Amsterdam, Netherlands.
HERKLOTS, G.A.C. 1972. Vegetables of South-East Asia. Hafner Press, New York.
KNOTT, J.E., and DEANON, J., JR. 1967. Vegetable Production in Southeast Asia. Univ. of Philippines Press, Manila.
MARTIN, F.W., and RUBERTE, R.M. 1975. Edible leaves of the tropics. Mayaguez Inst. Trop. Agric., Sci. Ed. Admin., US Dept. Agric., Mayaguez, Puerto Rico.
MARTIN, F.W., and TELEK, L. 1979. Vegetables of the hot humid tropics. Part 6. Amaranth and celosia, *Amaranthus* and *Celosia*. Mayaguez Inst. Trop. Agric., Sci. Ed. Admin., US Dept. Agric., Mayaguez, Puerto Rico.

Malabar Spinach *(Basella alba)*

HERKLOTS, G.A.C. 1972. Vegetables of South-East Asia. Hafner Press, New York.
KNOTT, J.E., and DEANON, J., JR. 1967. Vegetable Production in Southeast Asia. Univ. of Philippines Press, Manila.
MARTIN, F.W., and RUBERTE, R.M. 1975. Edible leaves of the tropics. Mayaguez Inst. Trop. Agric., Sci. Ed. Admin., US Dept. Agric., Mayaguez, Puerto Rico.
WINTER, H.F. 1963. Ceylon spinach (*Basella rubra* L.) Econ. Bot. *17*, 195–199.

New Zealand Spinach *(Tetragonia expansa)*

HERKLOTS, G.A.C. 1972. Vegetables in South-East Asia. Hafner Press, New York.
KAYS, S.J. 1975. Production of New Zealand spinach (*Tetagonia expansa*, Murr.) at high plant densities. J. Hort. Sci. *50*, 135–141.
KNOTT, J.E., and DEANON, J., JR. 1967. Vegetable Production in Southeast Asia. Univ. of Philippines Press, Manila.
MacGILLIVRAY, J.H. 1953. Vegetable Production. McGraw-Hill Book Co., New York.

Okra *(Hibiscus esculentus)*

ALBREGTS, E.E., and HOWARD, C.M. 1974. Responses of okra to plant density and fertilization. Hort Science *9*, 400.

ARULRAJAH, T., and ORMROD, D.P. 1973. Responses of Okra (*Hibiscus esculentia* L.) to photoperiod and temperature. Ann. Bot. (Rome) *37*, 331–340.
MARTIN, F.W., and RUBERTE, R. 1978. Vegetables for the hot humid tropics, Part 2. Okra, *Abelmoschus esculentus*. Mayaguez Inst. Trop. Agric., Sci. Ed. Admin., US Dept. Agric., Mayaguez, Puerto Rico.
PURSEGLOVE, J.W. 1968. Tropical Crops: Dicotyledons. John Wiley & Sons, New York.
SCHWEERS V.H., and SIMS, W.L. 1976. Okra production. Leaflet 2679, Div. of Agric. Sci., Univ. of California, Berkeley.

Purslane *(Portulaca oleracea)*

HERKLOTS, G.A.C. 1972. Vegetables of South-East Asia. Hafner Press, New York.
MARTIN F.W., and RUBERTE, R. 1975. Edible leaves of the tropics. Mayaguez Inst. Trop. Agric., Sci. Ed. Admin., US Dept. Agric., Mayaguez, Puerto Rico.
ZIMMERMAN, C.A. 1976. Growth characteristics of weediness in *Portulaca oleracea* L. Ecology *57*, 964–974.

Water Spinach *(Ipomoea aquatica)*

EDIE, H.H., and HO, B.W.C. 1969. *Ipomoea aquatica* as a vegetable crop in Hong Kong. Econ. Bot. *23*, 32–36.
KNOTT, J.E., and DEANON, J.R., JR. 1967. Vegetable Production in Southeast Asia. Univ. of Philippines Press, Manila.
MARTIN, F.W., and RUBERTE, R. 1975. Edible leaves of the tropics. Mayaguez Inst. Trop. Agric., Sci. Ed. Admin., US Dept. Agric., Mayaguez, Puerto Rico.

Mushrooms *(Agaricus bisporus)*

KNEEBONE, L.R., and MASON, E.C. 1962. Mushroom yields as influenced by degree of maturity at time of harvest. Proc. 5th Int. Conf. Scientific Aspects of Mushroom growing. Philadelphia, Pennsylvania.
LAMBERT, E.B. 1955. Mushroom growing in the United States. USDA Farmers Bull. No. 1875, US Dept. Agric., Washington, DC.
MURR, D.P., and MORRIS, L.L. 1975. Effect of storage temperature on postharvest changes in mushrooms. J. Am. Soc. Hort. Sci. *100*, 16–19.
SIMS, W.L., and HOWARD, F.D. 1979. Growing mushrooms. Leaflet 2640, Div. Agric. Sci., Univ. of California, Berkeley.
TAPE, N.W. 1965. How to grow mushrooms. Pub. No. 1205. Canada Dept. Agric. Ottawa, Ontario, Canada.

Straw Mushroom *(Volvariella volvacea)*

CHANG, S.-T. 1977. The origin and early development of straw mushroom cultivation. Econ. Bot. *31*, 374–376.
SAMARAWIRA, I. 1979. A classification of the stages in the growth cycle of the cultivated paddy straw mushrooms (*Volvariella volvacea* Singer) and its commercial importance. Econ. Bot. *33*, 163–171.

SAN ANTONIO, J.P., and FORDYCE, C., JR. 1972. Cultivation of the paddy straw mushroom. *Volvariella volvacea*. HortScience *7*, 461–464.

Skiitake *(Lentinus edodes)*

KUMAZAWA, S. 1965. Vegetable Crops. Yokendo Publishers, Tokyo, Japan. (*In* Japanese.)
SAN ANTONIO, J.P. 1981. Cultivation of the shiitake mushroom. HortScience *16*, 151–156.

Appendix

SOME USEFUL CONVERSIONS

Length

1 in. = 2.54 cm
1 ft = 0.305 meter (m)
1 mile = 1.61 km

1 cm = 0.394 in.
1 m = 3.28 ft
1 km = 0.62 mile

Area

1 acre = 43,560 ft^2 = 0.405 hectare (ha)
1 $mile^2$ = 640 acres = 2.59 km^2

1 ha = 100 are (a) = $10^4 m^2$ = 2.47 acres
1 km^2 = 100 ha = 0.386 mi^2

Volume (fluid)

1 gallon (US) = 3.785 liters = 231 $in.^3$

1 liter = 1.06 quart = 0.264 gallon

1 quart = 946 ml
1 bushel (US dry measure = 2150 $in.^3$ = 32 quarts (dry) = 35.24 liters

Weight

1 oz = 28.35 grams (g)
1 lb = 0.454 kg
1 ton (short) = 0.907 metric ton (MT)
1 bushel = 60 lb

1 g = 0.0353 oz
1 kg = 2.205 lb
1 MT = 1.1 tons
1 quintal = 100 kg = 220.5 lb

Yield or Rate

1 lb/acre = 1.12 kg/ha
100 lbs/acre = 1.12 quintal/ha

1 ton/acre = 2.24 MT/ha

1 kg/ha = 0.89 lb/acre
1 quintal/ha = 0.89 cwt (100 lbs)/acre = 1.487 bushels/acre
1 MT/ha = 0.445 ton/acre

Temperature

°F = 9/5°C + 32 °C = 5/9 (°F – 32)

Light (Radiation)

Quality: Wavelength measured in angstroms Å = 10^{-10}m; or nanometers (nm) = 10^{-9}m

Intensity: (point source) candle power (CP) or candela (Note: 1 CP emits 4 π lumens (luminous flux)

Illuminance: (luminous flux incident on a surface)
Ft candle (fc) = 1 CP at 1 ft distance = 1 lumen/ft^2
Meter candle = 1 CP at 1 meter distance = 1 lumen/m^2 = 1 lux
1 Ft candle = 10.76 meter candles = 10.76 lux

Energy: Langley (unit of solar radiation) = 1 gram-calorie/cm^2
1 g-calorie = 1.162 × 10^{-3} watt-hr = 4.184 Joules
1 watt/m^2 = 10 microeinsteins/m^2-sec.

Pressure and Soil Moisture Tension

1 atmosphere = 760 mm Hg = 1034 g/cm^2 = 14.7 lb/in.^{2} (psi) = 1.013 bar
1 bar = 10^6 dynes/cm^2 = 0.988 atmosphere = 10^5 Pascals (Pa) = 100 kPa

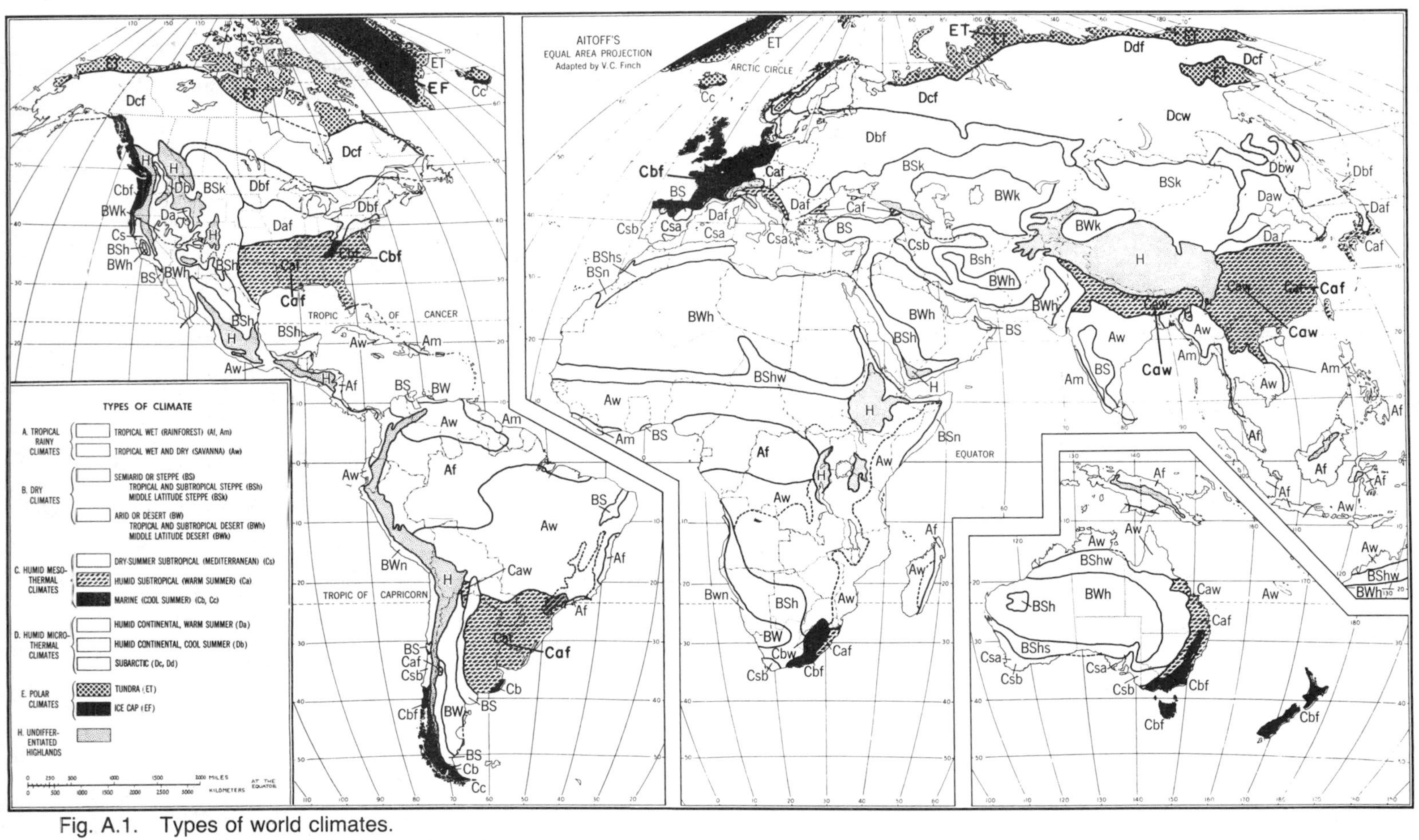

Fig. A.1. Types of world climates.

From: Elements of Geography by Finch, Trewartha, Robinson, and Hammond. Used with permission of McGraw-Hill Book Co. Copyright 1957.

TABLE A.1. BOTANICAL CLASSIFICATION, COMMON NAMES, CLIMATIC ADAPTATION, REGION GROWN, AND ENDEMIC REGIONS OF SOME VEGETABLES

Classification	Common names	Edible part[a]	Climatic zone adaptation[b]	Countries or regions grown[c]	Endemic regions[c]	Ref. page
I. Fungi and algae						
A. Agaricaceae—Mushroom family						
Agaricus bisporus	Mushroom	Fruiting body (sporangium)	C, Da, Db	Temp. World		373
Lentinus edodes (Cortinellus berkeleyanus)	Shitake	Fruiting body (sporangium)	C, Da, Db	E. Asia	E. Asia	376
Armillaria matsudake	Pine mushroom	Fruiting body (sporangium)	D, Da, Db	Temp. Asia	Pine forests	—
Volvariella volvacea	Straw mushroom	Fruiting body (sporangium)	A, C, Da	S.E. Asia,China Madagascar	S.E. Asia, China	374
Pholiota nameko	Namekotake	Fruiting body (sporangium)	C, Da	Japan	Japan	—
B. Algae						
Laminaria japonica	Kobu or kombu	Blade (lamina)	C, Da	Japan	Temp. Oceans and Seas	—
Porphyra umbilicalis	Laver	Blade (lamina)	C, Da	Brit., Iceland	Temp. Oceans and Seas	—
P. tenera	Nori	Blade (lamina)	C, Da	Japan	Temp. Oceans and Seas	—
Ulva lactuca	Sea lettuce	Blade (lamina)	C, Da	Brit., Iceland, Japan	Temp. Oceans and Seas	—
II. Ferns						
Pteridium aquilinum	Braken or brakefern	[Immature frond]	C, Da, Db	Japan, China, France		—
Osmunda japonica	Zen mai	[Immature frond]	C, Da, Db	Japan	Japan	—
III. Gymnosperms (cone-bearing plant)						
Ginkgoaceae—Ginkgo family						
Ginkgo biloba	Ginkgo	Mature nut	C, Da, Db	China, Japan	E. Asia	—
IV. Angiosperms (flowering plants)						
A. Monocotyledons						
1. Alismataceae—water plantain family (Alismaceae)						
Sagittaria sagittifolia	Old World arrowhead	Corm	Aw, Am, Ca	China, S.E. Asia	Eur., Asia	364

2. Amaryllidaceae—Amaryllis family						
Allium fistulosum	Welsh onion, Japanese bunching onion	Leaves	C, Da, Db	Temp. Orient	Prob. N. Asia	202
A. cepa	Onion, shallot	Leaves, bulb	C, Da, Db	Temp. World	Iran, W. Pakistan	184
A. schoenoprasum	Chive	Leaves	C, Da, Db	Eur. N. Amer.	N. Amer., Eur., N. Asia	202
A. ampeloprasum (A. porrum)	Leek, greathead garlic, kurrat	Bulb, leaves	C, Da, Db	Eur. Egypt	Medit., Middle East	201
A. sativum	Garlic	Bulb, leaves	C, Da, Db	Temp. World	Cent. Asia	195
A. chinense (A. bakerii)	Rakkyo	Bulb	C, Da, Db	Japan	E. Asia	202
A. tuberosum	Chinese chive	Leaves, flower, stalk	C, Da, Db	China, S.E. Asia	E. & S.E. Asia	203
3. Araceae—Arum family						
Colocasia esculenta	Taro, dasheen, old cocoyam	[corm, immature leaves]	A, (BSh), Ca, (Cs)	Orient, Cent. Afr., S. Pac. Is.	S.E. Asia, S. Pac. Is.	148
Xanthosoma sagittifolium	Yautia, tannia, new cocoyam	[Corm]	A, (BSh), Ca, (Cs)	Cent. Afr., Carrib., Trop. Amer.	Carrib.	148
X. brasiliense	Belembe, calalu	[Immature leaves]	A, Ca, (Cs)	Trop. Amer.	Brazil	156
Alocasia macrorrhiza	Alocasia, giant taro, ape	[Corm]	A, Ca, (Cs)	S.E. Asia, S. Pac. Is.	Ceylon, S.E. Asia	155
Cyrtosperma chamissonia	Giant swamp taro	[Corm]	A, Ca, (Cs)	S.E. Asia, S. Pac. Is.	S.E. Asia	156
4. Cannaceae—Canna family						
Canna edulis	Edible canna, Queensland arrowroot, tous-les-moi, achira	Rhizomes	A, Ca	Aust., Java, W. Indies, Peru, Argentina	S. Amer. (Andes)	158
5. Cyperaceae—Sedge family						
Eleocharis dulcis	Water chestnut, matai	Corms	A, Ca	China, S.E. Asia	Old World, tropics from Madagascar to India, China, S.E. Asia, and Fiji Is.	361

(continued)

TABLE A.1. *(cont.)*

Classification	Common names	Edible part[a]	Climatic zone adaptation[b]	Countries or regions grown[c]	Endemic regions[c]	Ref. page
6. Dioscoreaceae—Yam family						
Dioscorea batatas	Chinese yam, nagaimo	Tuber	A, Ca	Orient	E. Asia	139
D. alata	Greater yam, water yam, winged yam, ubi	Tuber	A	S.E. Asia	S.E. Asia	139
D. rotundata	White yam, Guinea yam, eboe yam	Tuber	A	W. Cent. Afr.	Ivory Coast	139
7. Graminae—Grass family						
Zea mays	Corn, maize	Immature seeds Immature cob	Am, Aw, C, Da, (BSh)	Warm Temp., Sub-trop. Trop.	Meso S. Amer.	167
Phyllostachys edulis	Bamboo shoot	[Shoot]	A, Ca	Orient, Brazil	Orient	358
Bambusa spp.	Bamboo shoot	[Shoot]	A, Ca	Orient	Orient	358
Dendrocalamus spp.	Bamboo shoot	[Shoot]	A, Ca	Orient	Orient	358
Sasa spp.	Bamboo shoot	[Shoot]	A, Ca, Cs	Orient	Orient	358
8. Liliaceae—Lily family						
Lilium tigrinum	Tiger lily	Bulb	C, Da	Temp. Orient	Orient	—
Asparagus officinalis	Asparagus	Shoot (spear)	C, Da, Db	Temp. N. Amer., Eur., Asia	Eur., C. Asia	356
9. Musaceae—Banana family						
Musa paradisiaca	Plantain, starchy banana	Fruit	A	Trop. World	India, Malaysia	176
M. acuminata (M. sapientum)		Green fruit, male blossom	A	Trop. World	S.E. Asia	176
10. Palmaceae—Palm family						
Euterpe oleraceae and *E. edulis*	Palm heart, palmito	Tender shoot	A, Ca	Brazil	Coastal S. Cent. Brazil	—
Many palm species	Palm cabbage	Tender shoot	A	Cent. Africa	Cent. Africa	—

11. Zingiberaceae—Ginger family						
Zingiber officinale	Ginger	Rhizome	A, Ca	Trop., Subtrop. World	Trop. Asia (India)	360
Z. mioga	Japanese ginger (mioga)	Rhizome	Ca	Japan	Japan	—
B. *Dicotyledons*						
1. Aizoaceae—Carpetweed family						
Tetragonia expansa	New Zealand spinach	Tender shoots, leaves	Aw, Ca, Cs	S.E. Asia, Brazil	N.Z., Aust.	368
2. Amaranthaceae—Amaranth family						
Amaranthus tricolor	Tampala, Chinese spinach, edible amaranth	Tender shoots, leaves	Aw, Am, Ca	Subtrop., Trop. Asia	Asian Trop.	364
3. Araliaceae—Aralia or Ginseng family						
Aralia cordata	Udo	Tender shoots	Da	Japan	E. Asia, Japan	—
4. Basellaceae—Basella family						
Basella alba (*B. ruba*: for red colored types)	Malabar nightshade, Malabar spinach	Leaves, tender shoots	Am, Aw, Ca	Trop., Subtrop.	Asian Trop.	367
5. Chenopodiaceae—Goosefoot family						
Beta vulgaris var. *vulgaris*	Beet, beetroot	Root, leaves	C, Da, Db	Eur., N. Amer.	Eur., N. Afr.	352
B. vulgaris var. *cicla*	Chard, Swiss	Leaves	C, Da, Db	Eur., N. Amer.	Eur., N. Afr.	353
Spinacia oleracea	Spinach	Leaves	Ca, Da, Db	Temp. World	S.W. Asia	350
Atriplex hortensis	Orach, mountain spinach	Leaves	Ca, Da, Db	Eur., N. Amer.	Prob. Asia	354
6. Compositae—Sunflower family						
Lactuca sativa	Lettuce	Leaves	C, Da, Db	Temp.	Medit.	207
Cichorium intybus	Chicory	Leaves, root	C, Da, Db	Eur., N. Amer.	Medit.	210
C. endiva	Endive	Leaves	C, Da, Db	Eur., N. Amer.	Prob. India	210
Tragopogon porrifolius	Salsify, vegetable oyster	Root	C, Da, Db	Eur., N. Amer.	S. Eur., N. Amer.	213

(continued)

TABLE A.1. *(cont.)*

Classification	Common names	Edible part[a]	Climatic zone adaptation[b]	Countries or regions grown[c]	Endemic regions[c]	Ref. page
Taraxicum offinale	Dandelion	Leaves	C, Da, Db	Eur., N. Amer.	Eurasia, N. Amer.	—
Helianthus tuberosus	Jerusalem	Tubers	C, Da, Db	Eur., N. Amer.	N.E. N. Amer.	163
Cynara scolymus	Artichoke	Immature flower bud	C, Da, Db	Medit., N. Amer. (Calif.)	Medit.	213
C. cardunculus	Cardoon	Petiole	C, Da, Db	Medit., N. Amer. (Calif.)	Medit.	213
Arctium lappa	Edible burdock, gobo	Root	Ca, Da, Db	Japan	China, Japan	215
Chrysanthemum coronarium	Garland chrysanthemum	Leaves, tender shoot	Ca, Da, Db	Japan	China, Japan	214
Petasites japonica	Fuki	Petiole	Ca, Da	Japan	Japan?	217
7. Convolvulaceae—Morning glory family						
Ipomoea batatas	Sweet potato, kumara	Roots, leaves	A, Ca, (BSh)	Trop., Subtrop., Warm Temp.	C. Amer.	123
I. aquatica (*I. reptans*)	Water convolvulus, water spinach, kangkong	Tender shoots, leaves	A, Ca	S.E. Asia	India	371
8. Cruciferae—Mustard family		Leaves				
Brassica napus (*B. napobrassica*)	Rutabaga, Siberian kale, Hanover salad		Cb, Da, Db	Eur., N. Amer.	Medit.	219
B. oleracea	Cole crops		C, Da, Db (BSk)	Temp. World	Medit., S.W. Eur., S. Brit.	219
Capitata gp.	Cabbage	Leaves				
Botrytis gp.	Cauliflower	Immature flower				
Italica gp.	Broccoli	Immature flower buds				
Gemmifera gp.	Brussels sprouts	Axillary buds				
Acephala gp.	Kale	Leaves				
Gongylodes gp.	Kohlrabi	Enlarged stem				
B. campestris	Turnip group		C, Da, Db (BSk)	Temp. World	Medit., E. Asia, India	227

Rapifera gp. (*B. rapa*)	Turnip	Enlarged root				
Chinensis gp.	Chinese mustard, pak choi	Leaves				
Pekinensis gp.	Chinese cabbage, nappa	Leaves				
Perviridis gp.	Mustard, brown, Chinese spinach	Leaves				
Ruvo gp.	Broccoli raab, rapa, Italian turnip	Leaves, roots				
B. juncea	Indian mustard	Leaves	C, Da, Db	Temp. Asia	Cent. Asia, Himalayas	219
Eruca sativa	Rocket salad	Leaves	C, Da, Db	Eur., N. Amer., Medit.	Eur., W. Asia	234
Raphanus sativus	Radish	Root, leaves	C, Da, Db	Temp. World	W. Asia	231
Lepidium sativum	Garden cress	Leaves	C, Da, Db	Temp. World	E. Medit.	237
Barbarea verna	Upland cress, winter cress	Leaves	C, Da, Db	Eur.	Eur., N. Amer.	—
Rorippa nasturtium-aquatica (Nasturtium officinale)	Watercress	Leaves	C, Da, Db	Temp. World	Brit., S. Cent. Eur., W. Asia	235
Eutrema wasabi	Japanese horseradish	Rhizomes	C, Da, Db	Japan	Prob. Japan	—
Armoracia rusticana	Horseradish	Rhizomes	C, Da, Db	Eur., N. Amer.	E. Eur., Turkey	236
Lepidium meyenii	Maca	Enlarged root	H	S. Amer. Andes	Andes (Peru)	236
9. Cucurbitaceae—Gourd family						
Cucurbita pepo	Pumpkin, summer squash, winter squash	Mature/ immature fruit	Am, Aw, Ca, (Cs), Da, (BSh)	Trop., Subtrop., Warm Temp. Temp.	N. Amer., Mex.	330
C. moshata	Pumpkin, winter squash	Mature fruit	Am, Aw, Ca, (Cs), Da, (BSh)	Trop., Subtrop., Warm Temp.	Meso, S. Amer.	330
C. mixta	Pumpkin, winter squash	Mature fruit	Am, Aw, Ca, (Cs), Da, (BSh)	Trop., Subtrop., Warm Temp.	Meso, Amer.	330

(continued)

TABLE A.1. *(cont.)*

Classification	Common names	Edible part[a]	Climatic zone adaptation[b]	Countries or regions grown[c]	Endemic regions[c]	Ref. page
C. maxima	Pumpkin, winter squash	Mature fruit	Am, Aw, Ca, (Cs), Da, (BSh)	Trop., Subtrop., Warm Temp.	Meso, S. Amer.	330
Citrullus lanatus (*C. vulgaris*)	Watermelon	Ripe fruit, seeds	Am, Aw, Ca, (Cs) (BSh)	Trop., Subtrop.	Trop. Afr.	327
Cucumis sativus	Cucumber	Immature fruit	Am, Aw, Ca, (Cb), Da, (BSh)	Trop., Subtrop. Warm Temp.	Trop., Subtrop. Afr.	317
C. melo	Muskmelon	Ripe fruit	Am, Aw, Ca, (Cs), (BSh)	Trop., Subtrop.	Trop., Subtrop. Afr.	322
C. anguria	West India gherkin	Immature fruit	Am, Aw, Ca, (Cs), (BSh)	Trop., Subtrop. Amer., Asia	Trop. Afr.	321
Sechium edule	Chayote	Immature fruit	A, Ca, (Cs), (BSh)	Trop., Subtrop. Amer., Asia	Meso Amer.	339
Luffa cylindrica	Dishcloth gourd, Sponge gourd, loofa	Immature fruit	A, Ca, (Cs), (BSh)	Trop., Subtrop. Asia	Trop. Asia (India)	342
L. acutangula	Chinese okra, vegetable gourd	Immature fruit	A, Ca, (Cs), (BSh)	Trop., Subtrop. Asia	Tropical Asia	342
Benincasa hispida	Wax gourd, Chinese winter melon, preserving melon	Immature/ mature fruit	A, Ca, (Cs), (BSh)	Trop., Subtrop. Asia	Java	336
Lagenaria siceraria	Calabash gourd, zucca melon, white flowering gourd	Immature fruit	A, Ca, (Cs), (BSh)	Trop., Subtrop.	Prob. Trop. Afr.	338
Tricosanthes anguina	Snake gourd, serpent gourd	Immature fruit	A, Ca, (Cs), (BSh)	Trop., Subtrop. Asia	India, S.E. Asia	344
Momordica charantia	Bitter melon, balsam pear	Immature fruit	A, Ca, (Cs), (BSh)	Trop., Subtrop. Asia	Tropical Asia	341
Cyclanthera pedata	Caihua, achoecha (Peruvian)	Immature fruit	A, Ca, (Cs), (BSh)	Trop. Amer.	Trop. S. Amer.	345
10. Euphorbiaceae—Spurge family (Castor bean)						
Manihot esculenta	Cassava, manioc, yuca, tapioca plant	[Roots] (bitter-type), leaves	A, Ca	Trop. World	N. Brazil	132

11. Labiatae—Mint family						
Perilla frutescens	Shiso	Leaves, seeds	C	Japan	S. Cent. & E. Asia	—
Stachys sieboldii (S. tuberifera)	Chinese artichoke	Tubers	Aw, C	Subtrop., China, Japan	China	165
12. Leguminosae—Pea or bean family						
Pisum sativum	Garden pea	Immature seeds, tender shoots	Cb, Cc, Da, Db	Eur., N. Amer., Argent., Temp. Oceania	Near East, Eur.	264
P. sativum (macrocarpon gp.)	Edible podded pea, sugar pea, China pea	Immature pods	Cb, Cc, Da, Db	N. Amer., Asia	Near East, Eur.	264
Phaseolus vulgaris	Bean, snap	Immature pods	A, C, Da, (BSh)	Trop., Subtrop., Warm Temp. World	Cent. S. Amer.	267
P. coccineus	Bean, scarlet runner	Immature pods	A, C	Cent. Amer.	Cent. Amer.	282
P. lunatus	Bean, lima	Immature pods	A, C	N. Amer., Trop. Amer.	Andes	270
P. aureus	Mungbean	Immature pods, sprouted seeds	A, C	Trop. Asia	India	281
P. aconitifolis	Mat bean	Immature pods/ seeds	A, C	Trop. and Subtrop. Asia	India	280
P. angularis	Adzuki bean	Immature pods/ seeds	A, C	Trop. and Subtrop. Asia	E. Asia	281
P. calcaratus	Rice bean	Immature pods/ leaves	A, C	Trop. and Subtrop. Asia	S.E. Asia	281
P. mungo	Black bean	Immature pods/ seeds	A, C	Trop. and Subtrop. Asia	India	282
Arachis hypogea	Peanut, ground nut	Immature/ mature seeds	A, C	Trop., Subtrop., Warm Temp.	S. Amer.	275
Canavalia gladiata	Sword bean	Immature pod/ [seed]	A, Ca, Cb	Trop., Subtrop., Warm Temp.	Asia	277
C. ensiformis	Jack bean	Immature pod/ [seed]		Trop., Subtrop., Warm Temp.	Trop. Amer.	277
Cyamposis tetragonobus	Cluster bean, guar	Immature pods	A, C	Trop., Subtrop., Warm Temp.	India	278
Lalab niger	Hyacinth bean	Immature seeds	A, C, Da	Trop., Subtrop., Warm Temp.	India	278
Lathyris sativus	Chickling pea	Immature pods/ [seed]	A, C, Da	Trop., Subtrop., Warm Trop.	S. Eur.	279

(continued)

TABLE A.1. *(cont.)*

Classification	Common names	Edible part[a]	Climatic zone adaptation[b]	Countries or regions grown[c]	Endemic regions[c]	Ref. page
Lens esculenta	Lentil	Immature pods	A, C	Trop., Subtrop., Warm Temp.	E. Eur.	279
Lupinus terminus	Egyptian lupine	[Seeds]	A, C	Trop., Subtrop., Warm Temp.	Medit.	280
Glycine max	Soybean, soya	Immature seeds, sprouted seeds	A, C, Da	N. Amer., Asia, S. Brazil	N.E. China	273
Pachyrrhizus erosus	Yam bean, jicama	Enlarged root	A, Ca	Mex., S.E. Asia	Mex.	164
P. tuberosus	Potato bean	Enlarged root	A, Ca	Trop. Amer.	W. S. Amer.	165
Psophocarpus tetragonolobus	Winged bean, goa bean	Immature pods, seeds, leaves, enlarged root	A, Ca	S.E. Asia, New Guinea	New Guinea	282
Trigonella foenum-graceum	Fenugreek	Leaves, mature seeds (condiment)	C, Da	Mideast, Eur., India	Medit.	284
Vicia faba	Broad bean, horse bean field bean, fava bean	Immature seeds/pods	Ca, Cb, Da, Db	N. Amer., Eur.	Near East	285
Vigna sesquipedalis	Asparagus bean	Immature pods	A, C	China, S.E. Asia	Ethiopia, W. Afr.	285
V. sinensis	Cow peas	Immature pods	A, C	Trop. Afr.	Afr., Asia	285
Voandzeia subterrancea	Bambara ground nut	Immature/mature seeds	A, Ca, Cs	Trop. Africa, India S.E. Asia	W. Africa	287
13. Malvaceae—Mallow or cotton family						
Hibiscus esculentus (Abelmoschus esculentus)	Okra	Immature fruit	A, Ca, (Cs)	Trop., Subtrop. World	Trop. Afr.	368
H. sabdariffa	Roselle, Jamaican sorrel	Calyx	A, C	W. Indies	W. Afr.	—
Malva paraviflora	Egyptian mallow	Leaves	A, C	Egypt, Sudan	Prob. N. Afr.	—
14. Moraceae—Mulberry family						
Artocarpus altilis	Breadfruit	Fruit	A	S. Pac. Is., Indonesia	S. Pac. Is.	183
A. heterophyla	Jackfruit	Immature fruit	A	S. Pac. Is., Indonesia	S. Pac. Is.	182
15. Nymphaeceae—Water lily family						
Nelumbo nucifera	Lotus root, East Indian lotus	Rhizomes, seeds	A, (BSh), Ca	Trop., Subtrop. Asia	Trop. Asia, N.E. Aust.	363

16. Polygonaceae—Buckwheat family						
Rheum rhaponticum	Rhubarb, pie-plant	Petiole	C, Da, Db	Eur., N. Amer.	Siberia	—
Rumex acetosa	Garden sorrel	Leaves	C, Da, Db	Eur., N. Amer.	Eur., Asia	—
R. scutatus	French sorrel	Leaves	C, Da, Db	Eur., N. Amer.	Eur., Asia	—
17. Portulacacea—Purslane family						
Portulaca oleracea	Purslane	Leaves	(BWh)	Egypt, Sudan, S.E. Asia	Old World Temp.	370
18. Solanaceae—Potato or nightshade family						
Solanum tuberosum	Potato	Tuber	Cb, Da, Db, H, (BSk)	Temp. World	Peru, Andes	111
S. melongena	Eggplant	Immature fruit	A, Ca, (BSh)	Trop., Subtrop.	Trop. Asia (India)	298
S. muricatum	Pepino	Ripe fruit	Aw, Ca	Trop., Subtrop. Amer.	S. Amer.	306
S. nigrum	Garden huckle-berry, wonder-berry	Mature fruit	Da	N. Amer.	N. Amer.	307
S. gilo	Jilo	Immature fruit	A, Ca	Brazil, W. Africa	W. Africa	307
S. quitoense	Naranjillo, lulo	Ripe fruit	A, Ca	Trop, Subtrop., S. Amer.	Trop. S. Amer.	307
S. macrocarpon	African eggplant	Immature fruit	A	West Cent. Africa	Trop. Afr.	306
Capsicum annuum	Bell pepper, chili pepper, pimento	Mature fruit	A, Ca, (Cs), (BSh)	Trop., Subtrop., Warm Temp.	Andes	303
C. frutescens	Tobasco pepper	Mature fruit	A, Ca, (Cs), (BSh)	Trop., Subtrop., Warm Temp.	Andes	303
Lycopersicon esculentum	Tomato	Ripe fruit	A, Ca, (Cs), Da, (BSh)	Trop., Subtrop., Warm Temp.	Andes	292
Physalis spp.	Husk tomato	Ripe fruit	A, Ca	Trop., Subtrop. Amer.	Mexico	308
Cyphomandra betacea	Tree tomato	Ripe fruit	A, Ca	Trop., Subtrop. Amer., N.A.	Subtrop. S. Amer.	309
19. Tiliaceae—Basswood family						
Corchorus oliforius	Jew's mallow	Leaves, tender shoot	Aw, Ca, (Cs), (BWh)	Egypt, Sudan	India	366
20. Umbelliferae—Parsley family						
Daucus carota	Carrot	Root	C, Da, Db, (BSk)	Temp. World	Eur., Asia	240

(continued)

TABLE A.1. *(cont.)*

Classification	Common names	Edible part[a]	Climatic zone adaptation[b]	Countries or regions grown[c]	Endemic regions[c]	Ref. page
Pastinaca sativa	Parsnip	Root	C, Da, Db	Eur., N. Amer.	Eur.	249
Apium graveolens	Celery, celeriac	Leaves, root	C, Da, Db	Eur., N. Amer.	Eurasia	246
Foeniculum vulgare	Florence fennel	Leaves	C, Da	S. Eur.	Medit.	239
Coriandrum sativum	Coriander	Leaves	C, Da	Temp. Subtrop. Asia	Medit.	239
Cryptotaenia canadensis	Mitsuba	Leaves	C, Da	Japan	E. Asia, Japan	249
Arracacia xanthorrhiza	Arracacha, zanahoria blanca apio, mandioquina-salsa	Enlarged root	Cb, H	Peru, Colombia, Brazil	S. Amer. Andes	160
Petroselinum	Parsley	Leaves	C, Da, Db	Eur., N. Amer.	S. Eur.	249

[a] Edible parts in square brackets contain toxic substances that must be treated by cooking or other means of removing or/inactivating toxins before ingestion.

[b] Climatic zones according to Trewartha's Modification of Köppen System. The listed adaptation zones in the table are approximate.
A = Tropical, coolest month > 18°C (64°F).
Am = Tropical wet (monsoon).
Aw = Tropical wet (savanna), dry period at low sun (winter), driest month < 60 mm (2.4 in.) precipitation.

B = Dry, evaporation > precipitation.
BSh = Low latitude, semiarid.
BSk = Middle latitude steppe, semiarid.
BWh = Low latitude steppe, arid.

C = Humid mesothermal (Mild winter temperate), coldest month between 18°C (64°F) and 0°C (32°F).
Ca = Humid subtropical, warmest month > 22°C (72°F).
Cb = Marine, warmest month < 22°C (72°F).
Cs = Dry summer subtropical (Mediterranean), precipitation driest month < 30 mm (1.2 in.); wettest month in winter > 90 mm (3.5 in.).
Cc = Marine, warmest month < 22°C (72°F); < 4 months > 10°C (50°F).

D = Humid microthermal (severe winter temperate), coldest month < 0°C (32°F); warmest month > 10°C (50°F).
Da = Humid continental, warm summer, warmest month > 22°C (72°F).
Db = Humid continental, cool summer, warmest month < 22°C (72°F).

H = Undifferentiated highlands.

( ) = Irrigation may be required e.g., (BSh), (BSk), (BWh), and (Cs). (Not Köppen or Trewartha's designation.)

[c] Abbreviations:

Afr. = Africa
Amer. = America
Aust. = Australia
Brit. = Britain
Carrib. = Carribbean
Cent. = Central
E. = East
Eur. = Europe
Medit. = Mediterranean
Mex. = Mexico
N. = North
New Z. = New Zealand
Prob. = Probably
S. = South
S. Pac. Is. = South Pacific Islands
Subtrop. = Subtropical
Temp. = Temperate
Trop. = Tropics
W. = West

TABLE A.2. ESSENTIAL AMINO ACIDS IN VEGETABLES

Vegetable	Moisture (g/100g)	Protein (g/100g)	Essential amino acid[a] (mg/100 g fresh wt)								Protein quality	
			Ile	Leu	Lys	Met + Cys	Phe + Tyr	Thr	Trp	Val.	Limiting amino acid[b]	Chemical score[c]%
Starchy, Roots, Tubers, and Fruits												
White potato *(Solanum tuberosum)*	78	2.0	76	121	96	38	135	75	33	93	SC	34
											Ile	57
Sweet potato *(Ipomoea batatas)*	70	1.3	48	71	45	36	81	50	22	59	SC	51
											Lys	53
Cassava (Manioc) — meal, roots	13.1	1.6	46	64	67	45	67	43	19	54	AR	41
(Manihot esculenta)											Ile	42
Cassava (Manioc) — leaves	71.7	7.0	339	900	437	195	661	327	102	401	SC	50
(Manihot esculenta)											Ile	73
Yam *(Dioscorea* spp.)	72.4	2.4	89	154	97	65	190	86	30	110	SC	50
											Ile	56
Taro *(Colocasia esculenta)*	72.5	1.8	64	133	70	72	158	74	26	111	Ile	53
											Ly	60
Yautia *(Xanthosoma sagittifolium)*	64.4	2.2	59	124	70	79	158	65	30	112	Ile	41
											Ly	50
Corn—immature seeds *(Zea mays)*	68	3.7	137	407	137	134	331	151	23	231	Tr	42
											Ly	53
Plantain (banana)—*(Musa* spp.)	71.0	1.1	32	53	46	52	73	38	13	45	Ile	44
											Leu	53
Breadfruit *(Arthocarpus atilis)*	25	1.4	94	104	81	17[d]	116	95	25	109	—	—
Alliums												
Onion, bulb *(Allium cepa)*	89	1.4	20	37	63	16[d]	38[e]	20	20	30	—	—
Composites												
Endive *(Cichorium endiva)*	93	1.8	72	123	78	34	132	71	30	81	SC	34
											Ile	60
Lettuce *(Lactuca sativa)*	95	1.3	50	83	50	24[d]	102	54	10	71	—	—
Crucifers												
Broccoli, leaves	86	4.3	186	236	218	108	177[e]	161	46	210	—	—
(Brassica oleracea, italica)												
Brussels sprouts	85	4.7	230	257	252	71	172[e]	199	58	228	—	—
(B. oleracea, gemmifera)												
Cabbage *(B. oleracea,* capitata)	92	1.6	50	86	50	35	79	61	17	68	SC	39
											Ile	46
Cauliflower *(B. oleracea,* botrytis)	92	2.8	136	196	160	44[d]	101[e]	119	39	156	—	—
Radish, root *(Raphanus sativus)*	94	1.1	54	75	48	10[d]	48[e]	42	4	71	—	—

(continued)

TABLE A.2. *(cont.)*

Vegetable	Moisture (g/100g)	Protein (g/100g)	Essential amino acid[a] (mg/100 g fresh wt)								Protein quality	
			Ile	Leu	Lys	Met + Cys	Phe + Tyr	Thr	Trp	Val.	Limiting amino acid[b]	Chemical score[c]%
Turnip, leaves (*B. campestris*, rapa)	89	3.1	106	210	157	57	228	127	42	136	SC Ile	33 51
Turnip, root (*B. campestris*, rapa)	91	0.9	22	36	17	11	31	25	11	22	SC Lys	23 31
Umbellifers												
Carrot, root *(Dacus carota)*	87	1.1	33	50	44	26	56	32	8	50	SC Trp	40 44
Celery *(Apium graveolens)*	93	1.1	44	76	27	25[d]	51[e]	38	14	54	—	—
Parsley, leaves *(Petroselium crispum)*	85	3.7	—	—	531	18[d]	—	—	74	—	—	—
Legumes												
Bean *(Phaseolus vulgaris)*	11.0	22.1	927	1685	1593	422	1713	873	223	1016	SC Trp, Val	34 63
Broad bean *(Vicia faba)*	11.0	23.4	936	1659	1513	359	1760	786	202	1030	SC Trp	28 54
Chick pea *(Cicer arietium)*	11.0	20.1	891	1505	1376	447	1740	756	174	913	SC Trp	40 54
Cow pea *(Vigna* spp.)	11.0	23.4	895	1647	1599	528	1820	842	254	1060	SC Ile	41 58
Hyacinth *(Lablab niger)*	11.0	22.8	934	1795	1591	339	1810	755	164	1073	SC Trp	27 45
Jack bean *(Canavalia ensiformis)*	10.7	24.5	980	1779	1348	627	2121	1078	—	1129	SC Ile	46 60
Lentil *(Lens culinaris)*	11.0	24.8	1045	1847	1739	415	2053	960	231	1211	SC Trp	31 60
Lima bean *(Phaseolus lunatus)*	11.0	19.7	977	1604	1466	444	1831	823	199	1015	SC Trp	41 63
Lupine (*Lupinus* spp.)	—	31.2	1369	2241	1652	669	2256	1138	314	1258	SC Val	39 55
Mung bean *(Phaseolus aureus)*	11.0	23.9	891	1686	1927	294	1767	799	191	990	SC Trp	22 50
Pea *(Pisum sativum)*	11.0	22.5	961	1530	1692	457	1649	914	202	1058	SC Trp	37 56
Pigeon pea *(Cajanus cajan)*	11.0	20.9	648	1316	1607	311	2148	608	117	751	SC Trp	27 35
Soybean *(Glycine max)*	8.0	38.0	1889	3232	2653	1077	3358	1603	532	1995	SC Val	47 66

Solanaceous fruits												
Eggplant *(Solanum melongena)*	93	1.2	52	72	63	19	95	44	12	61	SC	30
											Trp	64
Pepper *(Capsicum frutescens)*	74	4.1	—	—	252	40[d]	—	—	—	—	—	—
Tomato *(Lycopersicon esculentum)*	94	1.1	20	30	32	14	34	25	9	24	SC	21
											Ile	27
Cucurbits												
Calabash, leaves (*Lagenaria* spp.)	84	4.4	179	284	210	114	380	153	—	280	SC	47
											Ile	62
Cucumbers, fruit *(cucumis sativus)*	95	0.8	25	34	35	8[d]	19[e]	21	6	28	—	—
Pumpkin, leaves *(Cucurbita pepo)*	89	4.0	218	400	254	118	446	204	52	250	SC	53
											Trp	81
Pumpkin, fruit *(Cucurbita pepo)*	93	1.0	37	52	43	9[d]	33[e]	27	11	48	—	—
Others												
Amaranth (*Amaranthus* spp.)	84	4.6	218	359	234	170	427	197	57	256	SC	66
											Ile	70
Asparagus *(Asparagus officinalis)*	93	2.1	55	96	96	47	99	60	25	79	Ile	39
											SC	40
Beet, leaves *(Beta vulgaris)*	90	2.1	40	93	60	37	104	62	21	60	Ile	28
											SC	32
Beet, root *(Beta vulgaris)*	88	1.8	44	80	96	54	128	60	17	44	Val	33
											Ile	36
Malabar spinach (*Basella alba*)	93	1.8	54	103	89	47	136	56	—	67	—	—
Mallow, leaves (*Malva* spp.)	84	5.0	187	215	118	94	194	168	86	323	—	—
Mushroom (*Agaricus bisporus*)	92	3.7	83	136	165	36	148	100	38	34	SC	18
											Ile	34
Okra *(Hibiscus esculentus)*	87	2.1	55	81	70	49	79	49	12	66	Ile, Trp	36
											AR	37
Spinach *(Spinacia oleracea)*	92	2.2	106	208	159	82	244	116	34	133	SC	68
											Ile	73
Water spinach *(Ipomoea aquatica)*	83	4.0	116	208	144	100	304	132	52	160	Ile	44
											SC	45

Source: Data from F.A.O. Nutritional Studies No. 24 (1970), Rome, Italy.

[a] Ile = isoleucine, Leu = leucine, Lys = lysine, Met + Cys = methionine + cystine, AR = Phe + Tyr = phenylalanine + tyrosine, Thr = threonine, Trp = tryptophan, Val = valine.

[b] SC = sulfur-containing amino acids, AR, aromatic amino acids.

[c] Chemical score: relative content of limiting essential amino acid in the protein expressed as percentage of the content of the same amino acid in egg protein.

[d] Methionine only.

[e] Phenylalanine only.

Glossary

Abaxil: Away from axis.
Abscission: Separation of leaves, flowers, fruits, or other plant parts, usually following the formation of a separation layer.
Absorption: Process by which substances are taken in.
Acid Soil: Soil with reaction below pH 7.0 (usually less than pH 6.6). Soil with preponderance of hydrogen ions.
Adaxil: On side of or toward axis.
Adsorb: Physical or chemical attraction of particles or molecular ions on surfaces.
Adventitious: Plant organs, such as shoots or roots, produced in an abnormal position or at an unusual time of development.
Afterripening: Metabolic changes occurring in some dormant seeds before germination can occur.
Alkali Soil: Soil containing alkali salts (usually Na_2CO_3) with pH of 8.5 or higher.
Alkaline soil: Soil with pH above 7.0 (usually above pH 7.3).
Alkaloid: Nitrogen-containing organic compound produced by plants which constitutes the active part of many drugs, bitter in taste and often times poisonous; e.g., nicotine, solanine, caffeine, quinine, morphine, strychnine.
Amino acids: Organic acids with nitrogen derived from ammonia. Building units for proteins.
Androecious: All staminate flowers.
Androecium: Stamens and their appendages collectively.
Andromonoecious: Some perfect and some staminate flowers.
Angiosperm: Plants having seed in a closed ovary.
Annual: Plant that completes its life cycle in 1 year; i.e., plant that starts from seed, matures, produces seed, and dies in 1 year.
Anther: Pollen bearing part of stamen.
Anthesis: When flower first opens or dehiscence of anthers.
Anthocyanins: Water-soluble plant pigments, usually red, blue, or purple, found in leaves, stems, flowers, and fruits; also in roots of some plants.
Apical dominance: Control of lateral buds by presence of an apical meristem (shoot).
Apical meristem: Meristematic cells of the apex of a shoot or a root.
Arable: Land capable of producing crops requiring tillage.
Arcuate: Forming an arc.
Arid: Dry, low in rainfall, usually less than 25 cm (10 in.) of precipitation.
Asexual: Reproduction that does not involve germ or sexual cells.
Atmosphere: Mass of air surrounding earth; also, the gaseous phase of a soil mass; unit of pressure equivalent to 760 mm (30 in.) Hg.

Auxin: Growth substance naturally present in plants. Indoleacetic acid (IAA).
Axil: Location on stem at upper side of leaf attachment.
Axillary bud: Term applied to bud arising at the axil of leaf.
Bar: Unit of pressure, approximately equal to pressure expressed in atmosphere (bar = 0.987 atm).
Biennial: A plant that normally requires two growing seasons for its life cycle; vegetative the first year, produces seed the second year, then dies.
Binomial system: Application of two names to any plant: the generic and species names.
Blade: Broad expanded part of a leaf.
Blanching: Horticulture: Prevention of stems and leaves from turning green; to etiolate. Food Science: Inactivate enzymes by heating in hot water or steam.
Bolt: Emergence of a seed stalk.
Bud: Embryonic or undeveloped shoot.
Bulb: Underground storage organ composed mainly of enlarged fleshy leaf bases, called sheath leaves, and a very short stem.
Bulbil: Aerial stem bulblet (miniature bulb developed from buds at axil of leaf) and capable of regenerating a new plant.
Calyx: Outer perianth of flower, collective term for sepals.
Cambium: A zone of meristemic cells (dividing cells) which give rise to vascular (conductive) tissues.
Capsule: A dry dehiscent fruit.
Carbohydrate: Carbon compounds with $(CH_2O)_n$ formula produced by plants.
Carotene: Yellow or orange pigments produced by plants in roots, stems, leaves, flowers, and fruits. Depending on structure, it may have vitamin A value.
Cellulose: A carbohydrate that is the main component of plant cell walls.
Chilling injury: Physiological injury caused by low but nonfreezing temperature. For sensitive crops temperatures of 0°–10°C (32°–50°F), upon sufficient exposure, cause injury to tissues.
Chlorophyll: Green pigment in chloroplasts necessary for photosynthesis.
Chlorosis: Reduced development of chlorophyll resulting in yellowing or whitening of normally green tissues.
Clay: Soil classification: particles less than 0.02 mm (2 μm) diameter.
Climate: Summation of all weather conditions for a particular region of the earth.
Clone: A group of individuals of common origin, produced by vegetative means.
Colloid: A dispersion of particles in a medium that does not settle out by force of gravity alone.
Complete flower: Flower having sepals, petals, stamens, and pistil(s).
Controlled atmosphere (CA): Storage in which the atmospheric content of O_2, CO_2, H_2O, C_2H_4, and/or N_2 is regulated.
Corm: A short, thickened underground stem, upright in position, in which food reserves are stored.
Cormel: A daughter corm arising from a parent corm.
Cotyledon: A seed leaf of an embryo.
Cover crop: Crop seeded to provide cover for the soil, usually to prevent leaching and erosion and to increase soil organic matter when turned over.
Cross-pollination: Transfer of pollen from anther of one plant to stigma of flower on another plant.
Cultigen: A plant or group that is known only under cultivation with no known place of origin except having originated under cultivation. Not synonymous with cultivar.

Cultivar: A horticultural variety or race that originated under cultivation and not essentially referable to a botanical species; abbreviated cv.
Cutting: A plant part, usually a stem, used for propogation.
Cytokinin: A natural plant growth regulator, important in promotion of cell division.
Day-neutral: Flowering not controlled by photoperiod.
Deciduous: Parts fall at end of growing season, such as leaves in the fall of the year or fruits at maturity.
Determinate: When the shoot terminates with an inflorescence and growth of the axis is arrested.
Dicotyledon: Plant that produces embryos with two cotyledons.
Digitate: Leaf resembling fingers of the hand.
Dioecious: Plants that have male and female flowers on separate plants.
Dominant: One of two contrasted parental characters that appear in the individual of the first generation.
Dormancy: A physiological state of some plant parts, e.g., bud and seed, in which sprouting or rooting do not occur due to unfavorable conditions. Also known as *imposed dormancy*, *external dormancy*, and *quiesence* (see *Rest*).
Ecology: Science of relationships of organisms to the environment.
Edaphic: Pertaining to soil conditions.
Embryo: Rudimentary plant formed in a seed.
Emergence: Appearance of the first part of the plant from the soil.
Endemic: Native to a region or place; not introduced.
Endogenous: Developed or produced internally.
Endosperm: A nutrient tissue of angiosperm ovule which persists in some seeds and is used by the embryo upon germination.
Epicotyl: Part of embryo above the attachment of cotyledons, it consists of stem tips and several embryonic leaves.
Epidermis: Outermost layer of cells of the leaf and of young stems and roots.
Erosion: Movement of soil and other materials by natural forces, primarily water and wind.
Ethylene: Gas composed of carbon and hydrogen (C_2H_4) having growth regulating capabilities. Induces physiological responses. Produced by plant tissues, especially by many ripening fruits. Hastens fruit ripening and abscission.
Etiolate: Loss of color, a condition caused by withholding light to plants as in blanching.
Evapotranspiration: The total loss of water by evaporation from a given area from the soil surface and from transpiration of plants.
Evolution: History of development of a race; species that follow modifications distinguishing it from others.
Exogenous: Produced from without.
Family: Category of classification above genus and below order. Suffix of family name is usually "aceae."
Fats: Organic compounds of C, H, and O. Proportion of O to C is less than in carbohydrates.
Fertilization (sexual): Union of two gametes to form a zygote.
Fertilizers: Material added to soil to provide essential elements for plant growth, usually to increase crop growth.
Flower: Organ adapted for sexual reproduction in angiosperms.
Foliar: Pertaining to leaves.
Food: Organic compounds that can be metabolized for energy and used in assimilation.
Food chain: A group of plants and animals linked together by their food relationship.

Fruit: An ovary or group of ovaries with any adjacent parts that may be found with the ovary at maturity.
Gene: Unit of inheritance, located on the chromosome.
Genome: Set of chromosomes corresponding to the haploid set; the n set.
Genotype: Genetic constitution, latent or expressed, of an organism.
Genus: A group of closely related species clearly marked off from other groups.
Germination: Resumption of growth of an embryo or spore.
Gibberellin: An endogenous growth regulator responsible for cell elongation. Other activities include differentiation, flowering, and fruit set.
Glucose: A simple sugar, hexose; important in metabolism.
Grain: Fruit of plants belonging to the grass family.
Green manure: Crop grown and then plowed under to increase soil organic matter.
Group: Horticultural designation of botanical variety that includes many cultivars.
Growth regulator: Chemical hormone needed by living plants.
Gynoecious: All pistillate flowers.
Gynomonoecious: Some perfect and some pistillate flowers.
Hard pan: A hard layer of cemented soil beneath the tilled zone through which water and root penetrations are difficult.
Herb: A plant with no persistent woody system above ground.
Herbicide: Phytotoxic chemical used for killing or inhibiting growth of plants.
Hermaphrodite: Having both male and female parts on the same flower (perfect).
Hilum: Scar or mark on seed where seed was attached to ovary.
Horticulture: Science and culture of fruits and garden plants.
Hybridization: Cross-fertilization of plants belonging to different genotypes.
Hydroponics: Growing of plants in aqueous chemical solutions.
Hypocotyl: Stem of embryo or young seedling below attachment of cotyledons to root (radicle).
Indeterminate: Shoot axis remains vegetative; does not terminate with an inflorescence.
Indigenous: Native or natural to an area or region.
Inflorescence: Flowering part of plant flower cluster, its arrangement, and disposition of flowers on axis.
Inheritance: Character transmitted to progeny from parent(s).
Insecticide: A chemical for killing of insects.
Internode: Between two nodes on stem; region of stem between attachment of two adjacent leaves.
Ion: Atom or group of atoms carrying charge(s) generally in a solution.
Juvenile phase: Stage of plant in which vegetative growth proceeds and plant is insensitive to conditions which induce flowering.
Latex: A milky exudate from certain plants.
Layering: Stem covered with wet soil or moist peat moss while still attached to plant. Roots form at the nodes. Well-rooted stem is cut off and used for propagation.
Leaf area index (LAI): Leaf foliage density expressed as leaf area subtended per unit area of land.
Legume: Plant belonging to the Leguminoseae family (pea or bean family).
Lignin: A complex organic compound usually associated with cellulose in the cell wall.
Lipid: Organic compounds composed of fats and fat-like substances, insoluble in water.
Loam: Soil composed of high proportion of silt (75–95%), about half sand (40–50%), and low proportion of clay (5–25%).
Long-day: Plants that flower only when day length exceeds some minimum.

Macronutrient: Essential element required by plants in relatively large amounts (N, P, K, S).
Meristem: Undifferentiated cells capable of active cell division.
Metabolism: Biochemical process in a cell by which nutritive material is built into living matter or broken down into simple substances.
Micronutrient: Essential element required in relatively small amounts by plants.
Monocotyledon: A plant that produces seed with embryo having one cotyledon.
Monoecious: Both male and female flowers are produced on same plant.
Mosaic: Disease characterized by mottling of plant due to spots of light green, yellow, or dark green, usually by virus infection.
Mulch: Material used to protect root of plant from extremes in weather conditions.
Mutation: A sudden heritable change appearing as a result of change in genes or chromosomes.
Necrosis: Plant tissue turning dark or black due to death or disintegration of cells; usually caused by disease.
Nematode: Microscopic soil organism (roundworm), which may be parasitic in plant roots.
Nitrification: Change of ammonia nitrogen to nitrate by certain soil bacteria.
Node: Slightly enlarged portion of stem where leaves and buds arise.
Nodule: Enlarged swelling on roots of legumes where nitrogen-fixing bacteria colonize.
Nutrient: In plants, compounds that furnish energy and raw materials for growth; elements essential for plant growth.
Olericulture: Science and culture of vegetables.
Organ: A distinct and visibly different part of a plant: a root, leaf, flower, fruit.
Organic: Carbon compounds. Many may have had origin in plants or animals.
Ovary: Enlarged basal portion of pistil, containing egg cell(s).
Ovule: A rudimentary seed containing unfertilized seed gametophyte.
Palmate: Compound leaves with leaflets arising from the apex of the petiole.
Parasite: An organism deriving its nourishment from the living body of another organism.
Parthenocarpy: Development of fruit without fertilization.
Pathogen: Organism capable of inducing disease.
Pectin: Organic compound cementing adjacent plant cells.
Pedicel: Stalk or stem of flower or inflorescence.
Peduncle: A stalk or stem of flower borne singly or on main stem of inflorescence.
Peltate: Leaf with petiole attached to the under or lower surface and not at edge.
Perennial: Plant that lives many years; does not die after flowering once.
Perfect flower: Bisexual flower: flower with both stamens and pistil.
Periderm: Skin of storage organ such as roots and tubers; usually several cell layers thick. Cortical tissue from cork combium.
Pesticide: Chemical that kills pests, e.g., insects, fungi, weeds, and rodents.
Petiole: Stalk or stem-like structure of leaf, arising from the node of stem.
pH: Negative log of hydrogen ion concentration. Denotes neutrality (pH 7.0), acidity ($pH < 7.0$), or alkalinity ($pH > 7.0$).
Phenotype: Physical or external appearance of an organism.
Phloem: Conductive tissue that transports carbohydrates, growth regulators, and other synthesized substances from tissue where formed to other parts of plant.
Photoperiod: Length of day. Period of illumination required for normal growth to plant maturity.

Photosynthesis: Process in which CO_2 and H_2O with light are combined in chlorophyllous tissue to form carbohydrates.
Phototropism: The response of a plant to light stimulus. The growth toward light.
Phylogeny: Evolution of a group of related individuals.
Physiology: Science of the functions and activities of living organisms.
Phytotoxin: Poison or toxin produced by plants.
Pistil: Central organ of flower consisting of ovary, style, and stigma.
Pistillate flower: Flower with pistil and no stamens.
Pith: Large cells (paranchyma) of central portion of stem.
Pollen: Developed male gametophytes.
Polysaccharide: Long chain molecules of sugars, e.g., cellulose, starch.
Protein: Long-chain molecules of different amino acids with many functions.
Radicle: Portion of plant embryo which develop into primary root.
Receptacle: Enlarged end of pedicle or peduncle to which flower parts are attached.
Recessive: That character that is suppressed by a dominant character in heredity.
Relative humidity: Ratio of actual amount of water vapor in air to the maximum amount (saturation) air can hold at the same temperature, multiplied by 100.
Reproduction: Process of giving rise to offspring(s).
Respiration: Biological oxidation by enzymes of organic matter to CO_2 and H_2O to obtain energy.
Rest: A physiological condition of some plant parts (seed and bud) in which sprouting and/or rooting is inhibited even when the environment is conducive to growth. Also called *innate dormancy* or *internal dormancy* (see *Dormancy*).
Reticulate: A network or pattern.
Rhizome: Elongated and underground stem, usually horizontal, capable of producing new roots and shoots at the node.
Rogue: Remove inferior plants, diseased plants, or plants not true to type.
Rootstock: Portion of a true root or root with crown which is capable of producing stems or having stems grafted on them.
Sagittate: Arrowhead-shaped, usually a leaf.
Sand: Mineral particles in soil between 50 and 2000 μm in diameter.
Savanna: Grassland with scattered trees and xerophytic undergrowth (tropical and subtropical).
Scape: Flower stalk arising from the ground.
Scarify: To cut or scratch hard seed coat to hasten germination.
Scion: Portion of stem used in grafting; the aerial portion.
Seed: Mature ovule, capable of forming a new plant under favorable environmental conditions. Anything that may be sown, e.g., seed potatoes.
Seed coat: Outer protective layer of seed developed from integuments of the ovule.
Segregation: Separation of genes from different parents.
Self-pollination: Transfer of pollen from anther to stigma of same flower or of another flower on the same plant, or within a clone.
Sepal: A part of the calyx.
Seta: Bristle.
Sexual reproduction: Reproduction requiring reduction division of chromosomes and fertilization for zygote formation.
Shoot: Main portion of plant above ground, may be main stem of a branch.
Short-day: Plants that flower only when day length is less than some maximum.
Silt: Soil particles between 2 and 50 μm in diameter.
Soil: Uppermost stratum of earth's crust, modified by weathering.

Spadix: A spike with fleshy axis; flower of aroids.
Species: Class of individuals that interbreed freely and may have common characteristics.
Stamen: A structure of a flower made up of anther (pollen-bearing portion and stalk or filament).
Staminate flower: Pollen bearing flower with no pistil.
Starch: Polysaccharides of glucose insoluble in water. Food storage substance in plants.
Stock: Part of stem that receives scion in grafting; the root or lower portion.
Stolon: Stem growing horizontally along the ground surface, also called runner, capable of producing a shoot and root at the node.
Stoma: Specialized opening or pore with guard cells for gaseous exchange in leaves and stems.
Sucrose: A disaccharide composed of glucose and fructose.
Symbiosis: Association of two living organisms involving benefit to both.
Symptom: Response of organism or part of organism to disease or abnormal condition.
Tannin: Chemical substances produced by some plants which have an astringent and bitter taste.
Taproot: A stout main root.
Taxonomy: Science of classification into natural relationships.
Tendril: A modified leaf; a slender coiling organ of climbing plants.
Tiller: A shoot from the axis of lower leaves.
Toxin: Poisonous substance to living organisms; may be organic or inorganic.
Translocation: Movement of food or water and minerals from one part of a plant to another.
Transpiration: Loss of water from plant's tissues, usually from leaves.
Trichome: Hair or bristle from epidermis.
Trimonoecious: Three kinds of flowers (perfect, staminate, and pistillate) on same plant.
Tropism: Response of plants to stimulus.
Truss: A flower cluster, e.g., tomato flower cluster.
Tuber: Enlarged fleshy portion of rhizome bearing "eyes" or buds.
Umbel: An influorescence with individual pedicels arising from the apex of the peduncle.
Variety: A subdivision of species.
Vegetable: A horticultural food crop grown in most cases as annuals, and few as perennials. Most plants are herbaceous and not woody; parts used for food are roots, stems, leaves, immature floral parts, immature seeds, and immature and mature fruits. The edible part is high in water content, eaten raw or cooked, and can be stored only for short periods of time, rarely over 9 months even under ideal conditions.
Vegetative propagation: Reproduction by other than seeds.
Vernalization: Low temperature induction of floral initiation.
Virus: A filterable organic substance capable of self-replication in living organisms and may cause disease in the host.
Vitamin: Complex organic substance required by animals, obtained from other sources (usually plants).
Weed: Herbaceous plant or shrub of no value, hindering growth of other plants or crops.
Windrow: Row of harvested crop placed or stacked to dry.
Xerophyte: Plant resistant to droughty conditions.
Xylem: Plant conductive tissue that transports water and absorbed nutrients from roots to rest of plant.
Zygote: Cell resulting from fusion of gametes in sexual reproduction.

Index

C

Q

R

U

V